TENSOR ANALYSIS FOR PHYSICISTS

TENSOR ANALYSIS FOR PHYSICISTS

J. A. SCHOUTEN

Late Professor of Mathematics,
Amsterdam University

SECOND EDITION

Dover Publications, Inc., *New York*

Published in Canada by General Publishing Company, Ltd., 30 Lesmill Road, Don Mills, Toronto, Ontario.

Published in the United Kingdom by Constable and Company, Ltd.

This Dover edition, first published in 1989, is an unabridged republication of the 1959 corrected printing of the second (1954) edition of the work first published by the Clarendon Press, Oxford, in 1951. It is reprinted by special arrangement with Oxford University Press, 200 Madison Avenue, New York, N.Y. 10016.

Manufactured in the United States of America
Dover Publications, Inc., 31 East 2nd Street, Mineola, N.Y. 11501

Library of Congress Cataloging-in-Publication Data

Schouten, J. A. (Jan Arnoldus), b. 1883.
Tensor analysis for physicists / J.A. Schouten.
p. cm.
Reprint. Originally published: 2nd ed. Oxford : Clarendon Press, 1954.
Bibliography: p.
Includes index.
ISBN 0-486-65582-2
1. Calculus of tensors. 2. Mathematical physics. I. Title.
QC20.7.C28S36 1988
515′.63—dc19 87-36451
CIP

PREFACE TO THE SECOND EDITION

THIS second edition does not differ radically from the first. Little need be said about the changes because they speak for themselves. I have only to express my most grateful thanks to all those friends who have sent corrections and advice on improvements for the new edition. But I have something to say to my readers. Do not use this book in the wrong way. It is meant to be an introduction to tensor analysis not a kind of short cut that will give much knowledge of dimension theory, elasticity, classical mechanics, relativity, or matrix calculus without any hard work. The chapters on these topics were only written to provide enough interesting applications for showing how tensor analysis works in practice. It is advisable first to study Chapters I–V thoroughly, giving special care to exactness in mathematical expression. It is no use relying on mere fragments of tensor calculus picked up anyhow. When a good groundwork has been obtained it is then time to choose a congenial topic from those dealt with in Chapters VI–X. This book should then be used side by side with a good textbook on the special subject.

EPE, HOLLAND
May 1953

J. A. S.

PREFACE TO THE FIRST EDITION

THIS book has grown from lectures delivered before and during the war at Delft, and after the war at Amsterdam. The tensor algebra in E_n and R_n is developed in Chapters I and II, and the tensor analysis in X_n and L_n in Chapters IV and V. Chapter III belongs to the algebra and deals with the identifications of quantities in E_n after the introduction of a sub-group of the affine group. These five chapters contain as much of the theory of the calculus as is necessary for physical purposes.

Immediately after Chapter V there is a brief summary of the salient points of the theory. This was specially asked for by physicists and should prove valuable to experimentalists and others whose interests lie chiefly in the topics dealt with in the later chapters and by its means can avoid actually working through the whole of the earlier part.

In the next four chapters we give applications. Since there are sufficient topics to fill more than one book of this size, a selection had to be made. We have, of course, chosen only those applications that are both interesting in themselves and also good examples of the use and advantages of the calculus. In Chapter VI, intimately connected with Chapter III, we have shown that the dimensions of physical quantities depend on the choice of the underlying group. From the days of Voigt, who introduced the term 'tensor', some kind of tensor calculus has always been the best instrument for dealing with the properties of anisotropic media. In Chapter VII modern tensor calculus is applied both to some old and also some modern problems of elasticity and piezo-electricity. It is not so well known that classical dynamics can be treated in a very elegant way by using tensor calculus. In Chapter VIII we give some examples concerning anholonomic systems and the homogeneous treatment of the equations of Lagrange and Hamilton. The development of tensor calculus from its original form given by Ricci to the most modern form used here has been influenced strongly by the development of the theory of relativity. In Chapter IX we deal first with relativistic kinematics and dynamics and then give in the last section an exposition of modern treatment of relativistic hydrodynamics. None of these four chapters is meant to be a small text-book on its particular subject, but we have endeavoured to avoid an incoherent enumeration of interesting facts, and have tried to make each chapter a short but systematic introduction to some branch of theoretical physics.

Matrix calculus and tensor calculus are so intimately related that it is not possible in a book like this to say nothing about the former. In Chapter II we do the usual thing and give a brief sketch of the relations between both disciplines. But in view of the latest developments in matrix calculus for quantum mechanics due to Dirac we thought it necessary to give in Chapter X an exposition of his beautiful methods.

At the end of each chapter there are some exercises. Many of them are formulated as proofs in order to give the answer in advance. In nearly all the others sufficient references to literature are given.

The index at the end of the book should prove useful.

I owe many thanks to Prof. E. T. Davies and the officials of the Clarendon Press who did much to improve my English and also made other very valuable suggestions. My collaboration with the Clarendon Press has been most agreeable, and I wish to express my sincere thanks for all they have done.

EPE, HOLLAND J. A. S.

CONTENTS

I. SPACES DEFINED BY LINEAR GROUPS

II. GEOMETRIC OBJECTS IN E_n

III. IDENTIFICATIONS OF QUANTITIES IN E_n AFTER INTRODUCING A SUB-GROUP OF G_a

IV. GEOMETRIC OBJECTS IN X_n

V. GEOMETRY OF MANIFOLDS WHICH HAVE A GIVEN DISPLACEMENT

VI. PHYSICAL OBJECTS AND THEIR DIMENSIONS

VII. APPLICATIONS TO THE THEORY OF ELASTICITY

VIII. CLASSICAL DYNAMICS

IX. RELATIVITY

X. DIRAC'S MATRIX CALCULUS

TENSOR ANALYSIS FOR PHYSICISTS

I

SPACES DEFINED BY LINEAR GROUPS†

1. The Group G_a. Affine geometry

WE consider an n-dimensional space with coordinates x^κ, subject to transformations of the form

$$x^{\kappa'} = A^{\kappa'}_{\kappa} x^\kappa + a^{\kappa'}; \quad \Delta \overset{\text{def}}{=} \text{Det}(A^{\kappa'}_{\kappa}) \neq 0‡§ \tag{1.1 a}$$

with *constant* coefficients $A^{\kappa'}_{\kappa}$, $a^{\kappa'}$. The set of these transformations forms a *group*, i.e.

(1) the result of two transformations of the set applied after each other is a transformation of the set;

(2) the inverse of every transformation of the set belongs to the set;

(3) the set contains the identical transformation.

This group is called the *affine group* G_a. The coordinates are called *rectilinear*.

The inverse transformation of (1.1)

$$x^\kappa = A^{\kappa}_{\kappa'} x^{\kappa'} + a^\kappa \tag{1.1 b}$$

contains *constant* coefficients $A^{\kappa}_{\kappa'}$, a^κ connected with a^κ and $A^{\kappa'}_{\kappa}$ by the relations

$$A^{\kappa'}_{\kappa} A^{\kappa}_{\lambda'} = \begin{cases} 1 \text{ for } \kappa' = \lambda' \\ 0 \;\;,, \;\; \kappa' \neq \lambda', \end{cases} \qquad A^{\kappa}_{\kappa'} A^{\kappa'}_{\lambda} = \begin{cases} 1 \text{ for } \kappa = \lambda \\ 0 \;\;,, \;\; \kappa \neq \lambda \end{cases} \tag{1.2}$$

$$A^{\kappa}_{\kappa'} a^{\kappa'} = -a^\kappa, \qquad A^{\kappa'}_{\kappa} a^\kappa = -a^{\kappa'}. \tag{1.3}$$

To determine $A^{\kappa}_{\lambda'}$ for some definitely given values of κ and λ' (e.g. $\kappa = 1$; $\lambda' = 3'$) we write out the matrix‖ of $A^{\kappa'}_{\lambda}$,

$$\left\| \begin{matrix} A^{1'}_{1} & \cdots & A^{n'}_{1} \\ \vdots & & \\ A^{1'}_{n} & \cdots & A^{n'}_{n} \end{matrix} \right\| \tag{1.4}$$

and form the *minor* of the element $A^{\lambda'}_{\kappa}$ by *striking out* the κth row and

† General references: R.K. 1924. 1; Veblen and Whitehead 1932. 1; Ei. 1935. 1; Lichnerowicz, 1947. 1; R.C. Ch. I.

‡ The indices $\kappa, \lambda, \mu, \nu, \rho, \sigma, \tau$ (and sometimes ω) always take the values $1,\ldots, n$ (italic), the indices $\kappa', \lambda', \mu', \nu', \rho', \sigma', \tau'$ (and sometimes ω') the values $1',\ldots, n'$ (italic), etc. $\overset{\text{def}}{=}$ means: is defined to be. Det( ) stands for determinant of ( ).

§ If an index occurs twice in any one term, once as an *upper* and once as a *lower* index, summation has to be effected. The summation sign Σ is omitted (Einstein convention).

‖ In dealing with expressions with two indices and their corresponding matrices we agree that the first index numbers the columns and the second the rows and that if the two indices are vertical, one above the other, the upper index is considered as first.

the λth column and multiplying by $(-1)^{\kappa+\lambda}$. Then $A^{\kappa}_{\lambda'}$ is equal to the determinant of this minor divided by Δ. If $\mathrm{Det}(A^{\kappa'}_{\lambda})$ is written out this minor is the coefficient of the element $A^{\lambda'}_{\kappa}$. Hence

$$A^{\kappa}_{\lambda'} = \Delta^{-1}\frac{\partial \Delta}{\partial A^{\lambda'}_{\kappa}} = \frac{\partial \log \Delta}{\partial A^{\lambda'}_{\kappa}}. \tag{1.5}$$

Every coordinate system $x^{\kappa'}$ that can be formed from the x^{κ} by means of a transformation of G_a is called an *allowable coordinate system* and the space provided with all allowable coordinate systems an *affine space* or E_n. The theory of all properties of figures in E_n which are invariant under the group G_a is called *affine geometry*.

In all formulae we have *kernel letters* like A, x, *running indices* like κ, κ', and *fixed indices* like $1,\ldots, n$; $1',\ldots, n'$. Running indices can also be taken from another alphabet and fixed indices can also be taken from the row 1, 2, 3,... instead of $1, 2, 3,\ldots$. For example, a transformation of coordinates could be denoted by passing from x^{κ}; $\kappa = 1,\ldots, n$ to x^{h}; $h = 1,\ldots$, n. Then there is of course a difference between x^{1} and x^{1} just as between x^{1} and $x^{1'}$. Using roman running indices we generally use *vertical* figures for the corresponding fixed indices. A set of fixed indices always belongs to one and only one set of running indices. Every set of running or fixed indices belongs to one definite coordinate system, and this coordinate system is denoted by one of its running indices in round brackets, for instance (κ), (κ'), (h), (h'). Points and kernel letters do not change with a *coordinate* transformation. The change of coordinates is indicated by a new set of running, and a corresponding new set of fixed, indices. With a *point* transformation, however, the coordinate system, and consequently also the running and the fixed indices, do not change while the points and the kernel letters are changed as in the following example:

$$y^{\kappa} = p^{\kappa} + P^{\kappa}_{\cdot\lambda} x^{\lambda}, \tag{1.6}$$

which, for constant p^{κ} and $P^{\kappa}_{\cdot\lambda}$ represents an affine point transformation.†

It is not always convenient, however, to use a new letter when a change of kernel letter is indicated. We shall indicate a change of kernel letter by

(1) changing the letter itself;

(2) adjoining an accent or asterisk, preferably *to the left* of the kernel letter;

† This is the principle of the *kernel-index method* used in this book and in many modern publications on differential geometry and partial differential equations.

(3) adding an index directly *above* or *below* the kernel letter since the upper and lower places to the right of the kernel letter are generally reserved for running and fixed indices;

(4) by the use of undisplaced 'strangled' indices as will be explained later.

2. Sub-groups of G_a

Particular cases of affine geometry arise if we use a sub-group of G_a instead of G_a. The following cases occur later in this book:

1. Group G_{ho} of all *linear homogeneous transformations* with $\Delta \neq 0$. The *origin* or *centre* $x^\kappa = 0$ is invariant and the space is called a *centred* E_n. Allowable coordinate systems are the allowable systems of E_n with the same origin.

2. Group G_{eq} of all *equivoluminar linear homogeneous transformations*, i.e. transformations with $\Delta = \pm 1$. The space is called a *centred* E_n *with a given unit volume*. Allowable coordinate systems are the allowable systems of E_n with the same origin and the same unit volume. In the geometry of G_{eq}, volumes can be compared.

3. Group G_{sa} of all *special affine transformations*, i.e. the linear homogeneous transformations with $\Delta = +1$. The space is called an *oriented centred* E_n *with a given unit volume and a given screw-sense*. Allowable coordinate systems are the allowable systems of E_n with the same origin, the same unit volume, and the same *orientation* (in E_3 only right-handed systems or only left-handed systems). An *n-dimensional orientation* or *screw-sense* is fixed by n directions with sense (arrow) which are not contained in any E_{n-1} and which are given in a definite order.† For $n = 2$ we call the screw-sense *sense of rotation* and for $n = 1$ *sense*. Two such sets fix the same screw-sense if they can be transformed into each other by an affine point transformation with Δ positive. It follows from this that every coordinate system in E_n fixes a definite screw-sense by its n axes, their positive senses, and the order $1,\ldots, n$. This screw-sense changes under coordinate transformations if and only if Δ is negative.

4. Group G_{or} of all *orthogonal homogeneous transformations* (*rotations and reflections*). The space is called an R_n. $\Delta = \pm 1$. Allowable coordinate systems are the orthogonal systems with the same origin and the same unit volume. In an R_n the notions length and angle exist.

† Instead of n directions a part of a curve not contained in any E_{n-1} and provided with a sense (arrow) can be taken. If $\frac{1}{2}n(n+1)$ is even (e.g. for $n = 3$) the sense can be omitted.

5. Group G_{ro} of all *rotations*. $\Delta = +1$. The space is called an *oriented* R_n. Allowable coordinate systems are the orthogonal systems with the same origin, the same unit volume, and the same orientation.

To every group there belongs a geometry, viz. the theory of all those properties of figures that are invariant for all transformations of the given group.

The geometry of a group and the geometries of its sub-groups are connected in the following way (principle of F. Klein†).

If G is a group and G′ the sub-group of G consisting of all transformations of G that leave a set of figures A invariant, the geometry of a figure with respect to G′ is identical with the geometry of this figure together with A with respect to G.

In the preceding cases we can always take for the set A the set of all allowable coordinate systems or the figure from which they can be derived, for instance, the origin if we pass from G_a to G_{ho} or the unit hypersphere if we pass from G_a to G_{or}.

3. Flat sub-manifolds in E_n

In this section we give a brief account of the most important properties of flat sub-manifolds in an E_n.

We consider a system of $n-p$ linear equations

$$C_\lambda^x x^\lambda + C^x = 0 \quad (x = \mathrm{p}+1,\ldots,\mathrm{n}) \tag{3.1}$$

with *constant* coefficients C_λ^x, C^x and suppose that the matrix of the C_λ^x

$$\begin{Vmatrix} C_1^{\mathrm{p}+1} & \cdots & C_1^{\mathrm{n}} \\ \vdots & & \\ C_n^{\mathrm{p}+1} & \cdots & C_n^{\mathrm{n}} \end{Vmatrix} \tag{3.2}$$

has the highest possible rank $n-p$.‡ Then the equations (3.1) are linearly independent and $n-p$ of the variables x^κ can be solved from them as functions of the remaining ones. Hence these p variables can be used as coordinates in the sub-manifold of E_n consisting of all points satisfying (3.1). Since these variables are subject to the group of all transformations of G_a leaving the other $n-p$ variables invariant, the sub-manifold is an E_p. We call it the *null manifold* of (3.1) and (3.1) a *null form* of E_p. More generally the solution of (3.1) can be written in the form

$$x^\kappa = B_b^\kappa \eta^b + B^\kappa \quad (b = 1,\ldots,\mathrm{p}) \tag{3.3}$$

† 1872. 1.

‡ The rank of a matrix is r if there exists a non-vanishing r-rowed sub-determinant but no non-vanishing sub-determinant with more than r rows.

with *constant* coefficients B_b^κ, B^κ, containing p arbitrary parameters η^b. (3.3) is called a *parametric form* of E_p. Substituting (3.3) in (3.1) we get

$$C_\lambda^x B_b^\lambda = 0; \quad C_\lambda^x B^\lambda + C^x = 0 \quad (b = 1,..., \mathrm{p};\ x = \mathrm{p}+1,..., \mathrm{n}). \tag{3.4}$$

A point transformation of the form

$$'x^\kappa = x^\kappa + c^\kappa \tag{3.5}$$

(in which the change of the kernel letter is indicated by an accent to the *left* of the letter x) with constant c^κ is called a *translation* If we subject (3.5) to the transformation (1.1) of the group G_a, we have

$$'x^{\kappa'} = x^{\kappa'} + A_\kappa^{\kappa'} c^\kappa. \tag{3.6}$$

Hence translation in an E_n is invariant for allowable coordinate transformations. Two E_p's in E_n are called *parallel* if they can be transformed into each other by means of a translation. Parallel E_p's are said to have the same *p-direction*. A *1*-direction is called *direction*. This direction has not yet got a sense. Only if it is provided with a sense (e.g. an arrow) does it become identical with the direction in every-day language. In a centred E_n to every E_p there belongs one and only one parallel E_p through O whose null form can be derived from (3.1) by omitting C^x:

$$C_\lambda^x x^\lambda = 0 \quad (x = \mathrm{p}+1,..., \mathrm{n}). \tag{3.7}$$

Now consider an E_p and an E_q through O with the equations

$$\begin{matrix} \overset{1}{C}{}_\lambda^x x^\lambda = 0 \\ \overset{2}{C}{}_\lambda^y x^\lambda = 0 \end{matrix} \quad (x = \mathrm{p}+1,..., \mathrm{n};\ y = \mathrm{q}+1,..., \mathrm{n}). \tag{3.8}$$

If the rank of the combined matrix

$$\overset{\uparrow}{\underset{\downarrow}{n}} \left\| \overset{\leftarrow 2n-p-q \rightarrow}{\overset{1}{C}{}_\lambda^x \quad \overset{2}{C}{}_\lambda^y} \right\| \tag{3.9}$$

is $n-s$, there are just $n-s$ linearly independent equations in the set (3.8) and they therefore define an E_s, the *section*† of E_p and E_q. Hence an E_p and an E_q through O always intersect in an E_s; $0 \leqslant s \leqslant p$, $s \leqslant q$; $s \geqslant p+q-n$. This can easily be generalized for an arbitrary E_p and E_q if we consider the points of E_p 'at infinity' as an *improper* E_{p-1} (or E_{p-1} *'at infinity'*). Then two parallel E_p's intersect in an improper E_{p-1}, and from this we see that an improper E_{p-1} may be considered as a p-direction.

Using this expression, we may formulate the proposition

An E_p and an E_q in E_n either intersect in an E_s; $p+q-n \leqslant s \leqslant p$; $0 \leqslant s \leqslant q$ or they have no point in common ($s = -1$).

† Some authors use the terms *intersection* and *meet*.

An E_0 is a point, an E_1 is called a *straight line*, an E_2 a *plane*, and an E_{n-1} a *hyperplane*. If $u_\lambda x^\lambda = 0$ is the equation of a hyperplane through O, the u_λ are called its (homogeneous) *hyperplane coordinates*. To every allowable coordinate system in E_n there belong n E_1's, the *coordinate axes*, $\binom{n}{2}$ E_2's, the *coordinate planes*, each determined by two axes and $\binom{n}{p}$ E_p's, the *coordinate* E_p's; $1 \leqslant p \leqslant n-1$. By means of these it can be readily shown that an E_p in E_n is fixed by $(p+1)(n-p)$ numbers. If the coordinate system is not special, an E_p intersects each of the coordinate-E_{n-p}'s in exactly one point. The E_p is determined uniquely by $p+1$ of these points, and each of them can be fixed by $n-p$ numbers. In the same way it can be proved that an E_p in E_n through a given E_q; $q \leqslant p$, can be fixed by $(p-q)(n-p)$ numbers. Hence an E_p in a centred E_n through the origin can be fixed by $p(n-p)$ numbers.

If an E_p and an E_q are given, one may require the flat manifold E_t containing E_p and E_q and having the smallest dimension possible. This E_t is called the *join*† of E_p and E_q. Taking into account improper manifolds the following proposition can be proved:

The join of an E_p and an E_q intersecting in an E_s is an E_t; $t = p+q-s$.

For $n = 4$ we give here a table of intersections and joins in the general case:

Intersections

	E_0	E_1	E_2	E_3
E_0	×	×	×	×
E_1	×	×	×	E_0
E_2	×	×	E_0	E_1
E_3	×	E_0	E_1	E_2

Joins

	E_0	E_1	E_2	E_3
E_0	E_1	E_2	E_3	E_4
E_1	E_2	E_3	E_4	E_4
E_2	E_3	E_4	E_4	E_4
E_3	E_4	E_4	E_4	E_4

(3.10)

× = no intersection in general.

The table of intersections in E_4 can be verified by a simple experiment. In the E_4 of the ordinary coordinates x, y, z and the time t the diagram of a moving point is a curve, its *world line*. The world line is an E_1 in the case of uniform motion. A straight line moving uniformly describes a world-plane, and a plane moving uniformly describes a world-E_3. Now consider, for example, two E_1's moving uniformly. As they intersect at *one* instant, the two world-E_2's intersect in *one* point.

† In P.P. 1949, I we introduced the term *junction*. We have, however, since discovered that the term *join* had already been used in earlier publications.

In the same way an E_1 and an E_2, each moving uniformly, intersect in a point moving uniformly. Hence the world-E_2 and the world-E_3 intersect in an E_1.†

If a p-direction is given in E_n we may consider all E_p's with this p-direction as points of an E_{n-p}. This process is called the *reduction of the E_n with respect to the given p-direction.* If the coordinates are chosen in such a way that the parallel E_p's are represented by the equations

$$x^{p+1} = \text{const.}, \quad ..., \quad x^n = \text{const.} \tag{3.11}$$

the $x^{p+1}, ..., x^n$ can be used as coordinates in this E_{n-p}.

If in E_n, an E_p is given and also an $(n-p)$-direction that has no direction in common with E_p, every geometric figure in E_n can be subjected to the following processes:

(*a*) *Section with E_p*: the resulting figure consists of the common points of the original figure and E_p. Only the E_p is used.

(*b*) *Reduction with respect to the $(n-p)$-direction*: all points of the figure which lie in the same E_{n-p} with the given $(n-p)$-direction are identified. The figure that results is contained in the E_p that comes from the E_n by reduction with respect to the $(n-p)$-direction. Only the $(n-p)$-direction is used.

(*c*) *Projection on the E_p in the $(n-p)$-direction*: through every point of the figure an E_{n-p} is drawn with the given $(n-p)$-direction. The section of this E_{n-p} with the E_p is the projection of the point. Both the E_p and the $(n-p)$-direction are used.

In a given E_p in E_n an orientation or p-dimensional screw-sense may be fixed. Then the E_p is said to have an *inner orientation.* If an $(n-p)$-dimensional screw-sense is fixed in some E_{n-p} having no direction in common with the E_p, this screw-sense determines a screw-sense in *every* E_{n-p} of this kind by projection in the p-direction of the E_p. In this case the E_p is said to have an *outer orientation.* In an E_1 in E_3, for instance, an arrow *in* the E_1 and a sense of rotation *around* the E_1.

EXERCISES

I.1. Prove that the point transformation (cf. footnote on page 19)

$$y^\kappa = A^\kappa_{\kappa'} \delta^{\kappa'}_\lambda x^\lambda + a^\kappa \tag{1 α}$$

transforms every point x^κ into a point y^κ which has the same components with respect to (κ') as the point x^κ with respect to (κ). (Use 1.1 *a*.)

† This experiment can be performed for an auditorium with the aid of two sticks or a stick and a piece of cardboard.

I.2. If (3.8) are the equations of an E_p and an E_q through O in point coordinates, the parametric equations in hyperplane coordinates have the form

$$u_\lambda = \overset{1}{C}{}^x_\lambda \zeta_x, \qquad u_\lambda = \overset{2}{C}{}^y_\lambda \theta_y \quad (x = \mathrm{p}+1,\ldots,\mathrm{n};\ y = \mathrm{q}+1,\ldots,\mathrm{n}). \tag{2 α}$$

Since the rank of (3.9) is $n-s$ the n equations

$$\overset{1}{C}{}^x_\lambda \zeta_x - \overset{2}{C}{}^y_\lambda \theta_y = 0 \quad (x = \mathrm{p}+1,\ldots,\mathrm{n};\ y = \mathrm{q}+1,\ldots,\mathrm{n}), \tag{2 β}$$

have $n-p-q+s$ linearly independent sets of solutions ζ_x, θ_y, and the join of E_p and E_q is an E_{p+q-s}.

I.3. An E_p and an E_q through O can be given by the parametric equations

$$x^\kappa = \overset{1}{B}{}^\kappa_b \eta^b, \qquad x^\kappa = \overset{2}{B}{}^\kappa_c \chi^c \quad (b = 1,\ldots,\mathrm{p};\ c = 1,\ldots,\mathrm{q}) \tag{3 α}$$

in point coordinates and by the equations

$$\overset{1}{B}{}^\kappa_b u_\kappa = 0, \qquad \overset{2}{B}{}^\kappa_c u_\kappa = 0 \quad (b = 1,\ldots,\mathrm{p};\ c = 1,\ldots,\mathrm{q}) \tag{3 β}$$

in hyperplane coordinates. If E_s is the section of E_p and E_q the equations

$$\overset{1}{B}{}^\kappa_b \eta^b - \overset{2}{B}{}^\kappa_c \chi^c = 0 \quad (b = 1,\ldots,\mathrm{p};\ c = 1,\ldots,\mathrm{q}) \tag{3 γ}$$

have s linearly independent sets of solutions η^b, χ^c and the rank of the matrix (cf. (3.9))

$$p+q \left\{ \overbrace{\left\| \begin{matrix} \overset{1}{B}{}^\kappa_b \\ \\ \overset{2}{B}{}^\kappa_c \end{matrix} \right\|}^{n} \right. \tag{3 δ}$$

is $p+q-s$.

I.4. An E_p and an E_q in E_n which have no direction in common, and each of which has an inner orientation, always determine an inner orientation in the join E_{p+q} for pq even. For pq odd they only determine an inner orientation in the join if they are given in a definite order.

I.5. An E_p and an E_q in E_n which have an E_s in common, and each of which has an inner orientation, always determine an inner orientation in the join E_{p+q-s} for each choice of an inner orientation in E_s when $(p-s)(q-s)$ is even but only do this for $(p-s)(q-s)$ odd if they are given in a definite order.

I.6. If $u_\lambda x^\lambda = 1$ is the equation of an E_{n-1} not containing 0 the u_λ are called its hyperplane coordinates. Prove that an E_p has in these coordinates the nullform $B^\lambda_b u_\lambda = 0$; $B^\lambda u_\lambda = 1$, and the parametric form

$$u_\lambda = C^x_\lambda \zeta_x; \qquad -\zeta_x C^x = -1$$

with $n-p$ parameters ζ_x satisfying one condition.

II
GEOMETRIC OBJECTS IN E_n†

1. Definitions

If there is a correspondence between ordered sets of N numbers

$$\Phi_\Lambda \quad (\Lambda = 1,..., N)$$

and the allowable coordinate systems (κ) *in* E_n *such that*

(1) *to every* (κ) *there corresponds one and only one set* Φ_Λ;

(2) *the set* $\Phi_{\Lambda'}$ *corresponding to* (κ') *can be expressed in terms of* Φ_Λ, $A_\kappa^{\kappa'}$, *and* $a^{\kappa'}$ *only,*

then the Φ_Λ *are said to be the components of a geometric object with respect to* (κ) *in* E_n.

Geometric objects in E_n are classified according to the laws of transformation of their components. If in the condition (2) above the expression for the $\Phi_{\Lambda'}$ is linear and homogeneous in the Φ_Λ, is algebraically homogeneous in the $A_\kappa^{\kappa'}$, and does not involve the $a^{\kappa'}$, then the Φ_Λ are the components of a *geometric quantity in* E_n.

A simple example of a geometric object which is not a quantity is a point, for which the components Φ_Λ are the x^κ, whose law of transformation involves the $a^{\kappa'}$ (cf. I. 1.1 *a*).

Besides coordinate transformations and point transformations we may consider more general *object transformations*. An object transformation is a transformation by which an object is transformed into another object of the same kind, i.e. with the same law of transformation. Also with an object transformation the coordinate system, and consequently the running and the fixed indices, do not change while the kernel letters are changed.

In sections 2–8 of this chapter many examples of geometric objects will be dealt with.

2. Scalars and vectors

(*a*) *Scalars*. A scalar has one component, invariant for (I. 1.1).

(*b*) *Contravariant vectors*. A contravariant vector has n components v^κ with the transformation

$$v^{\kappa'} = A_\kappa^{\kappa'} v^\kappa. \tag{2.1}$$

Hence v^κ transforms like the difference of two radii vectores, and from this it follows that v^κ can be represented by a set of two points with

† General references: R.K. 1924. 1; Eisenhart 1926. 1; Levi Civita 1927. 2; Thomas 1931. 1; Veblen and Whitehead 1932. 1; E I. 1935. 1; Brillouin 1938.1; Lichnerowicz 1947. 1; Brandt 1947. 2; Michal 1947. 3; R.C. 1954. 1, Ch. I.

a sense fixing the order of these points. This sense can be given by an arrow (not necessarily straight) or by a positive and a negative sign or by the numbers *1* and *2*, etc. The point set is determined to within translations. The components with respect to (κ) are the projections on the axes, measured by the corresponding units on these axes. In a centred E_n a contravariant vector can be represented by a point. Here the radius vector x^κ is a quantity, viz. a contravariant vector. Hence if G_a defining an E_n is replaced by its sub-group G_{ho} defining a centred E_n, certain geometric objects which *differ* in their law of transformation with respect to the main group may have the *same* law of transformation with respect to the sub-group.

Addition of contravariant vectors is the process known in mechanics as the 'parallelogram of forces'. By multiplication of v^κ by a real scalar p we get a vector pv^κ that has $|p|$ times the magnitude of v^κ and the same or the opposite sense according as p is greater or less than zero.

(*c*) *Covariant vectors.* A covariant vector has n components w_λ with the transformation

$$w_{\lambda'} = A^{\lambda}_{\lambda'} w_\lambda. \tag{2.2}$$

In order to give a geometrical interpretation to the covariant vector, let us take two parallel hyperplanes with the equations

$$w_\lambda x^\lambda = c, \qquad w_\lambda x^\lambda = c+1, \tag{2.3}$$

where c is an arbitrary constant. These equations represent a set of ∞^1 pairs of parallel hyperplanes dependent on one parameter c. With the transformations (1.1, 2.2) they transform into

$$w_{\lambda'} x^{\lambda'} = c', \qquad w_{\lambda'} x^{\lambda'} = c'+1, \tag{2.4}$$

where

$$c' = c - w_\lambda a^\lambda. \tag{2.5}$$

Hence the form of the equations (2.3) is invariant for the coordinate transformations (1.1). The hyperplanes (2.3) intersect the *1*-axis in the points p^κ and q^κ defined by the equations

$$p^1 = \frac{c}{w_1}; \qquad q^1 = \frac{c+1}{w_1}, \tag{2.6}$$

hence

$$w_1 = \frac{1}{q^1 - p^1} \tag{2.7}$$

independent of the choice of c.

Consequently the w_λ can be represented by a set of two parallel hyperplanes provided with a sense fixing the order of these hyperplanes. This sense can be given by an arrow (preferably not straight)† or a positive and a negative sign, or by painting one hyperplane white and the other black, etc. The set of hyperplanes is determined to within translations. The components with respect to (κ) are the inverses of

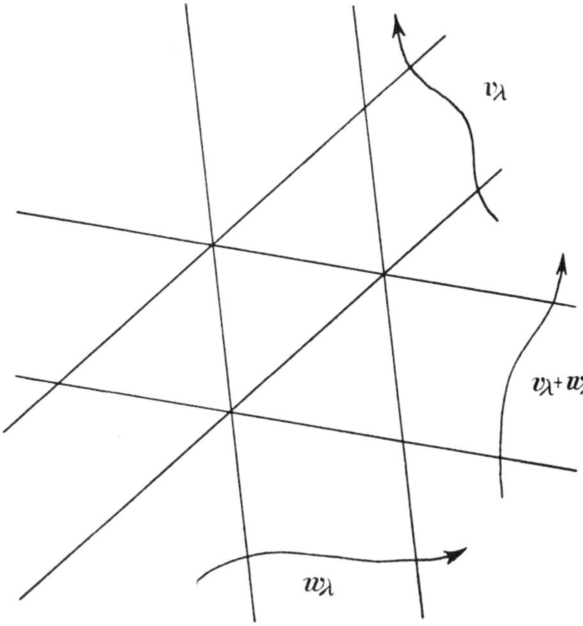

FIG. 1.

the segments cut out on the axes, measured by the corresponding units on these axes. In a centred E_n a covariant vector can be represented by one E_{n-1}. But in a general E_n the u_λ in the equation of an E_{n-1}

$$u_\lambda x^\lambda = 1 \tag{2.8}$$

transforms as follows: $u_{\lambda'} = u_\lambda A^\lambda_{\lambda'}/(1-u_\lambda a^\lambda).$ (2.8 a)

From this equation we see that a hyperplane is a geometric object in E_n but not a quantity because the law of transformation of the u_λ involves $a^\kappa = -A^\kappa_{\kappa'} a^{\kappa'}$.

Addition of covariant vectors is illustrated geometrically in Fig. 1, in which is shown the section of v_λ, w_λ, and $v_\lambda + w_\lambda$ with some E_2.

On multiplying w_λ by a real scalar p we get another covariant vector pw_λ whose magnitude is $|p|^{-1}$ times the magnitude of w_λ and which has the same or the opposite sense to w_λ according as p is positive or negative.

† A straight arrow could suggest that some direction was given.

p linearly independent contravariant vectors determine a p-direction, and the directions of all contravariant vectors linearly dependent on them are contained in this p-direction. Such a system of ∞^p contravariant vectors is called a *contravariant domain* and the p-direction its *support*. In a centred E_n, an E_p through the origin can be used as support instead. p is called the *dimension of the domain*. A contravariant domain and its support determine each other uniquely, and they are said to be *spanned* by the p vectors.

p linearly independent covariant vectors determine an $(n-p)$-direction, and the $(n-1)$-directions of all covariant vectors linearly dependent on them contain this $(n-p)$-direction. Such a system of ∞^p covariant vectors is called a *covariant domain* and the $(n-p)$-direction its *support*. In a centred E_n, an E_{n-p} through the origin can be used as support instead. p is called the *dimension of the domain*. A covariant domain and its support determine each other uniquely, and they are said to be *spanned* by the p vectors.

To each allowable coordinate system (κ) in E_n there belong n contravariant vectors $\underset{\lambda}{e}^\kappa$; $\lambda = 1, \dots, n$ with the components

$$\begin{array}{llll} \underset{1}{e}^1 = 1, & \underset{1}{e}^2 = 0, & \dots, & \underset{1}{e}^n = 0, \\ \underset{2}{e}^1 = 0, & \underset{2}{e}^2 = 1, & \dots, & \underset{2}{e}^n = 0, \\ \vdots & & & \\ \underset{n}{e}^1 = 0, & \underset{n}{e}^2 = 0, & \dots, & \underset{n}{e}^n = 1, \end{array} \tag{2.9}$$

where $\underset{\lambda}{e}$ is the kernel and κ the running index. In the same way n covariant vectors $\overset{\kappa}{e}_\lambda$; $\kappa = 1, \dots, n$ belong to (κ). These have the components

$$\begin{array}{llll} \overset{1}{e}_1 = 1, & \overset{1}{e}_2 = 0, & \dots, & \overset{1}{e}_n = 0, \\ \overset{2}{e}_1 = 0, & \overset{2}{e}_2 = 1, & \dots, & \overset{2}{e}_n = 0, \\ \vdots & & & \\ \overset{n}{e}_1 = 0, & \overset{n}{e}_2 = 0, & \dots, & \overset{n}{e}_n = 1. \end{array} \tag{2.10}$$

Here $\overset{\kappa}{e}$ is the kernel and λ the running index. $\underset{\lambda}{e}^\kappa$ and $\overset{\kappa}{e}_\lambda$ are called the *contravariant* and *covariant measuring vectors of* (κ). For every value of μ, the vector $\underset{\mu}{e}^\kappa$ has a direction contained in the $(n-1)$-direction of every vector $\overset{\nu}{e}_\lambda$ ($\nu \neq \mu$) and fits between the two E_{n-1}'s of $\overset{\mu}{e}_\lambda$. In Fig. 2 the vectors $\underset{\lambda}{e}^\kappa$ and $\overset{\kappa}{e}_\lambda$ are shown for $n = 3$.

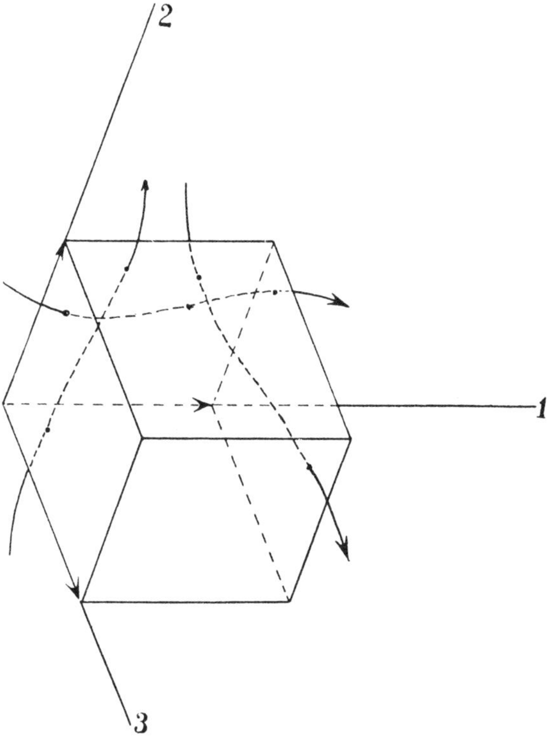

FIG. 2.

If another coordinate system (κ') is introduced the components of $\underset{\lambda}{e}^\kappa$ and $\overset{\kappa}{e}_\lambda$ with respect to (κ') are

$$\begin{aligned} \underset{\lambda}{e}^{\kappa'} &= A^{\kappa'}_{\kappa} \underset{\lambda}{e}^{\kappa}, \\ \overset{\kappa}{e}_{\lambda'} &= A^{\lambda}_{\lambda'} \overset{\kappa}{e}_{\lambda}. \end{aligned} \tag{2.11}$$

Of course these components are no longer *1* or *0*. In the transformation (2.11) we remark that the complete kernel consists of the letter e together with an index directly below (as $\underset{\lambda}{e}$) or above (as $\overset{\kappa}{e}$) in accordance with the convention (3) mentioned in I, § 1. Such indices, forming part of the kernel, are called *dead*, and indices subjected to the transformations of the underlying group *living*.

The measuring vectors belonging to (κ') are $\underset{\lambda'}{e}{}^{\kappa'}$, $\overset{\kappa'}{e}_{\lambda'}$. Their components with respect to (κ') are one or zero, and their components $\underset{\lambda'}{e}{}^{\kappa}$, $\overset{\kappa'}{e}_{\lambda}$ with respect to (κ) can have any other values.

The combination $v^\lambda w_\lambda$ is an invariant because of

$$v^{\lambda'} w_{\lambda'} = v^\kappa A^{\lambda'}_\kappa A^\mu_{\lambda'} w_\mu = v^\lambda w_\lambda. \tag{2.12}$$

This invariant is zero if and only if the direction of v^κ is contained in the $(n-1)$-direction of w_λ and 1 if v^κ fits between the E_{n-1}'s of w_λ and if the senses of v^κ and w_λ correspond. We call $v^\lambda w_\lambda$ the *transvection* of v^κ and w_λ.

If n linearly independent contravariant vectors $\underset{i}{v}{}^\kappa$ $(i = 1,\ldots, \mathrm{n})$ are given ($\underset{i}{v}$ is the kernel and i a dead index) a parallelepiped is determined by these vectors to within translations. The n pairs of parallel E_{n-1}'s of this parallelepiped, each having the sense of the only vector among the $\underset{i}{v}{}^\kappa$ whose direction is not contained in their $(n-1)$-direction, represent n covariant vectors $\overset{h}{w}_\lambda$ $(h = 1,\ldots, \mathrm{n})$ satisfying the conditions

$$\underset{i}{v}{}^\mu \overset{h}{w}_\mu = \delta^h_i \quad \left(\delta^h_i = \begin{cases} 1 \text{ for } h = i \\ 0 \text{ for } h \neq i \end{cases}\right).\dagger \tag{2.13}$$

The two systems $\underset{i}{v}{}^\kappa$ and $\overset{h}{w}_\lambda$ are called *reciprocal*. Hence the co- and contravariant measuring vectors belonging to the same coordinate system are reciprocal systems.

Co- and contravariant measuring vectors can be used to form the components of a given vector or to construct a vector from its given components. Take, for instance, a vector v^κ and form the n scalars

$$\overset{\kappa}{v} = v^\lambda \overset{\kappa}{e}_\lambda. \tag{2.14}$$

Obviously these n scalars are numerically equal to the v^κ. But with respect to the system (κ'), the equation (2.14) takes the form

$$\overset{\kappa}{v} = v^{\lambda'} \overset{\kappa}{e}_{\lambda'}. \tag{2.15}$$

The v^κ are transformed while the $\overset{\kappa}{v}$ remain invariant. This means that

† δ^h_i is the so-called *Kronecker symbol*. Its indices are dead ones and it would be more consistent to write $\underset{i}{\overset{h}{\delta}}$. But we prefer to avoid, as much as possible, the tower-like constructions arising if indices directly below and above a central letter are used at the same time.

we cannot write $v^\kappa = \overset{\kappa}{v}$ because the left-hand side is a vector and the right-hand side consists of n scalars. We therefore write

$$v^\kappa \overset{*}{=} \overset{\kappa}{v}. \tag{2.16}$$

The sign $\overset{*}{=}$ (read as 'equal cross') will always be used if *we wish to express the fact that an equation is valid with respect to the coordinate system or the systems used in the equation but is not necessarily valid with respect to all other coordinate systems* (see Note A, p. 272).

In (2.14) we get a dead index from a living one. This process occurs very often and will be called *strangling of an index.* When an index has been strangled *it then belongs to the kernel,* and the fact must be indicated either by putting the strangled index directly above or below the central letter or by putting a round bracket around it to show that the coordinate transformations of the underlying group no longer apply to it.

Conversely the vector v^κ can be constructed from the scalars $\overset{\kappa}{v}$ by using the contravariant measuring vectors:

$$v^\kappa = \overset{\lambda}{v} \underset{\lambda}{e^\kappa}, \tag{2.17}$$

or, with respect to (κ'),

$$v^{\kappa'} = \overset{\lambda}{v} \underset{\lambda}{e^{\kappa'}}. \tag{2.18}$$

This process is the inverse of the process of strangling. In the system (κ) the equations (2.14) and (2.17) mean the same thing numerically. In view of the restricted nature of the equality sign in (2.16), however, this is not true in any other system (κ'). The two equations have of course a different geometrical significance. (2.14) gives the construction of the components of a vector by means of the covariant measuring vectors and (2.17) gives the construction of a vector from its components by means of the contravariant measuring vectors.

Because of the numerical identity of v^κ and $\overset{\kappa}{v}$ we get from (2.15) and (2.18)

$$\begin{aligned} \overset{\kappa}{e}_{\lambda'} &\overset{*}{=} \underset{\lambda'}{e^\kappa} \overset{*}{=} A^\kappa_{\lambda'}, \\ \underset{\lambda}{e^{\kappa'}} &\overset{*}{=} \overset{\kappa'}{e}_\lambda \overset{*}{=} A^{\kappa'}_\lambda, \end{aligned} \tag{2.19}$$

and from this we get for the transformation of the measuring vectors of (κ) to those of (κ'):

$$\begin{aligned} \underset{\lambda'}{e^\kappa} &\overset{*}{=} A^\lambda_{\lambda'} \underset{\lambda}{e^\kappa}, \\ \overset{\kappa'}{e}_\lambda &\overset{*}{=} A^{\kappa'}_\kappa \overset{\kappa}{e}_\lambda. \end{aligned} \tag{2.20}$$

We give an example for $n = 3$. Suppose that

$$\begin{aligned} \underset{1'}{e}^{\kappa} &= \phantom{\underset{1}{e}^{\kappa}+{}} \underset{2}{e}^{\kappa} - 2\underset{3}{e}^{\kappa}, \\ \underset{2'}{e}^{\kappa} &= \underset{1}{e}^{\kappa} \phantom{{}+\underset{2}{e}^{\kappa}} + \underset{3}{e}^{\kappa}, \\ \underset{3'}{e}^{\kappa} &= \underset{1}{e}^{\kappa} + \underset{2}{e}^{\kappa}, \end{aligned} \tag{2.21}$$

as illustrated in Fig. 3.

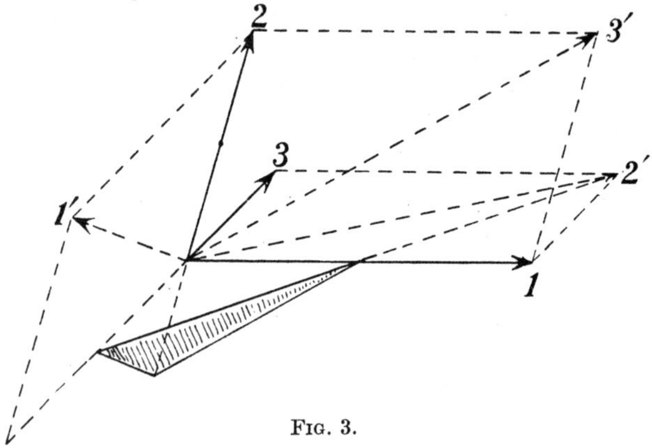

FIG. 3.

Then the matrix $A^{\kappa}_{\lambda'}$ is

$\lambda' \backslash \kappa$	1	2	3
1'	0	1	−2
2'	1	0	1
3'	1	1	0

$(\mathrm{Det}(A^{\kappa}_{\lambda'}) = -1).$ (2.22)

By the algebraic process described in I, § 1, we get the matrix of $A^{\kappa'}_{\lambda}$

$\lambda \backslash \kappa'$	1'	2'	3'
1	1	2	−1
2	−1	−2	2
3	−1	−1	1

(2.23)

and from this it follows that

$$\begin{aligned} \overset{1'}{e}_{\lambda} &= \overset{1}{e}_{\lambda} - \overset{2}{e}_{\lambda} - \overset{3}{e}_{\lambda}, \\ \overset{2'}{e}_{\lambda} &= 2\overset{1}{e}_{\lambda} - 2\overset{2}{e}_{\lambda} - \overset{3}{e}_{\lambda}, \\ \overset{3'}{e}_{\lambda} &= -\overset{1}{e}_{\lambda} + 2\overset{2}{e}_{\lambda} + \overset{3}{e}_{\lambda}. \end{aligned} \tag{2.24}$$

Now this can be verified in Fig. 3. For instance, the plane through the end-point of $\underset{2}{e}^{\kappa'}$ parallel to $\underset{1'}{e}^{\kappa}$ and $\underset{3'}{e}^{\kappa}$ has the equation

$$2x^1 - 2x^2 - x^3 = 1, \tag{2.25}$$

and this plane cuts off segments $\frac{1}{2}$, $-\frac{1}{2}$, -1 on the axes. Hence the components of $\overset{2'}{e}_\lambda$ with respect to (κ) are 2, -2, -1.

From the matrices (2.22) and (2.23) we get for the old measuring vectors in terms of the new ones

$$\begin{aligned} \underset{1}{e}^{\kappa} &= \underset{1'}{e}^{\kappa} + 2\underset{2'}{e}^{\kappa} - \underset{3'}{e}^{\kappa}, \\ \underset{2}{e}^{\kappa} &= -\underset{1'}{e}^{\kappa} - 2\underset{2'}{e}^{\kappa} + 2\underset{3'}{e}^{\kappa}, \\ \underset{3}{e}^{\kappa} &= -\underset{1'}{e}^{\kappa} - \underset{2'}{e}^{\kappa} + \underset{3'}{e}^{\kappa}, \end{aligned} \tag{2.26}$$

and

$$\begin{aligned} \overset{1}{e}_\lambda &= \overset{2'}{e}_\lambda + \overset{3'}{e}_\lambda, \\ \overset{2}{e}_\lambda &= \overset{1'}{e}_\lambda + \overset{3'}{e}_\lambda, \\ \overset{3}{e}_\lambda &= -2\overset{1'}{e}_\lambda + \overset{2'}{e}_\lambda. \end{aligned} \tag{2.27}$$

3. Affinors

A geometric quantity $P^{\kappa_1\ldots\kappa_p}{}_{\lambda_1\ldots\lambda_q}$ which has p upper indices and q lower indices and the law of transformation

$$P^{\kappa'_1\ldots\kappa'_p}{}_{\lambda'_1\ldots\lambda'_q} = A^{\kappa'_1\ldots\kappa'_p\,\lambda_1\ldots\lambda_q}_{\kappa_1\ldots\kappa_p\,\lambda'_1\ldots\lambda'_q}\, P^{\kappa_1\ldots\kappa_p}{}_{\cdot\ \ \cdot\ \lambda_1\ldots\lambda_q} \;\dagger\ddagger \tag{3.1}$$

is said to be an *affinor* of *contravariant valence* p, *covariant valence* q, and valence $p+q$.

If $q = 0$ the affinor is called *contravariant*, if $p = 0$ *covariant* and in the general case *mixed*; $p = 0$, $q = 0$ gives a scalar; $p = 1$, $q = 0$ a contravariant and $p = 0$, $q = 1$ a covariant vector. The place of the indices is important, so we must not write the same quantity sometimes as $P^{\kappa}{}_{\cdot\mu\lambda}$ and at other times as $P_{\mu}{}^{\kappa}{}_{\lambda}$.

If an affinor has more than one index there exist *intermediate components*, i.e. components belonging to two or more different coordinate systems, for example,

$$P^{\kappa'}{}_{\cdot\lambda} = A^{\kappa'}_{\kappa} P^{\kappa}{}_{\cdot\lambda} = A^{\lambda'}_{\lambda} P^{\kappa'}{}_{\cdot\lambda'}. \tag{3.2}$$

The indices of a quantity need not all belong to the same space. In

† $A^{\kappa'_1\ldots\kappa'_p\,\lambda_1\ldots\lambda_q}_{\kappa_1\ldots\kappa_p\,\lambda'_1\ldots\lambda'_q}$ stands for $A^{\kappa'_1}_{\kappa_1}\ldots A^{\kappa'_p}_{\kappa_p} A^{\lambda_1}_{\lambda'_1}\ldots A^{\lambda_q}_{\lambda'_q}$.

‡ Apart from a few exceptions co- and contravariant indices are never written in the same vertical line because this may easily lead to mistakes.

this case the quantity is called a *connecting quantity*. B_b^κ and C_λ^x in (I.3.3) and (I.3.1) are examples of such connecting quantities.

A mixed affinor with valence *2* represents a homogeneous linear transformation of contravariant vectors

$$'v^\kappa = P^\kappa_{.\lambda} v^\lambda. \tag{3.3}$$

If $\mathrm{Det}(P^\kappa_{.\lambda}) \neq 0$, the inverse transformation exists:

$$v^\kappa = \overset{-1}{P}{}^\kappa_{.\lambda}\, 'v^\lambda. \tag{3.4}$$

$\overset{-1}{P}{}^\kappa_{.\lambda}$ is called the *inverse* of $P^\kappa_{.\lambda}$ and can be derived from $P^\kappa_{.\lambda}$ in the same way as $A_{\lambda'}^\kappa$ from $A_\lambda^{\kappa'}$ in I, § 1. Hence, for instance, $\overset{-1}{P}{}^2_{.3}$ is equal to the minor of $P^3_{.2}$ in the matrix of the $P^\kappa_{.\lambda}$ divided by $\mathrm{Det}(P^\kappa_{.\lambda})$.

The affinor belonging to the identical transformation is called the *unity affinor* and is written A_λ^κ:

$$v^\kappa = A_\lambda^\kappa v^\lambda. \tag{3.5}$$

Obviously
$$A_\lambda^\kappa \overset{*}{=} \delta_\lambda^\kappa, \qquad A_\mu^\mu = n, \tag{3.6}$$

where the sign $\overset{*}{=}$ has to be used because the left-hand side is an affinor and the right-hand side is a system of n^2 scalars. Transforming κ in (3.5) we get

$$v^{\kappa'} = A_\kappa^{\kappa'} v^\kappa, \tag{3.7}$$

where the $A_\kappa^{\kappa'}$ are intermediate components of the unity affinor. This justifies the use of the same kernel A in (I.1.1) and (II.3.5).

Obviously
$$A_\lambda^\kappa \overset{*}{=} \overset{\kappa}{e}_\lambda \overset{*}{=} \underset{\lambda}{e}{}^\kappa, \qquad A_{\lambda'}^{\kappa'} \overset{*}{=} \overset{\kappa'}{e}_{\lambda'} \overset{*}{=} \underset{\lambda'}{e}{}^{\kappa'}. \tag{3.8}$$

Because these equations state that the n^2 components of an affinor with respect to a certain coordinate system happen to be equal to the n^2 components of n vectors with respect to this same system, the sign $\overset{*}{=}$ cannot be replaced by $=$.

For the affinors $P^\kappa_{.\lambda}$ and $\overset{-1}{P}{}^\kappa_{.\lambda}$ defined above, the relation

$$P^\kappa_{.\mu} \overset{-1}{P}{}^\mu_{.\lambda} = A_\lambda^\kappa \tag{3.9}$$

holds. In the same way inverses can be formed from a contra- or covariant affinor of valence *2* whose matrix has rank n, e.g.:

$$\begin{aligned} Q^{\kappa\mu} \overset{-1}{Q}_{\mu\lambda} &= A_\lambda^\kappa, \\ R_{\lambda\mu} \overset{-1}{R}{}^{\mu\kappa} &= A_\lambda^\kappa. \end{aligned} \tag{3.10}$$

4. Algebraic processes for quantities

Addition. Only affinors with the same valences can be added, e.g.:

$$R_{\mu\lambda}^{\cdot\cdot\kappa} = P_{\lambda\mu}^{\cdot\cdot\kappa} + Q_{\mu\cdot\lambda}^{\cdot\kappa}. \tag{4.1}$$

Multiplication. Affinors can always be multiplied. The valences of the product are the sums of the corresponding valences of the factors, e.g.:

$$S_{\mu\lambda\cdot\cdot\cdot\tau}^{\cdot\cdot\kappa\rho\sigma} = P_{\mu\lambda}^{\cdot\cdot\kappa} Q^{\rho\sigma} R_{\tau}. \tag{4.2}$$

This multiplication is not commutative, e.g.:

$$S^{\kappa\lambda} = v^{\kappa} w^{\lambda} \neq w^{\kappa} v^{\lambda} = T^{\kappa\lambda}. \tag{4.3}$$

Here $S^{\kappa\lambda}$ is a product of two vectors. In general an affinor is not a product of vectors. It is, however, always a sum of products of vectors, e.g.:

$$P^{\kappa\lambda}_{\cdot\cdot\mu} = A^{\kappa\lambda\tau}_{\rho\sigma\mu} P^{\rho\sigma}_{\cdot\cdot\tau} = \underset{\alpha}{e^{\kappa}}\underset{\beta}{e^{\lambda}}\overset{\gamma}{e}_{\mu}(\overset{\alpha}{e}_{\rho}\overset{\beta}{e}_{\sigma}\underset{\gamma}{e^{\tau}}P^{\rho\sigma}_{\cdot\cdot\tau}) = P^{(\alpha)(\beta)}_{\cdot\cdot(\gamma)}\underset{\alpha}{e^{\kappa}}\underset{\beta}{e^{\lambda}}\overset{\gamma}{e}_{\mu};$$

$$P^{(\alpha)(\beta)}_{\cdot\cdot(\gamma)} \overset{\text{def}}{=} P^{\rho\sigma}_{\cdot\cdot\tau}\overset{\alpha}{e}_{\rho}\overset{\beta}{e}_{\sigma}\underset{\gamma}{e^{\tau}} \quad (\alpha,\beta,\gamma = 1,\dots,n). \tag{4.4}$$

The affinor

$$T^{\kappa}_{\cdot\lambda} \overset{\text{def}}{=} \underset{1'}{e^{\kappa}}\overset{1}{e}_{\lambda} + \dots + \underset{n'}{e^{\kappa}}\overset{n}{e}_{\lambda}, \tag{4.5}$$

transforms the contravariant measuring vectors of (κ) into the contravariant measuring vectors of (κ'). Its intermediate components $T^{\kappa'}_{\cdot\lambda}$ are *0* or *1*:

$$T^{\kappa'}_{\cdot\lambda} = \delta^{\kappa'}_{\lambda}.\dagger \tag{4.6}$$

Transvection. The combination

$$R^{\kappa\rho} = P_{\lambda\mu}^{\cdot\cdot\kappa} Q^{\lambda\rho\mu} \tag{4.7}$$

is an affinor and called the *transvection* of $P_{\lambda\mu}^{\cdot\cdot\kappa}$ and $Q^{\lambda\rho\mu}$ over λ and μ. Transvections can be made up in different ways, but every summation must be effected with respect to an upper and a lower index. For instance, the expression $\sum\limits_{\lambda} v_{\lambda} P^{\kappa}_{\cdot\lambda}$ is not invariant and accordingly is not a vector.

Contraction. If the summation is effected with respect to indices belonging to the same affinor we get a *contraction*, e.g.:

$$Q^{\kappa} = P_{\lambda}^{\cdot\kappa\lambda}.\ddagger$$

The indices used in the summations in transvections and contractions are called *saturated*, the others *free indices*.§

† $\delta^{\kappa'}_{\lambda}$ is the generalized Kronecker symbol. $\delta^{\kappa'}_{\lambda} = 1$ if κ' and λ take corresponding values from the rows *1'*,..., n' and *1*,..., n and $= 0$ in all other cases.

‡ In the classical theory of invariants a process of this kind was originally called 'Faltung' (folding up). Physicists later introduced the name 'Verjüngung' (rejuvenation).

§ Eisenhart uses the term 'dummy indices' for saturated indices.

Building an isomer. An isomer is formed by interchanging the upper indices or the lower indices, or the upper and also the lower indices, e.g.:

$$Q^{\kappa\lambda\cdot\nu}_{\cdot\cdot\mu} = P^{\kappa\cdot\nu\lambda}_{\cdot\mu}. \tag{4.8}$$

Mixing. The process of mixing is always applied to a number of upper or lower indices. In order to mix over p indices we form $p!$ isomers by permuting these indices in all possible ways and take the sum of these isomers divided by $p!$. The mixing is always denoted by a pair of round brackets ( ). Indices to which the mixing process is not to be applied can be isolated by drawing vertical bars on each side of them. For example:

$$P_{\lambda}^{\cdot(\kappa\rho|\mu|\sigma)} = \tfrac{1}{6}(P_{\lambda}^{\cdot\kappa\rho\mu\sigma} + P_{\lambda}^{\cdot\rho\sigma\mu\kappa} + P_{\lambda}^{\cdot\sigma\kappa\mu\rho} + P_{\lambda}^{\cdot\rho\kappa\mu\sigma} + P_{\lambda}^{\cdot\sigma\rho\mu\kappa} + P_{\lambda}^{\cdot\kappa\sigma\mu\rho}). \tag{4.9}$$

If an affinor is invariant for mixing over p indices it is called *symmetrical* with respect to these indices. Then it is invariant for every interchanging of these indices. For instance

$$P_{\lambda}^{\cdot\kappa\mu\nu} = P_{\lambda}^{\cdot(\kappa\mu\nu)} = P_{\lambda}^{\cdot\mu\kappa\nu} = P_{\lambda}^{\cdot\nu\mu\kappa}, \text{ etc.} \tag{4.10}$$

Alternation. The process of alternation is effected in the same way as the process of mixing, but the $p!$ isomers are taken with the positive sign if the permutation is even and with the negative sign if it is odd. The alternation is denoted by a pair of square brackets [ ]. Indices to which the process of alternation is not to be applied can be isolated by vertical bars. For example:

$$P_{\lambda}^{\cdot[\kappa\rho|\mu|\sigma]} = \tfrac{1}{6}(P_{\lambda}^{\cdot\kappa\rho\mu\sigma} + P_{\lambda}^{\cdot\rho\sigma\mu\kappa} + P_{\lambda}^{\cdot\sigma\kappa\mu\rho} - P_{\lambda}^{\cdot\rho\kappa\mu\sigma} - P_{\lambda}^{\cdot\sigma\rho\mu\kappa} - P_{\lambda}^{\cdot\kappa\sigma\mu\rho}). \tag{4.11}$$

Alternating over more than n indices always leads to zero. If an affinor is invariant for alternating over p indices it is called *alternating* with respect to these indices. Then if two of these indices are interchanged it only changes sign. For instance,

$$P_{\lambda}^{\cdot\kappa\mu\nu} = P_{\lambda}^{\cdot[\kappa\mu\nu]} = -P_{\lambda}^{\cdot\mu\kappa\nu} = +P_{\lambda}^{\cdot\mu\nu\kappa} = -P_{\lambda}^{\cdot\kappa\nu\mu}. \tag{4.12}$$

The processes mentioned so far in this section are invariant with allowable coordinate transformations.

Strangling; rank with respect to certain indices. Transvection of an affinor with the measuring vectors leads to strangling of one or more indices. So we may obtain from $P^{\kappa\lambda}_{\cdot\cdot\mu\nu}$, for instance, a system of n^2 affinors of valence 2:

$$\overset{\kappa}{\underset{\nu}{P}}{}^{\lambda}_{\cdot\mu} = P^{(\kappa)\lambda}_{\cdot\cdot\mu(\nu)} \overset{\text{def}}{=} \overset{\kappa}{e}_{\rho}\, \underset{\nu}{e}^{\sigma} P^{\rho\lambda}_{\cdot\cdot\mu\sigma}. \tag{4.13}$$

According to § 2 the strangled indices are put here above or below the central letter or are marked in their own places by round brackets. The latter is more convenient if a large number of indices, or both upper and lower indices are strangled.

If the same indices are strangled on both sides of an invariant equation, the equation remains invariant. But if the strangled indices are not the same on both sides, the sign $\overset{*}{=}$ has to be introduced. For example:

$$A_\lambda^\kappa \overset{*}{=} A_\lambda^{(\kappa)} = \overset{\kappa}{e}_\lambda \overset{*}{=} A_{(\lambda)}^{(\kappa)} = \delta_\lambda^\kappa \overset{*}{=} A_{(\lambda)}^{\kappa} = \underset{\lambda}{e}^\kappa. \tag{4.14}$$

The *rank* of an affinor *with respect to certain indices* is the number of linearly independent quantities that are obtained if all *other* indices are strangled. The indices concerned need not be all upper or all lower indices nor need they belong to the same system of coordinates. *The rank is invariant for all allowable coordinate transformations* (see Note B, p. 272). From the definition it follows immediately that the rank with respect to some indices is the same as the rank with respect to all other indices, and that the rank with respect to all indices is *1*. In fact, suppose that the rank of an affinor of valence $s+t$ with respect to s indices is r. Then this affinor can be written as the sum of r linearly independent affinors of valence s each multiplied by an affinor of valence t. But from this it follows that the rank r' with respect to the t other indices can never be greater than r. The same reasoning holds if we start from these t indices and the rank r'. Hence $r = r'$.

An affinor with valence 2 has only one non-trivial rank and this is equal to the rank of its matrix.

If r is the rank of an affinor with respect to an index μ, by strangling all other indices exactly r linearly independent co- or contravariant vectors are obtained. The domain of these vectors (cf. II, § 2) is called the *μ-domain of the affinor.*

As an example we consider an E_p and an E_q $(p+q = n)$ through the origin, which have no direction in common. We wish to determine an affinor B_λ^κ of rank p which has E_p as support of the κ-domain and E_q as support of the λ-domain (cf. II, § 2) and which satisfies the condition

$$B_\mu^\kappa B_\lambda^\mu = B_\lambda^\kappa. \tag{4.15}$$

Let the coordinate system be taken in such a way that $\underset{1}{e}^\kappa,\dots, \underset{p}{e}^\kappa$ span E_p and that the other contravariant measuring vectors span E_q. Then E_q is also spanned by $\overset{1}{e}_\lambda,\dots, \overset{p}{e}_\lambda$ and E_p by the other covariant measuring vectors. From this we see that B_λ^κ must have the form

$$B_\lambda^\kappa = w_{(\beta)}^{(\alpha)} \overset{\beta}{e}_\lambda \underset{\alpha}{e}^\kappa \quad (\alpha, \beta = 1,..., p). \tag{4.16}$$

The matrix of the unknown coefficients $w_{(\beta)}^{(\alpha)}$ cannot have a rank less than p because the rank of B_λ^κ must be p. From (4.15) we get

$$w_{(\beta)}^{(\alpha)} w_{(\gamma)}^{(\beta)} = w_{(\gamma)}^{(\alpha)} \quad (\alpha, \beta, \gamma = 1,..., p). \tag{4.17}$$

Now this is only possible if $w_{(\beta)}^{(\alpha)} = \delta_\beta^\alpha$; hence

$$B_\lambda^\kappa = \overset{\gamma}{e}_\lambda \underset{\gamma}{e}^\kappa. \tag{4.18}$$

In the same way

$$C_\lambda^\kappa = \overset{\eta}{e}_\lambda \underset{\eta}{e}^\kappa \quad (\eta = p+1,..., n) \tag{4.19}$$

is the only affinor that has E_q as support of the κ-domain, E_p as support of the λ-domain, and that satisfies the condition

$$C_\mu^\kappa C_\lambda^\mu = C_\lambda^\kappa. \tag{4.20}$$

From (4.18, 19) it follows that

$$\begin{gathered} B_\lambda^\kappa + C_\lambda^\kappa = A_\lambda^\kappa, \qquad B_\mu^\mu = p, \qquad C_\mu^\mu = n-p; \\ B_\mu^\kappa C_\lambda^\mu = 0, \qquad B_\lambda^\mu C_\mu^\kappa = 0. \end{gathered} \tag{4.21}$$

The affinors B_λ^κ and C_λ^κ split up a contravariant vector v^κ into two parts in E_p and in E_q and a covariant vector w_λ into two parts through E_p and E_q:

$$\begin{aligned} v^\kappa &= B_\lambda^\kappa v^\lambda + C_\lambda^\kappa v^\lambda, \\ w_\lambda &= B_\lambda^\kappa w_\kappa + C_\lambda^\kappa w_\kappa. \end{aligned} \tag{4.22}$$

5. Tensors

A co- or contravariant affinor that is symmetrical in *all* indices is called a *tensor*:†

$$w_{\lambda_1 \ldots \lambda_p} = w_{(\lambda_1 \ldots \lambda_p)}. \tag{5.1}$$

Among the n^p components of a tensor there are exactly $\binom{n+p-1}{p}$ independent ones. In a centred E_n the equation

$$w_{\lambda_1 \ldots \lambda_p} x^{\lambda_1} \ldots x^{\lambda_p} = 1\,(0) \tag{5.2}$$

represents a hypersurface (cone) of *degree* p. Conversely the tensor $w_{\lambda_1 \ldots \lambda_p}$ is uniquely determined by its hypersurface and to within a scalar factor by its cone. In the same way a contravariant tensor $v^{\kappa_1 \ldots \kappa_p}$ in

† Many authors use the word 'tensor' for all affinors and introduce the term 'symmetric tensor' for what we call a tensor.

a centred E_n is connected with a hypersurface (cone) of *class* p with the equations

$$v^{\kappa_1\ldots\kappa_p}u_{\kappa_1}\ldots u_{\kappa_p} = 1\ (0) \tag{5.3}$$

in the E_{n-1}-coordinates u_λ If the tensor can be written as the mixed product of p vectors the cone degenerates into p hyperplanes or p straight lines through O.

The rank of a tensor is the same with respect to all indices. The support of the domain of a covariant tensor $w_{\lambda_1\ldots\lambda_p}$ of rank r is the E_{n-r} with the equations

$$w_{\lambda_1\ldots\lambda_p}x^{\lambda_p} = 0. \tag{5.4}$$

The hypersurface of $w_{\lambda_1\ldots\lambda_p}$ is a cylinder consisting of ∞^{r-1} parallel E_{n-r}'s. Its cone consists of ∞^{r-2} E_{n-r+1}'s, all of which contain the E_{n-r} (5.4).

6. Multivectors

A co- or contravariant affinor of valence p which is alternating in *all* indices is called a *p-vector*†

$$v^{\kappa_1\ldots\kappa_p} = v^{[\kappa_1\ldots\kappa_p]}. \tag{6.1}$$

p must be $\leqslant n$. Among the n^p components of a p-vector there are exactly $\binom{n}{p}$ independent ones. All components of a p-vector which have two equal indices are zero. An n-vector has only one independent component, for instance, $v^{1\ldots n}$. But since $v^{1\ldots n}$ is not invariant, an n-vector is by no means the same as a scalar.

If a p-vector can be written as the alternated product of p vectors

$$v^{\kappa_1\ldots\kappa_p} = p!\, \underset{1}{v}{}^{[\kappa_1}\ldots\underset{p}{v}{}^{\kappa_p]} = \begin{vmatrix} \underset{1}{v}{}^{\kappa_1} & \ldots & \underset{1}{v}{}^{\kappa_p} \\ \vdots & & \vdots \\ \underset{p}{v}{}^{\kappa_1} & \ldots & \underset{p}{v}{}^{\kappa_p} \end{vmatrix}, \tag{6.2}$$

it is called *simple*, and every factor of the product a *divisor* of $v^{\kappa_1\ldots\kappa_p}$.

We prove that a vector v^κ is a divisor of $v^{\kappa_1\ldots\kappa_p}$ if and only if

$$v^{[\kappa_1\ldots\kappa_p}v^{\kappa]} = 0. \tag{6.3}$$

The necessity of the condition is obvious. To prove the sufficiency let us take any vector w_λ satisfying the equation

$$w_\lambda v^\lambda = 1. \tag{6.4}$$

Then

$$\begin{aligned} 0 = (p+1)v^{[\kappa_1\ldots\kappa_p}v^{\kappa]}w_\kappa &= v^{\kappa_1\ldots\kappa_p}v^\kappa w_\kappa - v^{\kappa_1\ldots\kappa_{p-1}\kappa}v^{\kappa_p}w_\kappa - \ldots - v^{\kappa\kappa_2\ldots\kappa_p}v^{\kappa_1}w_\kappa \\ &= v^{\kappa_1\ldots\kappa_p} - pw_\kappa v^{\kappa[\kappa_2\ldots\kappa_p}v^{\kappa_1]}, \end{aligned} \tag{6.5}$$

and this proves the proposition.

† Also *multivector*, if the valence is not mentioned. Some authors use the term 'anti-symmetric tensor'.

A p-vector $v^{\kappa_1\ldots\kappa_p}$ is simple if and only if

$$v^{[\kappa_1\ldots\kappa_p}v^{\lambda_1]\ldots\lambda_p} = 0. \tag{6.6}$$

The necessity of the condition is obvious. To prove the sufficiency we remark that we can always interchange the indices so that $v^{1\ldots p} \neq 0$. It then follows from (6.6) that the p separate vectors $v^{\kappa 2\ldots p}$, $v^{1\kappa 3\ldots p}$,..., $v^{1\ldots(p-1)\kappa}$, obtained by keeping all the indices fixed except one, are all divisors of $v^{\kappa_1\ldots\kappa_p}$. Moreover, these vectors must be linearly independent because the first component of the first vector is $v^{1\ldots p} \neq 0$ and the first components of all other vectors are zero, and so on. Hence $v^{\kappa_1\ldots\kappa_p}$ has p linearly independent divisors and is therefore simple.

A less stringent necessary and sufficient condition

$$v^{[\kappa_1\ldots\kappa_p}v^{\lambda_1\lambda_2]\ldots\lambda_p} = 0 \tag{6.7}$$

was proved by Weitzenböck.† Givens‡ proved the theorem, already announced by Weitzenböck§ that for s odd we have

$$v^{[\kappa_1\ldots\kappa_p}v^{\lambda_1\ldots\lambda_s]\ldots\lambda_p} = 0 \tag{6.8}$$

if and only if

$$v^{[\kappa_1\ldots\kappa_p}v^{\lambda_1\ldots\lambda_{s+1}]\ldots\lambda_p} = 0 \quad \text{(cf. Ex. II.9).} \tag{6.9}$$

Now take a simple contravariant 2-vector $v^{\kappa\lambda}$ in E_3

$$v^{\kappa\lambda} = 2\underset{1}{v}^{[\kappa}\underset{2}{v}^{\lambda]}. \tag{6.10}$$

Here

$$|\underset{1}{v}^1\underset{2}{v}^2 - \underset{1}{v}^2\underset{2}{v}^1| \tag{6.11}$$

is the area of the projection of the parallelogram of $\underset{1}{v}^\kappa$ and $\underset{2}{v}^\kappa$ on the 12-plane, measured by the area of the parallelogram of $\underset{1}{e}^\kappa$, $\underset{2}{e}^\kappa$. The sign of $\underset{1}{v}^1\underset{2}{v}^2 - \underset{1}{v}^2\underset{2}{v}^1$ is positive or negative according as the sense from $\underset{1}{v}^\kappa$ to $\underset{2}{v}^\kappa$ corresponds to the sense from 1 to 2 in the 12-plane or not. Hence $v^{\kappa\lambda}$ is represented geometrically by a part of a plane with a definite area (but an arbitrary form) with an *inner orientation*, the whole figure being fixed only to within translations (Fig. 4).

The geometrical representation of a simple contravariant p-vector $v^{\kappa_1\ldots\kappa_p}$ in E_n can be obtained in the same way. We take an arbitrary E_p which has the p-direction of $v^{\kappa_1\ldots\kappa_p}$ and in this E_p we take a p-dimensional region whose projection in the $(n-p)$-direction of $\underset{p+1}{e}^{[\kappa_{p+1}}\ldots\underset{n}{e}^{\kappa_n]}$ on the E_p

† 1908. 1 for $p = 3$, $n = 5$; 1923. 3, p. 87, for the general case.
‡ 1937. 1, p. 364.
§ 1923. 3, p. 87.

of $\underset{1}{e}^{\kappa}, \ldots, \underset{p}{e}^{\kappa}$ measured by the parallelotope† of these latter p vectors has exactly the volume $|v^{1 \ldots p}|$. If this region has an inner orientation corresponding to the screw-sense of $\underset{1}{e}^{\kappa}, \ldots, \underset{p}{e}^{\kappa}$ in this order for $v^{1 \ldots p} > 0$ and in the opposite order for $v^{1 \ldots p} < 0$, we obtain the geometrical representation of $v^{\kappa_1 \ldots \kappa_p}$. If $v^{1 \ldots p} = 0$, another non-vanishing component can be

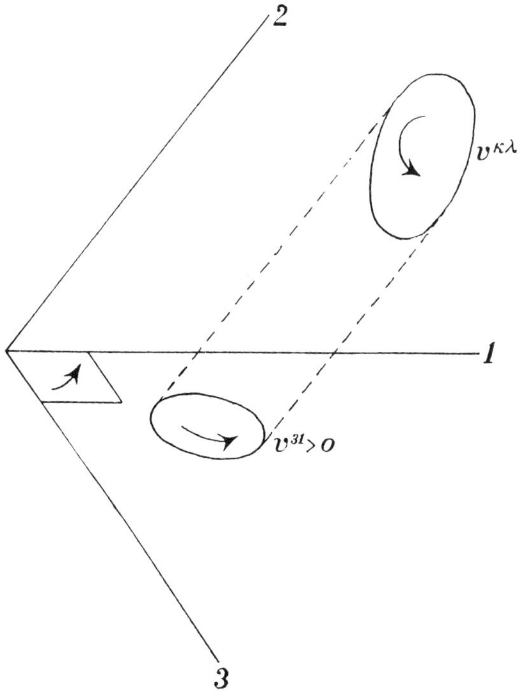

FIG. 4.

chosen from the $v^{\kappa_1 \ldots \kappa_p}$. Obviously the result is independent of the set of p measuring vectors chosen. The whole figure is fixed to within translations. We see that every simple contravariant p-vector provides its p-direction with an *inner* orientation.

If we take a simple covariant *2*-vector $w_{\lambda\kappa}$ in E_3

$$w_{\lambda\kappa} = \overset{1}{w}_{[\lambda} \overset{2}{w}_{\kappa]}, \tag{6.12}$$

the expression

$$|\overset{1}{w}_1 \overset{2}{w}_2 - \overset{1}{w}_2 \overset{2}{w}_1| \tag{6.13}$$

is the reciprocal of the area of the section of the tube of $\overset{1}{w}_{\lambda}$ and $\overset{2}{w}_{\lambda}$ with the *12*-plane, measured by the area of the parallelogram of $\underset{1}{e}^{\kappa}, \underset{2}{e}^{\kappa}$. The

† We use the term parallelotope instead of parallelepiped for dimensions > 3.

sign of $\overset{1}{w}_1 \overset{2}{w}_2 - \overset{1}{w}_2 \overset{2}{w}_1$ is positive if the sense fixed around the tube by the order of $\overset{1}{w}_\lambda$ and $\overset{2}{w}_\lambda$ corresponds to the sense from 1 to 2 in the 12-plane. Hence $w_{\lambda\kappa}$ can be represented geometrically by a cylinder with a definite direction, a definite section (but an arbitrary form), and an *outer* orientation, the whole figure being fixed to within translations (Fig. 5).

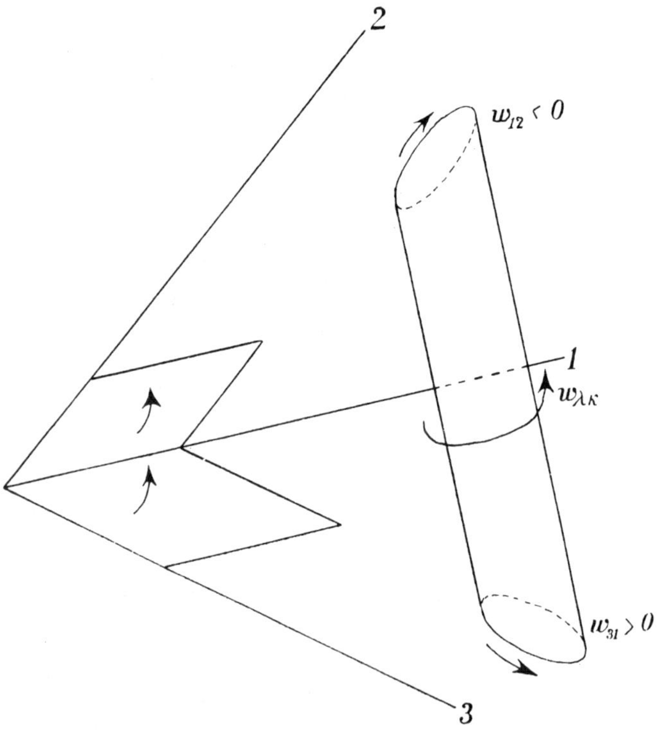

FIG. 5.

The geometrical representation of a simple covariant p-vector $w_{\lambda_1 \dots \lambda_p}$ in E_n can be obtained in the same way. We take an arbitrary n-dimensional cylinder, whose $(n-1)$-dimensional boundary consists of ∞^{p-1} E_{n-p}'s having the $(n-p)$-direction of $w_{\lambda_1 \dots \lambda_p}$ and whose section with the E_p of $\underset{1}{e}^\kappa, \dots, \underset{p}{e}^\kappa$ has the volume $1 : |w_{1 \dots p}|$ if measured with the parallelotope of these p vectors. If this cylinder has an outer orientation corresponding to the screw-sense of $\underset{1}{e}^\kappa, \dots, \underset{p}{e}^\kappa$ in this order for $w_{1 \dots p} > 0$ and in the opposite order for $w_{1 \dots p} < 0$, we obtain the geometric representation of

$w_{\lambda_1 \ldots \lambda_p}$. If $w_{1 \ldots p} = 0$ another set of p measuring vectors can be chosen from the $\underset{\lambda}{e^\kappa}$. Obviously the result is independent of the set of p measuring vectors chosen. The whole figure is fixed to within translations. We see that every simple covariant p-vector provides its $(n-p)$-direction with an *outer* orientation.

If a p-vector is not simple it is always the sum of a finite number of simple p-vectors because it is always possible to write the p-vector as a sum of simple p-vectors in the coordinate-E_p's or through the coordinate-E_{n-p}'s (cf. I, § 3) of an arbitrary coordinate system. We call this a *decomposition* of a p-vector *into blades*. The smallest number of blades into which a given p-vector can be decomposed is an invariant of the p-vector. For $2 < p < n-2$ not much is known about this invariant. But for $p = 2$ and $p = n-2$ a decomposition into a minimum number of blades can be obtained easily.

We prove this for $n = 4$ (cf. also R.C. I § 8). Let $F^{\kappa\lambda}$ be a bivector and

$$F^{[\kappa\lambda} F^{\mu]\nu} \neq 0. \tag{6.14}$$

From this it follows that

$$F^{12} F^{34} + F^{23} F^{14} + F^{31} F^{24} \neq 0. \tag{6.15}$$

Hence (cf. (6.7))

$$F^{[\kappa\lambda} F^{\mu\nu]} \neq 0. \tag{6.16}$$

Now take an arbitrary *simple* bivector $\underset{1}{F^{\kappa\lambda}}$ satisfying the condition

$$\underset{1}{F^{[\kappa\lambda}} F^{\mu\nu]} \neq 0. \tag{6.17}$$

Then we try to determine a simple bivector $\underset{2}{F^{\kappa\lambda}}$ such that

$$F^{\kappa\lambda} = \alpha \underset{1}{F^{\kappa\lambda}} + \underset{2}{F^{\kappa\lambda}}. \tag{6.18}$$

Since $\underset{2}{F^{\kappa\lambda}}$ must be simple we have the necessary and sufficient condition

$$F^{[12} F^{34]} - 2\alpha \underset{1}{F^{[12}} F^{34]} = 0, \tag{6.19}$$

and from this equation α can be solved uniquely. Hence, when $\underset{1}{F^{\kappa\lambda}}$ is chosen, $\underset{2}{F^{\kappa\lambda}}$ is uniquely determined. It can be proved in the same way that every bivector with rank 2ρ can be decomposed into ρ, and not less than ρ, simple bivectors. The decomposition is not uniquely determined. The decomposition of $(n-2)$-vectors can be reduced to the decomposition of bivectors as will be shown in § 7.

The rank of a p-vector is the same with respect to all its indices. Obviously the only p-vectors with rank p are the simple p-vectors.

7. n-vectors

An n-vector is represented by a volume of the E_n provided with an n-dimensional screw-sense. Its component $v^{1...n}$ ($w_{1...n}$) is this volume (the reciprocal volume), measured by the parallelotope of the measuring vectors and provided with a positive (negative) sign if the screw-sense is the same as (opposite to) the *1, 2,..., n*-sense. From this we see that two arbitrary co-(contra-)variant n-vectors differ only by a scalar factor.

Because

$$\begin{aligned} v^{1'...n'} &= A^{1'...n'}_{\kappa_1...\kappa_n} v^{\kappa_1...\kappa_n} \\ &= n!\, A^{1'...n'}_{[1...n]} v^{1...n} \\ &= \Delta v^{1...n}, \end{aligned} \tag{7.1}$$

the component $v^{1...n}$ is not a scalar. In the same way we can prove that

$$w_{1'...n'} = \Delta^{-1} w_{1...n}. \tag{7.2}$$

To every coordinate system there belong, for every value of $p \leqslant n$, $\binom{n}{p}$ contravariant and $\binom{n}{p}$ covariant p-vectors, for instance, $p!\, \underset{1}{e}^{[\kappa_1} \ldots \underset{p}{e}^{\kappa_p]}$.

The two n-vectors belonging to (κ) are written

$$\begin{aligned} \underset{(\kappa)}{E}{}^{\kappa_1...\kappa_n} &\overset{\text{def}}{=} n!\, \underset{1}{e}^{[\kappa_1} \ldots \underset{n}{e}^{\kappa_n]}, \\ \overset{(\kappa)}{e}_{\lambda_1...\lambda_n} &\overset{\text{def}}{=} n!\, \overset{1}{e}_{[\lambda_1} \ldots \overset{n}{e}_{\lambda_n]}. \end{aligned} \tag{7.3}$$

Obviously

$$\underset{(\kappa)}{E}{}^{1...n} = 1, \qquad \overset{(\kappa)}{e}_{1...n} = 1. \tag{7.4}$$

Both n-vectors are represented by the parallelotope of the measuring vectors with the *1, 2,..., n*-sense.

By means of these n-vectors a one-to-one correspondence can be established between contravariant p-vectors and covariant $(n-p)$-vectors. If we define

$$w_{\lambda_1...\lambda_p} \overset{\text{def}}{=} \frac{1}{(n-p)!} \overset{(\kappa)}{e}_{\lambda_1...\lambda_p\kappa_{p+1}...\kappa_n} v^{\kappa_{p+1}...\kappa_n}, \tag{7.5}$$

it is easily proved that

$$v^{\kappa_{p+1}...\kappa_n} = \frac{1}{p!} w_{\lambda_1...\lambda_p} \underset{(\kappa)}{E}{}^{\lambda_1...\lambda_p\kappa_{p+1}...\kappa_n}. \tag{7.6}$$

Both equations express the fact that

$$w_{\mu_1...\mu_p} \overset{*}{=} v^{\mu_{p+1}...\mu_n} \tag{7.7}$$

if $\mu_1...\mu_n$ is an even permutation of *1...n*. But this correspondence is not invariant because it depends on the choice of the system (κ). If (κ)

is changed $w_{\lambda_1 \dots \lambda_p}$ can acquire a scalar factor. The correspondence can be used to prove that an $(n-2)$-vector can always be decomposed into at most $\frac{1}{2}n$ blades. To do this we have only to decompose the corresponding bivector.

8. Densities†

As we have seen, the difference between $v^{1 \dots n}$ and a scalar is that in the transformation formula of $v^{1 \dots n}$ a factor Δ occurs. We define a *scalar-Δ-density of weight* $\mathfrak{k}$ as a quantity with one component $\tilde{\mathfrak{p}}$ and the transformation

$$\tilde{\mathfrak{p}}[\kappa'] = \Delta^{-\mathfrak{k}} \tilde{\mathfrak{p}}[\kappa] \quad \text{or} \quad \overset{(\kappa')}{\tilde{\mathfrak{p}}} = \Delta^{-\mathfrak{k}} \overset{(\kappa)}{\tilde{\mathfrak{p}}}.\ddagger \tag{8.1}$$

Then a contravariant n-vector can be considered as a scalar-Δ-density of weight -1 and a covariant n-vector as a scalar-Δ-density of weight $+1$. This is not a new geometrical conception, only a new notation enabling us to get rid of a lot of indices.

Since an n-vector has a screw-sense, every scalar-Δ-density with an odd positive or negative weight also has a screw-sense. This is the screw-sense of some coordinate system with respect to which the component is positive.

Any scalar, tensor, multivector, or affinor can be multiplied by a scalar-Δ-density of weight $\mathfrak{k}$. Then we get a *scalar-*, *tensor-*, *multivector-*, or *affinor-Δ-density of weight* $\mathfrak{k}$. Hence an affinor-Δ-density of weight $\mathfrak{k}$, contravariant valence p, and covariant valence q has ∞^{p+q} components $\tilde{\mathfrak{P}}^{\kappa_1 \dots \kappa_p}{}_{\lambda_1 \dots \lambda_q}$ and the transformation

$$\tilde{\mathfrak{P}}^{\kappa'_1 \dots \kappa'_p}{}_{\lambda'_1 \dots \lambda'_q} = \Delta^{-\mathfrak{k}} A^{\kappa'_1 \dots \kappa'_p \lambda_1 \dots \lambda_q}_{\kappa_1 \dots \kappa_p \lambda'_1 \dots \lambda'_q} \tilde{\mathfrak{P}}^{\kappa_1 \dots \kappa_p}{}_{\lambda_1 \dots \lambda_q}. \tag{8.2}$$

We shall generally use a gothic letter with a sign $\sim$ over it for the kernel of these densities.

All that has been said about indices, addition, isomers, multiplication, contraction, transvection, mixing, alternating, rank, domain, and support holds *mutatis mutandis* for affinor-Δ-densities (and also for affinor densities to be defined hereafter). Addition is only possible if the quantities have the same valences and the same weight.

The equations

$$\tilde{\mathfrak{E}}^{1 \dots n} = 1, \qquad \tilde{\mathfrak{e}}_{1 \dots n} = 1, \tag{8.3}$$

valid with respect to *every* coordinate system, define two n-vector-Δ-densities of weight $+1$ and -1 respectively. In fact,

$$\tilde{\mathfrak{E}}^{1' \dots n'} = \Delta^{-1} A^{1' \dots n'}_{\kappa_1 \dots \kappa_n} \tilde{\mathfrak{E}}^{\kappa_1 \dots \kappa_n} = n!\, \Delta^{-1} A^{1' \dots n'}_{[1 \dots n]} \tilde{\mathfrak{E}}^{1 \dots n} = \tilde{\mathfrak{E}}^{1 \dots n} = 1. \tag{8.4}$$

† Cf. also for literature Schouten 1938. 2; Schouten and v. Dantzig 1940. 1; Dorgelo and Schouten 1946. 1; R.C. 1954. 1, I § 2.

‡ Since a scalar density has no index, the coordinate system used has to be indicated in some other way.

These n-vector-Δ-densities are independent of the choice of the coordinate system. They can be used to establish an invariant one-to-one correspondence between contra-(co-)variant p-vectors and co-(contra-)variant $(n-p)$-vector-Δ-densities of weight -1 $(+1)$ by means of the formulae

$$\begin{aligned}
&(a) \quad \tilde{\mathfrak{v}}_{\lambda_1\ldots\lambda_p} = \frac{(-1)^{\alpha_p}}{(n-p)!}\,\tilde{\mathfrak{e}}_{\lambda_1\ldots\lambda_p\kappa_{p+1}\ldots\kappa_n}\,v^{\kappa_{p+1}\ldots\kappa_n},\\
&(b) \quad v^{\kappa_{p+1}\ldots\kappa_n} = \frac{(-1)^{-\alpha_p}}{p!}\,\tilde{\mathfrak{v}}_{\lambda_1\ldots\lambda_p}\,\tilde{\mathfrak{E}}^{\lambda_1\ldots\lambda_p\kappa_{p+1}\ldots\kappa_n},\\
&(c) \quad \tilde{\mathfrak{w}}^{\kappa_{p+1}\ldots\kappa_n} = \frac{(-1)^{\beta_p}}{p!}\,w_{\lambda_1\ldots\lambda_p}\,\tilde{\mathfrak{E}}^{\lambda_1\ldots\lambda_p\kappa_{p+1}\ldots\kappa_n},\\
&(d) \quad w_{\lambda_1\ldots\lambda_p} = \frac{(-1)^{-\beta_p}}{(n-p)!}\,\tilde{\mathfrak{e}}_{\lambda_1\ldots\lambda_p\kappa_{p+1}\ldots\kappa_n}\,\tilde{\mathfrak{w}}^{\kappa_{p+1}\ldots\kappa_n},
\end{aligned} \tag{8.5}$$

or, according to the values of the components of $\tilde{\mathfrak{E}}^{\kappa_1\ldots\kappa_n}$ and $\tilde{\mathfrak{e}}_{\lambda_1\ldots\lambda_n}$,

$$\begin{aligned}
\tilde{\mathfrak{v}}_{\mu_1\ldots\mu_p} &= (-1)^{\alpha_p} v^{\mu_{p+1}\ldots\mu_n},\\
\tilde{\mathfrak{w}}^{\mu_{p+1}\ldots\mu_n} &= (-1)^{\beta_p} w_{\mu_1\ldots\mu_p},
\end{aligned} \tag{8.6}$$

where $\mu_1\ldots\mu_n$ is an even permutation of $1\ldots n$. The coefficients α_p and β_p can be chosen arbitrarily. The choice will be postponed so that it can be made to agree with the identifications of quantities, made possible after introduction of a sub-group of G_a. As we shall see in Chapter III, it is only possible to obtain these identifications in the right way by choosing suitable values for α_p and β_p. But if no sub-group of G_a is introduced we may take $\alpha_p = \beta_p = 0$ for all values of p.

(8.6) shows that the components of a contra-(co-)variant vector may also be considered as components of a co-(contra-)variant $(n-p)$-vector-Δ-density of weight -1 $(+1)$. Hence the geometrical meanings of corresponding quantities do not differ. There is only a difference in notation. $\tilde{\mathfrak{E}}^{\kappa_1\ldots\kappa_n}$ and $\tilde{\mathfrak{e}}_{\lambda_1\ldots\lambda_n}$ correspond to the scalar $+1$. The use of Δ-densities is sometimes convenient; for instance, if a trivector $v^{\kappa\lambda\mu}$ in E_4 is written $\tilde{\mathfrak{v}}_\lambda$, the formulae contain less indices.

We can also prove in a direct way that the $v^{\kappa_1\ldots\kappa_p}$ transform like components of a covariant $(n-p)$-vector-Δ-density of weight -1. Take for example v^{23}, v^{31}, and v^{12} in an E_3. We know that

$$v^{2'3'} = A^{2'}_{\kappa} A^{3'}_{\lambda} v^{\kappa\lambda} = (A^{2'}_{2} A^{3'}_{3} - A^{2'}_{3} A^{3'}_{2}) v^{23} + (A^{2'}_{3} A^{3'}_{1} - A^{2'}_{1} A^{3'}_{3}) v^{31} + \\ + (A^{2'}_{1} A^{3'}_{2} - A^{2'}_{2} A^{3'}_{1}) v^{12}. \tag{8.7}$$

Now $A^{2'}_{2} A^{3'}_{3} - A^{2'}_{3} A^{3'}_{2}$ is the minor of the element $A^{1'}_{1}$ in the matrix of $A^{\kappa'}_{\lambda}$. Hence

$$A^{1}_{1'} = \frac{A^{2'}_{2} A^{3'}_{3} - A^{2'}_{3} A^{3'}_{2}}{\Delta}. \tag{8.8}$$

Working out $A^{2}_{1'}$ and $A^{3}_{1'}$ in the same way we get the equation

$$v^{2'3'} = \Delta A^{1}_{1'} v^{23} + \Delta A^{2}_{1'} v^{31} + \Delta A^{3}_{1'} v^{12}, \tag{8.9}$$

which proves the proposition.

The following table gives a summary of the quantities dealt with so far in E_3.

Figure	*Ordinary notation*	*Second notation*	*Weight*	*Relations (cycl.)*	*Number of components*	*Orientation*
	p	$\tilde{\mathfrak{p}}_{\mu\lambda\kappa}$	-1	$\tilde{\mathfrak{p}}_{123} = (-1)^{\alpha_3} p$	1	
		$\tilde{\mathfrak{p}}^{\kappa\lambda\mu}$	$+1$	$\tilde{\mathfrak{p}}^{123} = (-1)^{\beta_0} p$		
	v^{κ}	$\tilde{\mathfrak{v}}_{\lambda\kappa}$	-1	$\tilde{\mathfrak{v}}_{23} = (-1)^{\alpha_2} v^1$	3 (proj.)	inner
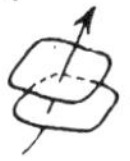	w_{λ}	$\tilde{\mathfrak{w}}^{\kappa\lambda}$	$+1$	$\tilde{\mathfrak{w}}^{23} = (-1)^{\beta_1} w_1$	$3\left(\frac{1}{\text{sect.}}\right)$	outer
	$f^{\kappa\lambda}$	$\tilde{\mathfrak{f}}_{\lambda}$	-1	$\tilde{\mathfrak{f}}_1 = (-1)^{\alpha_1} f^{23}$	3 (proj.)	inner
	$h_{\kappa\lambda}$	$\tilde{\mathfrak{h}}^{\kappa}$	$+1$	$\tilde{\mathfrak{h}}^1 = (-1)^{\beta_2} h_{23}$	$3\left(\frac{1}{\text{sect.}}\right)$	outer
	$p^{\kappa\lambda\mu}$	$\tilde{\mathfrak{p}}$	-1	$\tilde{\mathfrak{p}} = (-1)^{\alpha_0} p^{123}$	1 (vol.)	inner
	$q_{\mu\lambda\kappa}$	$\tilde{\mathfrak{q}}$	$+1$	$\tilde{\mathfrak{q}} = (-1)^{\beta_3} q_{123}$	$1\left(\frac{1}{\text{vol.}}\right)$	

(8.10)

If we replace Δ by $|\Delta|$ in the transformation formulae of Δ-densities, we get another kind of density. These new densities are called *ordinary densities* or briefly *densities*. If only coordinate transformations with positive Δ are used, *the difference between Δ-densities and ordinary densities vanishes*. This is another example of the identification of quantities after replacing G_a by a sub-group. Generally a Gothic letter without $\sim$ is used for the kernel of ordinary densities. But there are exceptions. For instance, a mass density, i.e. the mass per parallelepiped of the measuring vectors, is always denoted by a Greek letter.

In order to find the geometrical representation of contra- and covariant ordinary p-vector densities of weight $+1$ and -1 respectively we form the transvection of a contravariant vector-Δ-density of weight $+1$ and a covariant ordinary vector density of weight -1. This transvection has one component and the transformation

$$\tilde{p}[\kappa'] = \frac{|\Delta|}{\Delta} \tilde{p}[\kappa]. \tag{8.11}$$

Such a quantity is called a *W-scalar*.† A *W*-scalar changes its sign if

† In physical publications it is sometimes called *pseudoscalar*.

the coordinate system is transformed into a system with the opposite screw-sense. The product of an affinor with a W-scalar is called a *W-affinor*. Take, for example, a simple contravariant p-vector $v^{\kappa_1 \ldots \kappa_p}$ representing a part of an E_p with an *inner* orientation. Then this inner orientation together with the $1 \ldots n$ sense of the coordinate system defines an *outer* orientation of the E_p uniquely. This latter orientation changes if a coordinate transformation with negative Δ is applied. From $v^{\kappa_1 \ldots \kappa_p}$ and a W-scalar we can form a new quantity

$$\tilde{v}^{\kappa_1 \ldots \kappa_p} \stackrel{\text{def}}{=} \tilde{p} v^{\kappa_1 \ldots \kappa_p}; \quad \tilde{p}[\kappa] = 1 \tag{8.12}$$

called a *W-p-vector*. This quantity has the same components as $v^{\kappa_1 \ldots \kappa_p}$ with respect to every coordinate system having the same screw-sense as (κ), but it has the same components as $-v^{\kappa_1 \ldots \kappa_p}$ with respect to all other coordinate systems. This means that the *outer* orientation constructed above is invariant for all coordinate transformations. Hence $\tilde{v}^{\kappa_1 \ldots \kappa_p}$ is represented by the same part of an E_p as $v^{\kappa_1 \ldots \kappa_p}$ but with an outer orientation instead of an inner one. If we start from a covariant p-vector (outer orientation) we get in the same way a *covariant W-p-vector* with an inner orientation. All these W-quantities transform like the corresponding ordinary quantities but with an extra factor $|\Delta|/\Delta$. Generally a Roman letter with the sign $\sim$ is used for the kernel letter of W-quantities.

The quantities $\tilde{\mathfrak{E}}^{\kappa_1 \ldots \kappa_n}$ and $\tilde{\mathfrak{e}}_{\lambda_1 \ldots \lambda_n}$ can be used to establish a one-to-one correspondence between contra-(co-)variant W-p-vectors and co-(contra-)variant ordinary $(n-p)$-vector densities.

$$\begin{aligned}
&(a) \qquad \mathfrak{v}_{\lambda_1 \ldots \lambda_p} = \frac{(-1)^{\gamma_p}}{(n-p)!} \tilde{\mathfrak{e}}_{\lambda_1 \ldots \lambda_p \kappa_{p+1} \ldots \kappa_n} \tilde{v}^{\kappa_{p+1} \ldots \kappa_n}, \\
&(b) \qquad \tilde{v}^{\kappa_{p+1} \ldots \kappa_n} = \frac{(-1)^{-\gamma_p}}{p!} \mathfrak{v}_{\lambda_1 \ldots \lambda_p} \tilde{\mathfrak{E}}^{\lambda_1 \ldots \lambda_p \kappa_{p+1} \ldots \kappa_n}, \\
&(c) \qquad \mathfrak{w}^{\kappa_{p+1} \ldots \kappa_n} = \frac{(-1)^{\delta_p}}{p!} \tilde{w}_{\lambda_1 \ldots \lambda_p} \tilde{\mathfrak{E}}^{\lambda_1 \ldots \lambda_p \kappa_{p+1} \ldots \kappa_n}, \\
&(d) \qquad \tilde{w}_{\lambda_1 \ldots \lambda_p} = \frac{(-1)^{-\delta_p}}{(n-p)!} \tilde{\mathfrak{e}}_{\lambda_1 \ldots \lambda_p \kappa_{p+1} \ldots \kappa_n} \mathfrak{w}^{\kappa_{p+1} \ldots \kappa_n},
\end{aligned} \tag{8.13}$$

or, according to the values of the components of $\tilde{\mathfrak{E}}^{\kappa_1 \ldots \kappa_n}$ and $\tilde{\mathfrak{e}}_{\lambda_1 \ldots \lambda_n}$

$$\begin{aligned}
\mathfrak{v}_{\mu_1 \ldots \mu_p} &= (-1)^{\gamma_p} \tilde{v}^{\mu_{p+1} \ldots \mu_n}, \\
\mathfrak{w}^{\mu_{p+1} \ldots \mu_n} &= (-1)^{\delta_p} \tilde{w}_{\mu_1 \ldots \mu_p},
\end{aligned} \tag{8.14}$$

where $\mu_1 \ldots \mu_n$ is an even permutation of $1 \ldots n$. The coefficients γ_p and δ_p can be chosen arbitrarily. The choice will be postponed so that it

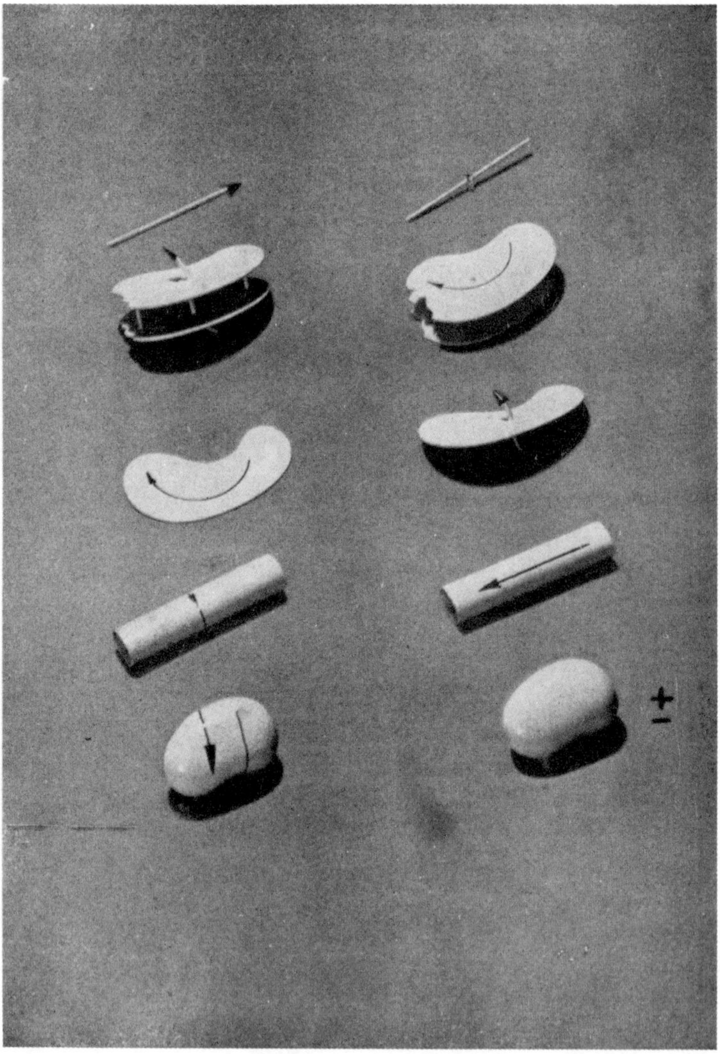
v^{κ}
w_{λ}
$f^{\kappa\lambda}$
$h_{\lambda\kappa}$
$p^{\kappa\lambda\mu}$
$q_{\mu\lambda\kappa}$
$\mathfrak{v}_{\lambda\kappa}$
$\mathfrak{w}^{\kappa\lambda}$
$\tilde{\mathfrak{f}}_{\lambda}$
$\mathfrak{h}^{\kappa}$
$\pm$
$\mathfrak{p}$
$\mathfrak{q}$

Fig. 6

can be made to agree with the identifications made possible after introduction of a sub-group of G_a. This is to be discussed in Chapter III. If no sub-group of G_a is introduced we may take $\gamma_p = \delta_p = 0$ for all values of p.

(8.14) shows that the components of a contra-(co-)variant W-p-vector may also be considered as components of a co-(contra-)variant $(n-p)$-vector density of weight $-1(+1)$. Hence the geometrical meanings of corresponding quantities do not differ. There is only a difference in notation.

The following table gives a summary of these quantities:

Figure	*Second notation*	*Ordinary notation*	*Weight*	*Relations (cycl.)*	*Number of components*	*Orientation*
	$\tilde{p}$	$\mathfrak{p}_{\mu\lambda\kappa}$	-1	$\mathfrak{p}_{123} = (-1)^{\gamma_3}\tilde{p}$	1	
		$\mathfrak{p}^{\kappa\lambda\mu}$	$+1$	$\mathfrak{p}^{123} = (-1)^{\delta_0}\tilde{p}$		
	$\tilde{v}^\kappa$	$\mathfrak{v}_{\lambda\kappa}$	-1	$\mathfrak{v}_{23} = (-1)^{\gamma_2}\tilde{v}^1$	3 (proj.)	outer
	$\tilde{w}_\lambda$	$\mathfrak{w}^{\kappa\lambda}$	$+1$	$\mathfrak{w}^{23} = (-1)^{\delta_1}\tilde{w}_1$	$3\left(\frac{1}{\text{sect.}}\right)$	inner
	$\tilde{f}^{\kappa\lambda}$	$\mathfrak{f}_\lambda$	-1	$\mathfrak{f}_1 = (-1)^{\gamma_1}\tilde{f}^{23}$	3 (proj.)	outer
	$\tilde{h}_{\lambda\kappa}$	$\mathfrak{h}^\kappa$	$+1$	$\mathfrak{h}^1 = (-1)^{\delta_2}\tilde{h}_{23}$	$3\left(\frac{1}{\text{sect.}}\right)$	inner
	$\tilde{p}^{\kappa\lambda\mu}$	$\mathfrak{p}$	-1	$\mathfrak{p} = (-1)^{\gamma_0}\tilde{p}^{123}$	1 (vol.)	outer
	$\tilde{q}_{\mu\lambda\kappa}$	$\mathfrak{q}$	$+1$	$\mathfrak{q} = (-1)^{\delta_3}\tilde{q}_{123}$	$1\left(\frac{1}{\text{vol.}}\right)$	

(8.15)

An ordinary scalar density has no screw-sense. This is the kind of density occurring in physics. For instance a mass density is an ordinary scalar density of weight $+1$. Fig. 6 shows models of the quantities occurring in (8.10) and (8.15).

9. Quantities of valence 2 and matrices (cf. R.C. I § 8)

It is easily proved that

$$\operatorname{Det}(P^\kappa_{.\lambda}) = n!\,P^{[1}_{.[1}\cdots P^{n]}_{.n]} = n!\,P^{[1}_{.1}\cdots P^{n]}_{.n} = n!\,P^{1}_{.[1}\cdots P^{n}_{.n]} \tag{9.1}$$

and that the components of the quantity

$$s!\,P^{[\kappa_1}_{.[\lambda_1}\cdots P^{\kappa_s]}_{.\lambda_s]} = s!\,P^{[\kappa_1}_{.\lambda_1}\cdots P^{\kappa_s]}_{.\lambda_s} = s!\,P^{\kappa_1}_{.[\lambda_1}\cdots P^{\kappa_s}_{.\lambda_s]} \tag{9.2}$$

are the s-rowed sub-determinants in the matrix of $P^\kappa_{.\lambda}$. From this it follows that the rank of $P^\kappa_{.\lambda}$ is r if and only if

$$P^{[\kappa_1}_{.[\lambda_1}\cdots P^{\kappa_s]}_{.\lambda_s]} \begin{cases} \neq 0 & \text{for } s \leq r, \\ = 0 & \text{for } s > r. \end{cases} \tag{9.3}$$

The same holds *mutatis mutandis* for all quantities of valence 2 and their matrices.

It can be proved for a bivector $v^{\kappa\lambda}$ that

$$(2\rho)!\, v^{[\kappa_1[\lambda_1}\dots v^{\kappa_{2\rho}]\lambda_{2\rho}]} = \left(\frac{(2\rho)!}{2^\rho \rho!}\right)^2 v^{[\kappa_1\kappa_2}\dots v^{\kappa_{2\rho-1}\kappa_{2\rho}]} v^{[\lambda_1\lambda_2}\dots v^{\lambda_{2\rho-1}\lambda_{2\rho}]},\dagger \tag{9.4}$$

and from this it follows that for $n = 2\rho$

$$\mathrm{Det}(v^{\kappa\lambda}) = \left\{\frac{(2\rho)!}{2^\rho \rho!} v^{[12}\dots v^{n-1,n]}\right\}^2, \tag{9.5}$$

and that $v^{\kappa\lambda}$ has rank r (always even) if and only if

$$v^{[\kappa_1\kappa_2}\dots v^{\kappa_{2\rho-1}\kappa_{2\rho}]} \begin{cases} \neq 0 & \text{for } 2\rho = r, \\ = 0 & \text{for } 2\rho > r. \end{cases} \tag{9.6}$$

For every definite choice of $\kappa_1,\dots,\kappa_{2\rho}$ the expression

$$(2\rho)!\, 2^{-\rho}(\rho!)^{-1} v^{[\kappa_1\kappa_2}\dots v^{\kappa_{2\rho-1}\kappa_{2\rho}]} \tag{9.7}$$

is a 'Pfaff aggregate' of order 2ρ;‡ viz. the sum of all essentially different terms of the form $v^{\nu_1\nu_2}v^{\nu_{2\rho-1}\nu_{2\rho}}\dots$, where $\nu_1,\dots,\nu_{2\rho}$ is an even permutation of $\kappa_1,\dots,\kappa_{2\rho}$. All terms which are equal according to $v^{\kappa\lambda} = -v^{\lambda\kappa}$ are considered as not being essentially different. The number of all essentially different terms is $(2\rho)!2^{-\rho}(\rho!)^{-1}$, for instance 3 for $\rho = 2$:

$$v^{\kappa_1\kappa_2}v^{\kappa_3\kappa_4} + v^{\kappa_2\kappa_3}v^{\kappa_1\kappa_4} + v^{\kappa_3\kappa_1}v^{\kappa_2\kappa_4}. \tag{9.8}$$

The same holds *mutatis mutandis* for all alternating quantities of valence 2.

If we are dealing with general quantities of valence 2, for instance $p^{\kappa\lambda}$, we may ask whether it is possible to choose $P^{\kappa}_{.\lambda}$ in such a way that the matrix $P^{\kappa}_{.\rho} P^{\lambda}_{.\sigma} p^{\rho\sigma}$ takes a simple form. The answer to this question is given fully in the theory of *elementary divisors*.§ In the two following sections we only give the results for symmetrical and alternating quantities.

10. Normal forms of a tensor of valence 2

If a tensor $h_{\lambda\kappa}$ is given in a centred E_3 the equation of its surface (cf. II, § 5) is

$$h_{\lambda\kappa} x^\lambda x^\kappa = 1. \tag{10.1}$$

If the rank of $h_{\lambda\kappa}$ is 3, the same surface can be represented by the equation

$$\overset{-1}{h}{}^{\kappa\lambda} u_\kappa u_\lambda = 1 \tag{10.2}$$

† In expressions of this kind alternation has to be effected separately over the indices κ and over the indices λ.

‡ Cf. v. Weber 1900. 1, p. 21.

§ Cf. for literature Schouten and Struik 1935, 1, pp. 39 ff.

in plane coordinates. Now if $h_{\lambda\kappa}$ is *real* it is well known from analytic geometry that there always exists a *real* coordinate transformation such that the only non-vanishing components $h_{\lambda'\kappa'}$ are

$$h_{1'1'} = \pm 1; \qquad h_{2'2'} = \pm 1 \text{ or } 0; \qquad h_{3'3'} = \pm 1 \text{ or } 0. \tag{10.3}$$

The *index*, that is the number of negative signs in (10.3), is an invariant of $h_{\lambda\kappa}$.

The same holds for $n > 3$. If r is the rank of $h_{\lambda\kappa}$ and if $h_{\lambda\kappa}$ is *real*, a *real* coordinate transformation $(\kappa) \rightarrow (\kappa')$ always exists such that the matrix of $h_{\lambda'\kappa'}$ takes the form

$$\begin{array}{c} \overleftrightarrow{\quad s \quad}\overleftrightarrow{\quad r-s \quad}\overleftrightarrow{\quad n-r \quad} \\ \left\| \begin{array}{ccccccc} -1 & \cdots & \cdots & \cdots & 0 & 0 & \cdots & 0 \\ \vdots & \ddots & & & \vdots & \vdots & & \vdots \\ \vdots & & -1 & & \vdots & \vdots & & \vdots \\ & & & +1 & \vdots & \vdots & & \vdots \\ \vdots & & & & \ddots & \vdots & & \vdots \\ 0 & \cdots & \cdots & \cdots & +1 & 0 & \cdots & 0 \\ 0 & \cdots & \cdots & \cdots & 0 & 0 & \cdots & 0 \\ \vdots & & & & \vdots & \vdots & & \vdots \\ 0 & \cdots & \cdots & \cdots & 0 & 0 & \cdots & 0 \end{array} \right\| \end{array} \tag{10.4}$$

This can be stated in another way. If $h_{\lambda\kappa}$ is *real* there is always a *real* affinor $P^{\kappa}_{\cdot\lambda}$ such that the matrix of $P^{\rho}_{\cdot\lambda} P^{\sigma}_{\cdot\kappa} h_{\rho\sigma}$ has the form (10.4). We call the sequence $-\ldots-+\ldots+$ occurring in (10.4) the *signature* and the number s the *index* of $h_{\lambda\kappa}$. s is an invariant for *real* coordinate transformations. The signature is said to be *even* if s is even and *odd* if s is odd.

If $r = n$ the signatures of $h_{\lambda\kappa}$ and $\overset{-1}{h}{}^{\kappa\lambda}$ are the same. In this case $h_{\lambda\kappa}$ and $h^{\kappa\lambda}$ are said to be *positive definite* if $s = 0$, *negative definite* if $s = n$, and *indefinite* in all other cases. The cases $r < n$, $s = 0$ and $r < n$, $s = r$ are often called *semi-definite*. Signature and index of a *real* contravariant tensor of valence 2 are defined in the same way.

11. Normal forms of a bivector

A bivector $f^{\kappa\lambda}$ in E_3 is always simple and accordingly its rank r is 2. If $f^{\kappa\lambda}$ is *real* and if

$$f^{\kappa\lambda} = 2v^{[\kappa}w^{\lambda]}, \tag{11.1}$$

we may choose a *real* coordinate system in such a way that

$$\underset{1'}{e^{\kappa}} = w^{\kappa}, \qquad \underset{2'}{e^{\kappa}} = v^{\kappa}. \tag{11.2}$$

Then the matrix of the $f^{\kappa'\lambda'}$ takes the form

$$\left\| \begin{matrix} 0 & 1 & 0 \\ -1 & 0 & 0 \\ 0 & 0 & 0 \end{matrix} \right\|. \tag{11.3}$$

Using the decomposition of a *real* bivector $f^{\kappa\lambda}$ in E_n, of rank 2ρ, into ρ *real* blades (cf. II, § 6) it can be proved in the same way that there is always a real coordinate system (κ') such that the matrix of $f^{\kappa'\lambda'}$ takes the form

$$\overbrace{\hphantom{0 \quad 1 \quad \cdots \quad 0}}^{2\rho}\overbrace{\hphantom{0 \quad \cdots \quad 0}}^{n-2\rho}$$
$$\left\| \begin{matrix} 0 & 1 & \cdots & 0 & 0 & \cdots & 0 \\ -1 & 0 & & \vdots & \vdots & & \vdots \\ \vdots & & 0 & 1 & \vdots & & \vdots \\ 0 & \cdots & -1 & 0 & 0 & \cdots & 0 \\ 0 & \cdots & \cdots & 0 & 0 & \cdots & 0 \\ \vdots & & & \vdots & \vdots & & \vdots \\ 0 & \cdots & \cdots & 0 & 0 & \cdots & 0 \end{matrix} \right\| \tag{11.4}$$

This can be stated in another way. If $f^{\kappa\lambda}$ is *real* there is always a *real* affinor $P^{\kappa}_{\cdot\lambda}$ such that the matrix of $P^{\kappa}_{\cdot\rho}\, P^{\lambda}_{\cdot\sigma} f^{\rho\sigma}$ has the form (11.4). The same holds *mutatis mutandis* for covariant bivectors.

12. The fundamental tensor

In E_n we introduce a *real* tensor $g_{\lambda\kappa}$ of rank n with *constant* components. Let the inverse of $g_{\lambda\kappa}$ be denoted by $g^{\kappa\lambda}$:

$$g^{\kappa\rho} g_{\rho\lambda} = A^{\kappa}_{\lambda}. \tag{12.1}$$

$g_{\lambda\kappa}$ and $g^{\kappa\lambda}$ are called the *covariant* and *contravariant fundamental tensor*. An E_n with a fundamental tensor is called an R_n. The scalar

$$g_{\lambda\kappa} v^{\lambda} v^{\kappa} \tag{12.2}$$

is an invariant of the vector v^{κ} in R_n. The invariant

$$|\sqrt{g_{\lambda\kappa} v^{\lambda} v^{\kappa}}| \tag{12.3}$$

is called the *length* of v^κ.† If $g_{\lambda\kappa}$ is indefinite and v^κ real, $g_{\lambda\kappa} v^\lambda v^\kappa$ can be positive, negative, or zero. A vector of zero length is called a *null vector* and a vector of length *1* a *unit vector*. The null vectors fill the *null cone* with the equation

$$g_{\lambda\kappa} x^\lambda x^\kappa = 0. \tag{12.4}$$

The null cone is real if and only if $g_{\lambda\kappa}$ is indefinite. In this case the centred R_n is split up into two parts, the *positive region* filled by the real vectors with $g_{\lambda\kappa} v^\lambda v^\kappa > 0$ and the *negative region* filled by the real vectors with $g_{\lambda\kappa} v^\lambda v^\kappa < 0$.

The hypersphere

$$g_{\lambda\kappa} x^\lambda x^\kappa = \mp 1 \tag{12.5}$$

is called the *unit hypersphere* in the negative and positive region respectively.

Two vectors u^κ and v^λ satisfying the condition

$$g_{\lambda\kappa} u^\lambda v^\kappa = 0 \tag{12.6}$$

are said to be *mutually perpendicular*. Hence a null vector is perpendicular to itself and only null vectors have this property.

Using a normal form of $g_{\lambda\kappa}$ (cf. II, § 10) a coordinate system x^h $(h = 1,..., \mathrm{n})$ may be introduced such that the non-vanishing components g_{ih}, g^{hi} are

$$\begin{array}{llll} g_{11} = -1, \quad ..., & g_{\mathrm{ss}} = -1, & g_{\mathrm{s+1s+1}} = +1, \quad ..., & g_{\mathrm{nn}} = +1, \\ g^{11} = -1, \quad ..., & g^{\mathrm{ss}} = -1, & g^{\mathrm{s+1s+1}} = +1, \quad ..., & g^{\mathrm{nn}} = +1. \end{array} \tag{12.7}$$

A coordinate system of this kind is called a *cartesian system in* R_n; the indices are always in roman letters and vertical figures. The measuring vectors of a cartesian system will be denoted by $\underset{i}{i}^h, \overset{h}{i}_i$. The $\underset{i}{i}^h$ are mutually perpendicular unit vectors. The components of a quantity with respect to a cartesian coordinate system are called *cartesian components*.

In an analogous way to the definition in ordinary space we define the angle between two vectors u^κ and v^κ, *both* lying in the negative region or *both* in the positive region by

$$\cos\phi = \frac{\mp g_{\lambda\kappa} u^\lambda v^\kappa}{|\sqrt{\{(g_{\nu\rho} u^\nu u^\rho)(g_{\sigma\tau} v^\sigma v^\tau)\}}|} \quad (\mp \text{ region}). \tag{12.8}$$

The fundamental tensor establishes a one-to-one correspondence between co- and contravariant vectors:

$$v^\kappa g_{\kappa\lambda} = w_\lambda, \qquad w_\lambda g^{\lambda\kappa} = v^\kappa. \tag{12.9}$$

† In the special theory of relativity space-time is considered as an R_4 with signature $---+$. If $g_{\lambda\kappa} v^\lambda v^\kappa > 0$ the term *duration* is used here instead of length.

From this it follows that v^κ and w_λ can be considered as two different representations of the same quantity which can be represented geometrically equally well by an arrow or by a system of two parallel hyperplanes. In these representations the distance between the two hyperplanes of w_λ is equal to the inverse of the length of v^κ. In both cases we will use the *same kernel letter* for this quantity. For example,

$$v^\kappa g_{\kappa\lambda} = v_\lambda, \qquad v_\lambda g^{\lambda\kappa} = v^\kappa. \tag{12.10}$$

This process is called the *raising* and *lowering* of indices. The equation

$$g_{\lambda\kappa} = g_{\lambda\rho} g_{\kappa\sigma} g^{\rho\sigma} \tag{12.11}$$

justifies the use of the same kernel letter in $g_{\lambda\kappa}$ and $g^{\lambda\kappa}$. By using this process, different kinds of components can be formed for every affinor or affinor density, e.g. $P_{\kappa\lambda}^{\cdot\cdot\nu}$,

$$P^{\mu\cdot\nu}_{\cdot\lambda} = g^{\mu\kappa} P_{\kappa\lambda}^{\cdot\cdot\nu}, \qquad P_{\kappa\lambda\mu} = P_{\kappa\lambda}^{\cdot\cdot\nu} g_{\kappa\mu}. \tag{12.12}$$

All these components have the same kernel letter.

The relations between the co- and contravariant measuring vectors belonging to the same cartesian coordinate system are

$$\underset{1}{i}^\kappa = -\overset{1}{i}{}^\kappa, \quad \ldots, \quad \underset{\mathrm{s}}{i}^\kappa = -\overset{\mathrm{s}}{i}{}^\kappa, \quad \underset{\mathrm{s}+1}{i}^\kappa = +\overset{\mathrm{s}+1}{i}{}^\kappa, \quad \ldots, \quad \underset{\mathrm{n}}{i}^\kappa = +\overset{\mathrm{n}}{i}{}^\kappa, \tag{12.13}$$

and for the cartesian components of a vector we have

$$v^h = \begin{cases} -v_h & (h = 1,\ldots,\mathrm{s}), \\ +v_h & (h = \mathrm{s}+1,\ldots,\mathrm{n}). \end{cases} \tag{12.14}$$

The length of a vector v^κ in cartesian components is

$$|\sqrt{(-v^1v^1-\ldots-v^{\mathrm{s}}v^{\mathrm{s}}+v^{\mathrm{s}+1}v^{\mathrm{s}+1}+\ldots+v^{\mathrm{n}}v^{\mathrm{n}})}|, \tag{12.15}$$

and the equation of the null cone is

$$-x^1x^1-\ldots-x^{\mathrm{s}}x^{\mathrm{s}}+x^{\mathrm{s}+1}x^{\mathrm{s}+1}+\ldots+x^{\mathrm{n}}x^{\mathrm{n}} = 0. \tag{12.16}$$

In ordinary space we use the signature $+++$ or $---$. The latter signature is preferred in physical investigations if the relativistic point of view has to be considered.

After introducing a fundamental tensor a unit volume is fixed, viz. the volume of the parallelotop of n mutually perpendicular unit vectors. The n-vectors

$$\begin{aligned} &(a) \qquad \underset{(h)}{I}^{\kappa_1\ldots\kappa_n} \overset{\text{def}}{=} n!\, \overset{1}{i}{}^{[\kappa_1}\ldots\overset{\mathrm{n}}{i}{}^{\kappa_n]}, \qquad \underset{(h)}{I}^{1\ldots\mathrm{n}} = +1, \\ &(b) \qquad \overset{(h)}{i}_{\lambda_1\ldots\lambda_n} \overset{\text{def}}{=} n!\, \overset{1}{i}_{[\lambda_1}\ldots\overset{\mathrm{n}}{i}_{\lambda_n]}, \qquad \overset{(h)}{i}_{1\ldots\mathrm{n}} = +1 \end{aligned} \tag{12.17}$$

with the volume $+1$ are invariant for orthogonal transformations of the cartesian coordinate system with $\Delta = +1$ and change sign for $\Delta = -1$. A scalar Δ-density of weight $+1$ corresponds to $\overset{(h)}{\dot{i}}_{\lambda_1\ldots\lambda_n}$:

$$\overset{(h)}{\tilde{\mathfrak{i}}}[h] = \overset{(h)}{i}_{1\ldots n} = +1. \tag{12.18}$$

The Δ-density 1: $\overset{(h)}{\tilde{\mathfrak{i}}}$ of weight -1 corresponds to $\underset{(h)}{I}{}^{\kappa_1\ldots\kappa_n}$. Hence the component of $\overset{(h)}{\tilde{\mathfrak{i}}}$ is $+1$ with respect to all cartesian coordinate systems with the same screw-sense as (h) and -1 with respect to all other cartesian systems. Accordingly $g_{\lambda\kappa}$ fixes $\overset{(h)}{\tilde{\mathfrak{i}}}$ only to within a factor ± 1, and this means *that $g_{\lambda\kappa}$ fixes a unit volume only and not a screw-sense.*

If $\overset{(h)}{\tilde{\omega}}$ is a W-scalar with the component $+1$ with respect to all coordinate systems with the same screw-sense as (h) and -1 with respect to all other systems, the quantity

$$\mathfrak{i} \overset{\text{def}}{=} \overset{(h)}{\tilde{\omega}}\, \overset{(h)}{\tilde{\mathfrak{i}}} \tag{12.19}$$

is an ordinary density of weight $+1$ with a component $+1$ with respect to *all* cartesian systems. This density is closely connected with $g_{\lambda\kappa}$. In fact, if

$$\mathfrak{g}[\kappa] \overset{\text{def}}{=} |\mathrm{Det}(g_{\lambda\kappa})| = n!\,|g_{[1[1}\cdots g_{n]n]}| \tag{12.20}$$

we have

$$\begin{aligned} \mathfrak{g}[\kappa'] &= n!\,|g_{[1'[1'}\cdots g_{n']n']}| = n!\,|A^{\lambda_1\ldots\lambda_n}_{1'\ldots n'}\,{}^{\kappa_1\ldots\kappa_n}_{1'\ldots n'}\, g_{[\lambda_1[\kappa_1}\cdots g_{\lambda_n]\kappa_n]}| \\ &= (n!)^3 A^{1\ldots n}_{1'\ldots n'}\,{}^{1\ldots n}_{1'\ldots n'}\,|g_{[1[1}\cdots g_{n]n]}| \\ &= \Delta^{-2}\mathfrak{g}[\kappa]. \end{aligned} \tag{12.21}$$

Hence $\mathfrak{g}$ is an ordinary scalar density of weight 2 with a component $+1$ with respect to *every* cartesian system and accordingly

$$\mathfrak{i} = |\mathfrak{g}^{\frac{1}{2}}|, \tag{12.22}$$

from which equation we see that $\mathfrak{i}$ is a concomitant of $\mathfrak{g}$.

13. Matrix calculus in E_n and R_n

The equations

$$P^{\kappa}_{\cdot\rho}\, Q^{\rho}_{\cdot\lambda} = R^{\kappa}_{\cdot\lambda} \tag{13.1}$$

can also be written in the abbreviated form

$$PQ = R, \tag{13.2}$$

in which P, Q, and R may be considered as symbols independent of any system of coordinates for the three affinors occurring in (13.1). But we may look upon P, Q, and R also as symbols for the matrices of the components of these affinors with respect to some arbitrarily but definitely given coordinate system, e.g. (κ). Then (13.2) defines a *multiplication of matrices,* provided that we agree about the connexion between the indices and the rows and columns. In what follows we always assume that the *right* index numbers the *rows* and the *left* index the *columns* and that if a lower index stands directly under the upper index, the *upper* is to be considered as *left* and the *lower* as *right*. Then the element in the κth row and the λth column of the product is found by multiplying, for all values of ρ, the element in the κth row and the ρth column of the left factor with the element in the ρth row and the λth column of the right factor and taking the sum of these n products.

Multiplication and addition of matrices are the constituents of the *matrix calculus*. The calculus can be extended by writing the equations

$$'v^\kappa = P^\kappa_{.\lambda} v^\lambda, \qquad 'w_\lambda = w_\kappa P^\kappa_{.\lambda}, \qquad p = w_\lambda v^\lambda \tag{13.3}$$

in the abbreviated form

$$'v = Pv, \qquad 'w = wP, \qquad p = wv = vw \tag{13.4}$$

and looking on v, $'v$ as symbols for matrices with one row and on w, $'w$ as symbols for matrices with one column.

Tensor calculus of quantities of valences up to 2 and matrix calculus are not quite the same. In tensor calculus the group G_a is given from the beginning and we are only interested in properties and operations which are invariant for this group. For example, we know of every quantity whether it is covariant, contravariant, mixed ${}^\kappa_{.\lambda}$ or mixed ${}_\lambda^{.\kappa}$, and multiplications are only effected if a transvection can be formed over one upper and one lower index. Also the properties of being symmetric or alternating have a sense only for co- or for contravariant quantities. On the other hand in matrix calculus no transformation group is given at first. Hence a matrix is a more general conception, and it is not necessary to consider it always as a set of components of some co- or contravariant or mixed quantity. Of course at any moment homogeneous linear transformations of rows and of columns, either co- or contravariant with respect to each other or just quite arbitrary, may be introduced and their invariants may be investigated. But in the matrix calculus as it was developed by Cayley and many others long before the discovery of tensor calculus, there are great parts of the theory and many applications where transformation of rows and columns is not discussed at all.†

By introducing the isomers

$$\overset{|}{P}{}_\lambda^{.\kappa} \overset{\text{def}}{=} P^\kappa_{.\lambda}, \qquad \overset{|}{Q}{}_\lambda^{.\kappa} \overset{\text{def}}{=} Q^\kappa_{.\lambda}, \qquad \overset{|}{R}{}_\lambda^{.\kappa} \overset{\text{def}}{=} R^\kappa_{.\lambda}, \tag{13.5}$$

† Cf. W. Givens, *Math. Rev.* **9** (1948), p. **324,** in his review of Duschek 1947. **6.** The reviewer criticizes the 'failure' of many authors 'to consider matrices other than those introduced to describe linear transformations of one set of variables into a covariant set' and the 'similar failure to note the effectiveness of the tensor notation and algebra in dealing with a number of variables, quite apart from the prior presence of a group of transformations'.

corresponding to the *transposed* matrices of P, Q, and R, in (13.2) and (13.4) we get

$$\overset{|}{Q}\overset{|}{P} = \overset{|}{R}, \qquad 'v = v\overset{|}{P}, \qquad 'w = \overset{|}{P}w.\dagger \tag{13.6}$$

When this notation is used we always have to remember whether a symbol like P has contravariant character to the left and covariant character to the right or vice versa, and whether a symbol like v stands for a co- or contravariant vector. For example, if v, w, and P have the meanings defined in (13.4), vP and Pw do not make sense.

The abbreviated notation can be extended further by introducing symbols for co- and contravariant quantities of valence *2*. For example, denoting $h_{\lambda\kappa}$ and $f^{\kappa\lambda}$ by h and f we get

$$\begin{gathered} hv \quad \text{or} \quad v\overset{|}{h} \quad \text{for} \quad h_{\lambda\kappa}v^{\kappa}; \\ Pf\overset{|}{Q},\ Q\overset{|}{f}\overset{|}{P} \quad \text{for} \quad P^{\kappa}_{\cdot\rho}Q^{\lambda}_{\cdot\sigma}f^{\rho\sigma},\ P^{\lambda}_{\cdot\rho}Q^{\kappa}_{\cdot\sigma}f^{\rho\sigma}; \\ hf,\ f\overset{||}{h} \quad \text{for} \quad h_{\lambda\rho}f^{\rho\kappa},\ \ f^{\rho\kappa}h_{\lambda\rho}; \quad \text{etc.} \end{gathered} \tag{13.7}$$

For the inverses of P, h, and f, which only exist if their rank is n, we write P^{-1}, h^{-1}, and f^{-1}. Then

$$\begin{aligned} PP^{-1} &= P^{-1}P = A, \\ hh^{-1} &= h^{-1}h = A, \\ ff^{-1} &= f^{-1}f = A, \end{aligned} \tag{13.8}$$

and from (13.1) we may derive

$$\begin{aligned} Qv &= P^{-1}Rv, \\ v &= Q^{-1}P^{-1}Rv, \\ R^{-1}PQv &= v. \end{aligned}$$

When this extension is used the covariant and contravariant character of all unmixed quantities of valence *2* must always be memorized.

All this memory-work is very inconvenient, and this causes the abbreviated method to be in general less fool-proof than a notation with indices. But there are two exceptions. In an R_n the difference between co- and contravariant *quantities* vanishes, and there are no longer any

† Many authors write P' or P^* instead of $\overset{|}{P}$. The accent or asterisk to the right is inconvenient if we wish to go back at some time to the notation with indices. So in this case it would be better to use $'P$ or $^*P$. But the vertical bar directly above the central letter has the advantage that if we use the notation $\bar{P}$ for the complex conjugate of P the two bars can be combined and give the notation $\overset{+}{P}$ for the frequently occurring conjugate of the transposed, for which other authors had to introduce a new symbol $\tilde{P}$ or $P^{\dagger}$.

operations of matrix calculus that are not invariant for the group G_{or}. If, moreover, the fundamental tensor is positive definite and if only orthogonal Cartesian coordinate systems are used, there is no longer any difference between co- and contravariant *components*. Here, matrix calculus is really the ideal method in all cases where only quantities of valences up to *2* are concerned and abbreviation is required.

The other exception where matrix calculus can be (still more) recommended is the U_n, i.e. the E_n with the group G_{un} of all unitary transformations, especially if the (hermitian) fundamental tensor is positive definite and the coordinate system is (unitary) cartesian.†

14. Orthogonal normal forms of tensors and bivectors of valence *2*

If a quantity $T^{\kappa}_{.\lambda}$ is given we may require a vector v^κ satisfying the equation

$$T^{\kappa}_{.\lambda} v^\lambda = \sigma v^\kappa, \tag{14.1}$$

where σ is a suitable scalar factor. σ is one of the roots of the equation

$$\operatorname{Det}(T^{\kappa}_{.\lambda} - \sigma A^{\kappa}_{\lambda}) = 0 \tag{14.2}$$

of degree n. The equation

$$w_\kappa T^{\kappa}_{.\lambda} = \sigma w_\lambda \tag{14.3}$$

leads to the same equation (14.2). Every root of (14.2) is called an *eigenvalue* of $T^{\kappa}_{.\lambda}$. A vector v^κ (w_λ) satisfying (14.1 (3)) for this value of σ is called a contravariant (covariant) *eigenvector* belonging to this eigenvalue.

Here we are especially interested in the case when there is a *definite* fundamental tensor $g_{\lambda\kappa}$ and when $T_{\mu\lambda}$ is *real* and *symmetric*. In this case (14.2) can be written

$$\operatorname{Det}(T_{\mu\lambda} - \sigma g_{\mu\lambda}) = 0, \tag{14.4}$$

and it can be proved that the roots of this equation are always *real*.‡ If $\overset{1}{\sigma}$ and $\overset{2}{\sigma}$ are two unequal roots of this equation and $\underset{1}{v}^\kappa$ and $\underset{2}{v}^\kappa$ two solutions of (14.4) for $\sigma = \overset{1}{\sigma}$ and $\sigma = \overset{2}{\sigma}$, we have

$$\underset{1}{v}^\mu T_{\mu\lambda} \underset{2}{v}^\lambda = \overset{1}{\sigma} \underset{1}{v}^\mu g_{\mu\lambda} \underset{2}{v}^\lambda = \overset{2}{\sigma} \underset{1}{v}^\mu g_{\mu\lambda} \underset{2}{v}^\lambda, \tag{14.5}$$

and this is only possible if the two eigenvectors $\underset{1}{v}^\kappa$ and $\underset{2}{v}^\kappa$ are perpendicular to each other and

$$\underset{1}{v}^\mu T_{\mu\lambda} \underset{2}{v}^\lambda = 0. \tag{14.6}$$

† Cf. Chapter X dealing with Dirac's matrix calculus.

‡ We omit the proof as a more general proposition is proved in Chapter X.

If $\overset{1}{\sigma}$ is an m-fold root of (14.4) it can be proved that the eigenvectors belonging to $\overset{1}{\sigma}$ are the vectors of a definite real R_m. In this R_m we can always choose m mutually perpendicular *real* eigenvectors. Dealing in this way with all roots we get in the end n different mutually perpendicular *real* eigenvectors of $T_{\mu\lambda}$. If the unit vectors of these eigenvectors are $\underset{i}{i}^{\kappa}$ $(i = 1,..., \mathrm{n})$, and if these vectors are chosen as measuring vectors we get

$$T_{ji} \overset{*}{=} \underset{j}{i}^{\mu}\underset{i}{i}^{\lambda}T_{\mu\lambda} = 0 \quad (i,j = 1,..., \mathrm{n};\ j \neq i). \tag{14.7}$$

Hence:

THEOREM OF PRINCIPAL AXES OF A TENSOR OF VALENCE 2. *If the fundamental tensor is definite and if $T_{\mu\lambda}$ is a real tensor, it is always possible to find a real cartesian system (h) such that*

$$T_{ji} = 0 \quad (i,j = 1,..., \mathrm{n};\ j \neq i).$$

There is a corresponding theorem for bivectors:

THEOREM OF PRINCIPAL BLADES OF A BIVECTOR. *If the fundamental tensor is definite and if $f_{\mu\lambda}$ is a real bivector of rank 2ρ it is always possible to decompose $f_{\mu\lambda}$ into ρ mutually perpendicular real blades.*

Note that neither the theorem of principal axes nor the theorem of principal blades holds if the fundamental tensor is indefinite. This is due to the fact that in this case a real tensor or bivector may have a singular position with respect to the null cone. The theorem holds however for tensors and bivectors that are not in such a singular position.

EXERCISES

II.1. If $P^{\kappa_1...\kappa_p}Q_{\kappa_1...\kappa_q}$, $q < p$, is an affinor for every choice of the affinor $Q_{\kappa_1...\kappa_q}$, then $P^{\kappa_1...\kappa_p}$ is an affinor.

II.2. If $v_{\lambda\kappa}$ is an affinor and if the equation

$$av_{\lambda\kappa}+bv_{\kappa\lambda} = 0$$

holds with respect to a coordinate system (κ), the same equation holds with respect to every other coordinate system. If $v_{\lambda\kappa} \neq 0$, either $a = b$ or $a = -b$.

II.3. Prove that

$$\underset{1}{v}^{[\kappa_1}...\underset{p}{v}^{\kappa_p]} = \underset{|1}{v}^{[\kappa_1}...\underset{p]}{v}^{\kappa_p]} = \underset{[1}{v}^{\kappa_1}...\underset{p]}{v}^{\kappa_p},$$

$$P^{[\kappa_1...\kappa_p]}_{\lambda_1...\lambda_p} = P^{[\kappa_1...\kappa_p]}_{[\lambda_1...\lambda_p]} = P^{\kappa_1...\kappa_p}_{[\lambda_1...\lambda_p]}.$$

II.4. Prove that

$$\underset{(\kappa)}{E}{}^{\tau_1\ldots\tau_m\kappa_{m+1}\ldots\kappa_n}\overset{(\kappa)}{e}{}_{\tau_1\ldots\tau_m\lambda_{m+1}\ldots\lambda_n} = m!(n-m)!\,A^{[\kappa_{m+1}\ldots\kappa_n]}_{[\lambda_{m+1}\ldots\lambda_n]}$$

and that the same equation holds for $\mathfrak{E}^{\kappa_1\ldots\kappa_n}$ and $\mathfrak{e}_{\lambda_1\ldots\lambda_n}$.

II.5. If p is the rank of $P^{\kappa}_{.\lambda}$ and q the rank of $Q^{\kappa}_{.\lambda}$, the rank r of $P^{\kappa}_{.\mu}Q^{\mu}_{.\lambda}$ is $\leqslant p$ and $\leqslant q$. If $q = n$, r is equal to p.

II.6. If $\underset{1}{i}{}^{\kappa}$ is a unit vector in the $-$ region and $\underset{2}{i}{}^{\kappa}$ a unit vector in the $+$ region of an R_n, perpendicular to each other, $\underset{1}{i}{}^{\kappa}+\underset{2}{i}{}^{\kappa}$ and $\underset{1}{i}{}^{\kappa}-\underset{2}{i}{}^{\kappa}$ are null vectors,

$$\underset{1'}{i}{}^{\kappa} \stackrel{\text{def}}{=} \underset{1}{i}\cosh\phi \pm \underset{2}{i}\sinh\phi$$

is a unit vector in the $-$ region and

$$\underset{2'}{i}{}^{\kappa} \stackrel{\text{def}}{=} \pm\underset{1}{i}\sinh\phi + \underset{2}{i}\cosh\phi$$

a unit vector in the $+$ region perpendicular to $\underset{1'}{i}{}^{\kappa}$.

II.7. If $g_{\lambda\rho}P^{\rho}_{.\kappa}$ and $g_{\lambda\rho}Q^{\rho}_{.\kappa}$ are symmetric, prove that after lowering the first index $PQ+QP$ is symmetric and $PQ-QP$ alternating.

II.8. The covariant vector v_λ has the same or the opposite orientation as the contravariant vector v^κ according as v^κ lies in the $+$ region or in the $-$ region.

II.9. Prove that, for every trivector $v^{\kappa_1\kappa_2\kappa_3}$, $v^{[\kappa_1\kappa_2\kappa_3}v^{\lambda_1]\lambda_2\lambda_3} = 0$ is a consequence of $v^{[\kappa_1\kappa_2\kappa_3}v^{\lambda_1\lambda_2]\lambda_3} = 0$ by writing out $v^{[\kappa_1\kappa_2\kappa_3}v^{\lambda_1\lambda_2]\lambda_3} - v^{[\kappa_1\kappa_2\kappa_3}v^{\lambda_1\lambda_3]\lambda_2}$.

III

IDENTIFICATIONS OF QUANTITIES IN E_n AFTER INTRODUCING A SUB-GROUP OF G_a†

1. Introduction of a unit volume (sub-group G_{eq})

A UNIT volume, represented by an ordinary scalar density $\mathfrak{q}$ of weight $+1$, establishes a one-to-one correspondence between contra-(co-)variant p-vectors (W-p-vectors) and contra-(co-)variant p-vector densities (p-vector-Δ-densities) of weight $+1$ and -1 respectively.

$$\mathfrak{v}^{\kappa_1\ldots\kappa_p} = \mathfrak{q} v^{\kappa_1\ldots\kappa_p}, \qquad \tilde{\mathfrak{v}}^{\kappa_1\ldots\kappa_p} = \mathfrak{q}\tilde{v}^{\kappa_1\ldots\kappa_p},$$

$$\mathfrak{w}_{\lambda_1\ldots\lambda_p} = \mathfrak{q}^{-1} w_{\lambda_1\ldots\lambda_p}, \qquad \tilde{\mathfrak{w}}_{\lambda_1\ldots\lambda_p} = \mathfrak{q}^{-1}\tilde{w}_{\lambda_1\ldots\lambda_p} \tag{1.1}$$

$$(\mathfrak{q}[\kappa] = \pm 1).$$

For example, for $n = 3$ (see Note C, p. 272):

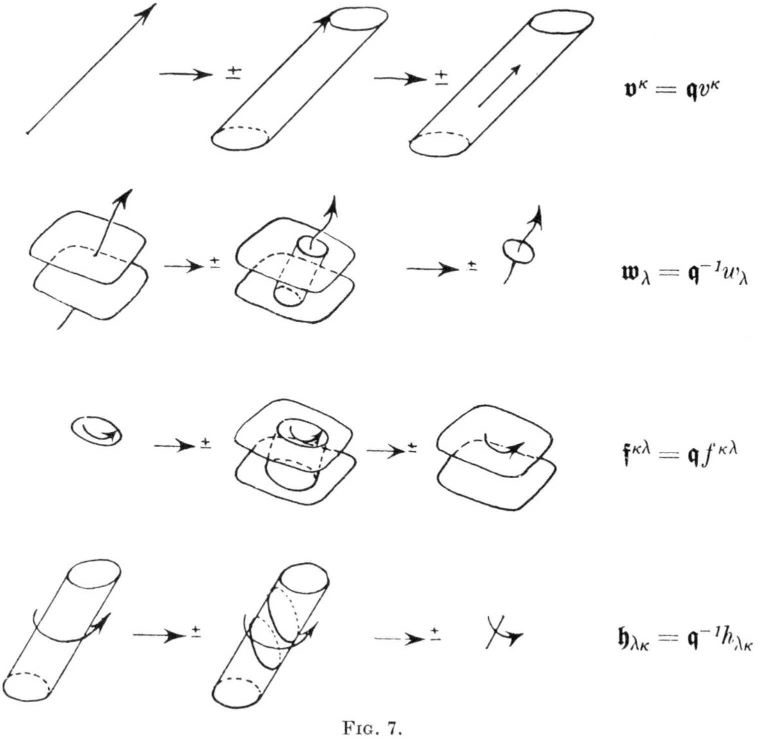

FIG. 7.

† Cf. E I, p. 30, 50; Givens 1937. 1; Schouten 1938. 2; Schouten and v. Dantzig 1940. 1; Dorgelo and Schouten, 1946. 1.

From (III.1.1), (II.8.6), and (II.8.14) it follows that

$$\boxed{\gamma_p \equiv -\beta_p \pmod{2}, \qquad \delta_p \equiv -\alpha_p \pmod{2}.} \tag{1.2}$$

If the unit volume belongs to a fundamental tensor $g_{\lambda\kappa}$ we have $\mathfrak{q} = \mathfrak{i}$, and accordingly in cartesian components

$$\begin{aligned} &\mathfrak{v}^{h_1 \dots h_p} \stackrel{*}{=} v^{h_1 \dots h_p}, && \tilde{\mathfrak{v}}^{h_1 \dots h_p} \stackrel{*}{=} \tilde{v}^{h_1 \dots h_p}, \\ &\mathfrak{w}_{i_1 \dots i_p} \stackrel{*}{=} w_{i_1 \dots i_p}, && \tilde{\mathfrak{w}}_{i_1 \dots i_p} \stackrel{*}{=} \tilde{w}_{i_1 \dots i_p} \\ & && (h_1, \dots, h_p, i_1, \dots, i_p = 1, \dots, \mathrm{n}). \end{aligned} \tag{1.3}$$

2. Introduction of a fundamental tensor (sub-group G_{or})

A fundamental tensor $g_{\lambda\kappa}$ establishes a one-to-one correspondence between contra-(co-)variant and co-(contra-)variant p-vectors, W-p-vectors, p-vector densities, and p-vector-Δ-densities

$$\begin{aligned} v^{\kappa_1 \dots \kappa_p} &= g^{\kappa_1 \lambda_1} \dots g^{\kappa_p \lambda_p} w_{\lambda_1 \dots \lambda_p}, & \tilde{v}^{\kappa_1 \dots \kappa_p} &= g^{\kappa_1 \lambda_1} \dots g^{\kappa_p \lambda_p} \tilde{w}_{\lambda_1 \dots \lambda_p}, \\ \mathfrak{v}^{\kappa_1 \dots \kappa_p} &= \mathfrak{g} g^{\kappa_1 \lambda_1} \dots g^{\kappa_p \lambda_p} \mathfrak{w}_{\lambda_1 \dots \lambda_p}, & \tilde{\mathfrak{v}}^{\kappa_1 \dots \kappa_p} &= \mathfrak{g} g^{\kappa_1 \lambda_1} \dots g^{\kappa_p \lambda_p} \tilde{\mathfrak{w}}_{\lambda_1 \dots \lambda_p}, \end{aligned} \tag{2.1}$$

or in cartesian coordinates, if t_p is the number of negative signs in the first p signs of the signature

$$\begin{aligned} v^{1 \dots \mathrm{p}} &= (-1)^{t_p} w_{1 \dots \mathrm{p}}, & \tilde{v}^{1 \dots \mathrm{p}} &= (-1)^{t_p} \tilde{w}_{1 \dots \mathrm{p}}, \\ \mathfrak{v}^{1 \dots \mathrm{p}} &= (-1)^{t_p} \mathfrak{w}_{1 \dots \mathrm{p}}, & \tilde{\mathfrak{v}}^{1 \dots \mathrm{p}} &= (-1)^{t_p} \tilde{\mathfrak{w}}_{1 \dots \mathrm{p}}. \end{aligned} \tag{2.2}$$

From (III.2.1) or (III.2.2) and (II.8.6), (II.8.14) it follows that for odd/even signature

$$\boxed{\begin{aligned} \alpha_{n-p} - \beta_p &\equiv p(n-p) \begin{cases} +1 \\ +0 \end{cases} \pmod{2}, \\ \gamma_p - \delta_{n-p} &\equiv p(n-p) \begin{cases} +1 \\ +0 \end{cases} \pmod{2}. \end{aligned}} \tag{2.3}$$

The congruences (1.2) hold in this case also because a fundamental tensor fixes a unit volume $\mathfrak{i}$.

3. Introduction of a screw-sense

A screw-sense, represented by a W-scalar $\tilde{\omega}$, establishes a one-to-one correspondence between contra-(co-)variant p-vectors and contra-(co-)variant W-p-vectors and also between contra-(co)-variant p-vector densities of weight $+1$ (-1) and contra-(co-)variant p-vector-Δ-densities of weight $+1$ (-1)

$$\begin{aligned} \tilde{v}^{\kappa_1 \dots \kappa_p} &= \tilde{\omega} v^{\kappa_1 \dots \kappa_p}, & \tilde{\mathfrak{v}}^{\kappa_1 \dots \kappa_p} &= \tilde{\omega} \mathfrak{v}^{\kappa_1 \dots \kappa_p}, \\ \tilde{w}_{\lambda_1 \dots \lambda_p} &= \tilde{\omega} w_{\lambda_1 \dots \lambda_p}, & \tilde{\mathfrak{w}}_{\lambda_1 \dots \lambda_p} &= \tilde{\omega} \mathfrak{w}_{\lambda_1 \dots \lambda_p} \quad (\tilde{\omega}\tilde{\omega} = +1). \end{aligned} \tag{3.1}$$

From (III.3.1), (II.8.6), and (II.8.14) it follows that

$$\boxed{\gamma_p \equiv \alpha_p \,(\mathrm{mod}\, 2), \qquad \delta_p \equiv \beta_p \,(\mathrm{mod}\, 2).} \tag{3.2}$$

4. Simultaneous identifications (sub-group G_{ro})

If all identifications are performed simultaneously we have the diagram

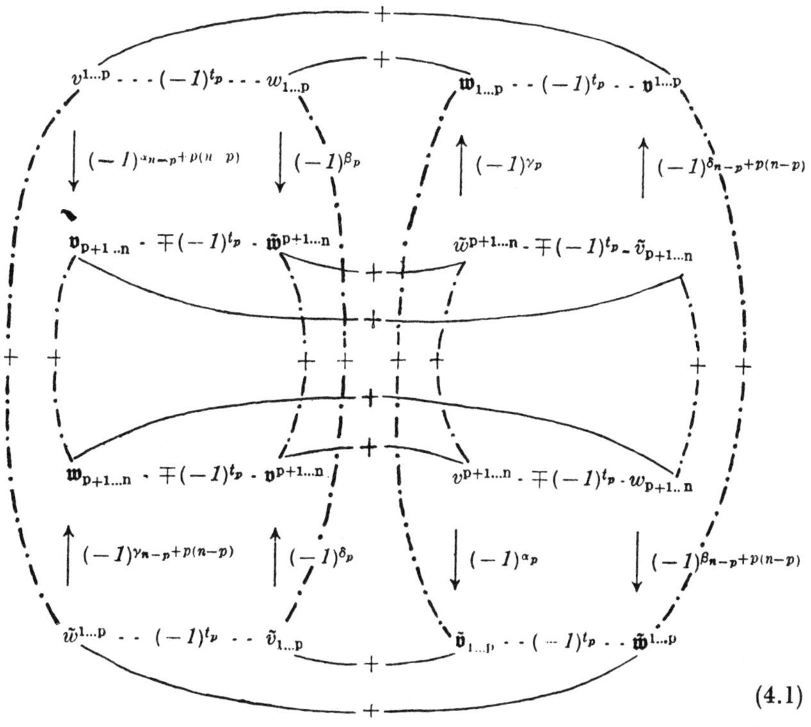

(4.1)

and accordingly the following twelve congruences mod 2 hold

$$\begin{array}{lll}
\gamma_p \equiv -\beta_p, & \alpha_p \equiv \gamma_p, & \alpha_{n-p}-\beta_p \equiv p(n-p)\begin{cases}+1\\+0\end{cases} \\[2ex]
(\gamma_{n-p} \equiv -\beta_{n-p}), & \alpha_{n-p} \equiv \gamma_{n-p}, & \alpha_p-\beta_{n-p} \equiv p(n-p)\begin{cases}+1\\+0\end{cases} \\[2ex]
\delta_p \equiv -\alpha_p, & (\beta_p \equiv \delta_p), & \left(\gamma_p-\delta_{n-p} \equiv p(n-p)\begin{cases}+1\\+0\end{cases}\right) \\[2ex]
\delta_{n-p} \equiv -\alpha_{n-p}, & (\beta_{n-p} \equiv \delta_{n-p}), & \left(\gamma_{n-p}-\delta_p \equiv p(n-p)\begin{cases}+1\\+0\end{cases}\right)
\end{array} \tag{4.2}$$

for the eight unknowns $\alpha_p, \alpha_{n-p}, \beta_p, \beta_{n-p}, \gamma_p, \gamma_{n-p}, \delta_p, \delta_{n-p}$ for odd/even signature. Seven of these congruences are independent (the others are put in round brackets). Expressing all unknowns in α_p we get

$$\begin{aligned} \alpha_p &\equiv \alpha_p, & \beta_p &\equiv -\alpha_p, \\ \alpha_{n-p} &\equiv -\alpha_p+p(n-p)\begin{cases}+1\\+0\end{cases} & \beta_{n-p} &\equiv \alpha_p+p(n-p)\begin{cases}+1\\+0\end{cases} \\ \gamma_p &\equiv \alpha_p, & \delta_p &\equiv -\alpha_p, \\ \gamma_{n-p} &\equiv -\alpha_p+p(n-p)\begin{cases}+1\\+0\end{cases} & \delta_{n-p} &\equiv -\alpha_p+p(n-p)\begin{cases}+1\\+0\end{cases} \end{aligned} \tag{4.3}$$

There are four cases:

(1) n odd, $p(n-p) \equiv 0$, odd signature:

$$\begin{aligned} \alpha_p &\equiv -\beta_p \equiv \gamma_p \equiv -\delta_p, \\ \alpha_{n-p} &\equiv -\beta_{n-p} \equiv \gamma_{n-p} \equiv -\delta_{n-p} \equiv -\alpha_p+1. \end{aligned} \tag{4.4}$$

(2) n odd, $p(n-p) \equiv 0$, even signature:

$$\begin{aligned} \alpha_p &\equiv -\beta_p \equiv \gamma_p \equiv -\delta_p, \\ \alpha_{n-p} &\equiv -\beta_{n-p} \equiv \gamma_{n-p} \equiv -\delta_{n-p} \equiv -\alpha_p. \end{aligned} \tag{4.5}$$

(3) n even, $p(n-p) \equiv p$, odd signature:

$$\begin{aligned} \alpha_p &\equiv -\beta_p \equiv \gamma_p \equiv -\delta_p, \\ \alpha_{n-p} &\equiv -\beta_{n-p} \equiv \gamma_{n-p} \equiv -\delta_{n-p} \equiv -\alpha_p+p+1. \end{aligned} \tag{4.6}$$

(4) n even, $p(n-p) \equiv p$, even signature:

$$\begin{aligned} \alpha_p &\equiv -\beta_p \equiv \gamma_p \equiv -\delta_p, \\ \alpha_{n-p} &\equiv -\beta_{n-p} \equiv \gamma_{n-p} \equiv -\delta_{n-p} \equiv -\alpha_p+p. \end{aligned} \tag{4.7}$$

For n odd, p and $n-p$ are always different (mod 2). The simplest solutions are

(1) n odd, $p(n-p) \equiv 0$, odd signature:

$$\begin{aligned} \alpha_p &= -\beta_p = \gamma_p = -\delta_p = p, \\ \alpha_{n-p} &= -\beta_{n-p} = \gamma_{n-p} = -\delta_{n-p} = n-p. \end{aligned} \tag{4.8}$$

(2) n odd, $p(n-p) \equiv 0$, even signature:

$$\begin{aligned} \alpha_p &= -\beta_p = \gamma_p = -\delta_p = 0, \\ \alpha_{n-p} &= -\beta_{n-p} = \gamma_{n-p} = -\delta_{n-p} = 0. \end{aligned} \tag{4.9}$$

For $n = 2\nu$ we get $2\alpha_\nu = \begin{cases}\nu+1\\ \nu\end{cases}$ for odd/even signature, and from this we see that we get imaginary coefficients for p even/odd. The simplest solutions are

(3) n even, $p(n-p) \equiv p$, odd signature:

$$\begin{aligned} \alpha_p = \alpha_{n-p} = \gamma_p = \gamma_{n-p} &= \frac{p(n-p)+1}{2}, \\ \beta_p = \beta_{n-p} = \delta_p = \delta_{n-p} &= -\frac{p(n-p)+1}{2}, \end{aligned} \tag{4.10}$$

(4) n even, $p(n-p) \equiv p$, even signature:

$$\begin{aligned} \alpha_p = \alpha_{n-p} = \gamma_p = \gamma_{n-p} &= \frac{p(n-p)}{2}, \\ \beta_p = \beta_{n-p} = \delta_p = \delta_{n-p} &= -\frac{p(n-p)}{2}. \end{aligned} \tag{4.11}$$

The fractional values in (4.10) for even values of p and in (4.11) for odd values of p are sometimes inconvenient because they give rise to imaginary coefficients. The only way to get rid of them is to return from G_{ro} to the sub-group G_{or}, that is, to give up the identification of quantities which only differ by their orientation (second column in (4.2)). Then the identifications marked with –·– in (4.1) drop out. Instead of $v^{1\ldots p} = +\tilde{w}^{1\ldots p}$, etc. we now have $v^{1\ldots p} = \pm\tilde{w}^{1\ldots p}$, and the sign depends on the orientation of the coordinate system. The remaining six congruences are

$$\begin{aligned} \gamma_p &\equiv -\beta_p, & \alpha_{n-p}-\beta_p &\equiv p\begin{cases}+1\\+0\end{cases} \\ \gamma_{n-p} &\equiv -\beta_{n-p}, \\ \delta_p &\equiv -\alpha_p, & \alpha_p-\beta_{n-p} &\equiv p\begin{cases}+1\\+0\end{cases} \\ \delta_{n-p} &\equiv -\alpha_{n-p}. \end{aligned} \tag{4.12}$$

The simplest solutions are

$$\begin{aligned} \alpha_p = \alpha_{n-p} = \delta_p = \delta_{n-p} &= \begin{cases}1\\0\end{cases} \\ \beta_p = \beta_{n-p} = \gamma_p = \gamma_{n-p} &= p. \end{aligned} \tag{4.13}$$

These solutions are most convenient for use in the theory of relativity. But the solutions (4.10, 11) of the full congruences (4.2) are interesting because of their connexion with the theory of spinors. For $p = \nu$; $2\nu = n$ we get from them the following diagram for G_{ro}:

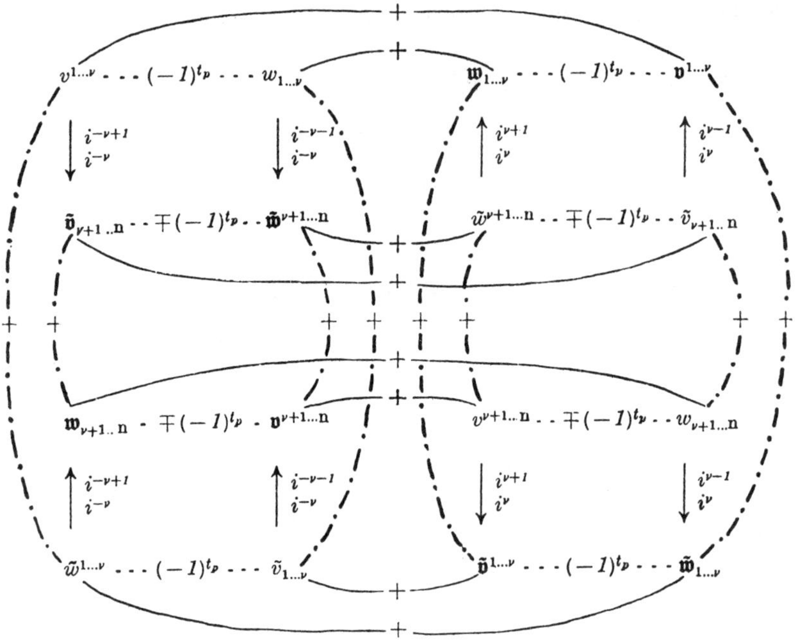

for odd/even signature. Accordingly we have for $n = 4$ and signature $---+$

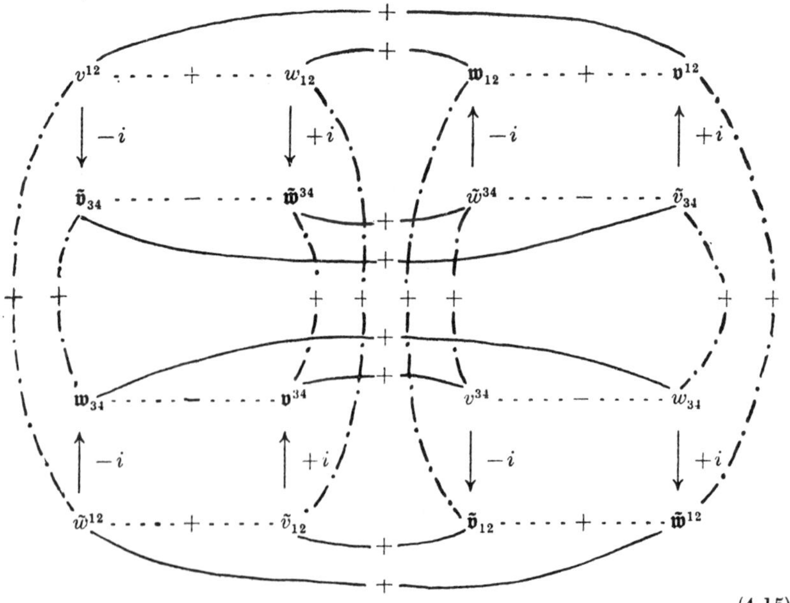

(4.15)

For G_{or} the identifications marked with -·-· drop out. Instead of $+$ we get $\pm$ according to the orientation of the coordinate system. From (4.15) we see that v^{23}, v^{31}, v^{12}, v^{14}, v^{24}, v^{34} transform in the same way as $-iv^{14}$, $-iv^{24}$, $-iv^{34}$, iv^{23}, iv^{31}, iv^{12}. Accordingly the six sums (differences)

$$\begin{array}{llll} (a) & v^{23}\pm iv^{14}, & v^{31}\pm iv^{24}, & v^{12}\pm iv^{34}; \\ (b) & v^{14}\mp iv^{23}, & v^{24}\mp iv^{31}, & v^{34}\mp iv^{12} \end{array} \tag{4.16}$$

transform with G_{ro} as the components v^{23}, v^{31}, v^{12}, v^{14}, v^{24}, v^{34}. But the components (4.16 *b*) only differ from the components (4.16 *a*) by a factor i and from this it follows that each of the sets of *three* sums (differences)

$$v^{23}\pm iv^{14}, \qquad v^{31}\pm iv^{24}, \qquad v^{12}\pm iv^{34} \tag{4.17}$$

form a geometric quantity with three components with respect to the group G_{ro}. We call these quantities *special bivectors of the first and second kind.*

The same holds for $n = 2\nu$. In this case we have according to (4.1) and (4.10, 11)

$$v^{1\ldots\nu} = (-1)^{t_\nu} i^{\left\{ {\nu^2+1 \atop \nu^2} \right.} v^{\nu+1\ldots n} \text{ (for odd/even signature)}, \tag{4.18}$$

and from this we see that among the $\binom{n}{\nu}$ sums

$$v^{1\ldots\nu} + (-1)^{t_\nu} i^{\left\{ {\nu^2+1 \atop \nu^2} \right.} v^{\nu+1\ldots n} \text{ (for odd/even signature)} \tag{4.19}$$

there are exactly $\frac{1}{2}\binom{n}{\nu}$ linearly independent ones which form a geometric quantity, viz. a *special ν-vector of the first kind*, with $\frac{1}{2}\binom{n}{\nu}$ components with respect to G_{ro}. The same holds for the $\binom{n}{\nu}$ differences

$$v^{1\ldots\nu} - (-1)^{t_\nu} i^{\left\{ {\nu^2+1 \atop \nu^2} \right.} v^{\nu+1\ldots n} \text{ (for odd/even signature)}. \tag{4.20}$$

They form a *special ν-vector of the second kind.*

Every ν-vector can be split up in one and only one way into two special ν-vectors of different kinds.

For $n = 2\nu+1$ we have $4n+2$ different quantities for G_a:

Scalar	contr. vector cov. vector	contr. bivector cov. bivector	...	contr. n-vector cov. n-vector
W-scalar	contr. W-vector cov. W-vector	contr. W-bivector cov. W-bivector	...	contr. W-n-vector cov. W-n-vector.

The first row can also be interpreted as Δ-densities and the second as ordinary densities. For G_{eq} (introduction of a unit volume) we have only *2n* different quantities:

Scalar	contr. vector cov. vector	...	contr. ν-vector cov. ν-vector
W-scalar	contr. W-vector cov. W-vector	...	contr. W-ν-vector cov. W-ν-vector.

For G_{or} (introduction of a fundamental tensor) we have only

$$2\nu+2 = n+1$$

different quantities:

Scalar	vector	...	ν-vector
W-scalar	W-vector	...	W-ν-vector.

For G_{ro} (introduction of a fundamental tensor and a screw-sense) we have only $\nu+1$ different quantities:

Scalar vector ... ν-vector.

As an example we take vectors and bivectors in E_3. For signature $\mp\mp\mp$ we have, using the values (4.8) (cf. 4.1),

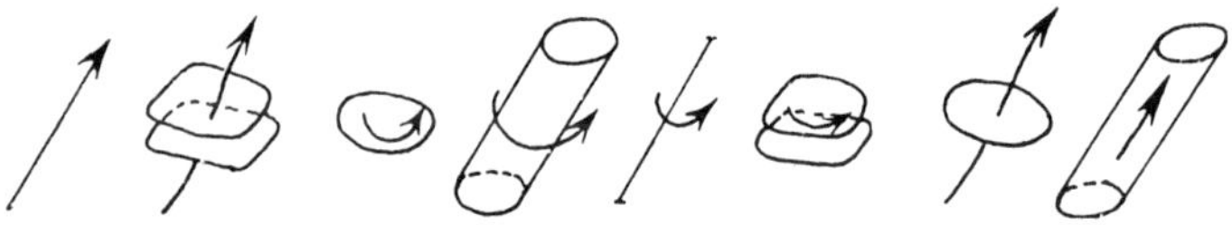

$$\overset{+}{\underset{+}{}}v^1 \quad \overset{+}{\underset{+}{}}w_1 \quad \overset{+}{\underset{+}{}}f^{23} \quad \overset{+}{\underset{+}{}}h_{23} \quad \overset{+}{\underset{+}{}}\mathfrak{v}_{23} \quad \overset{-}{\underset{+}{}}\mathfrak{w}^{23} \quad \overset{-}{\underset{+}{}}\mathfrak{f}_1 \quad \overset{+}{\underset{+}{}}\mathfrak{h}^1$$

$$\overset{+}{\underset{+}{}}\tilde{\mathfrak{v}}_{23} \quad \overset{-}{\underset{+}{}}\tilde{\mathfrak{w}}^{23} \quad \overset{-}{\underset{+}{}}\tilde{\mathfrak{f}}_1 \quad \overset{+}{\underset{+}{}}\tilde{\mathfrak{h}}^1 \quad \overset{+}{\underset{+}{}}\tilde{v}^1 \quad \overset{+}{\underset{+}{}}\tilde{w}_1 \quad \overset{+}{\underset{+}{}}\tilde{f}^{23} \quad \overset{+}{\underset{+}{}}\tilde{h}_{23}.$$

FIG. 8.

According to (4.1) we have the diagram (4.21) shown opposite, for the signature $\mp\mp\mp$.

After introduction of the unit volume $\mathfrak{q}$ we get (group G_{eq})

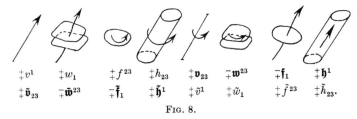

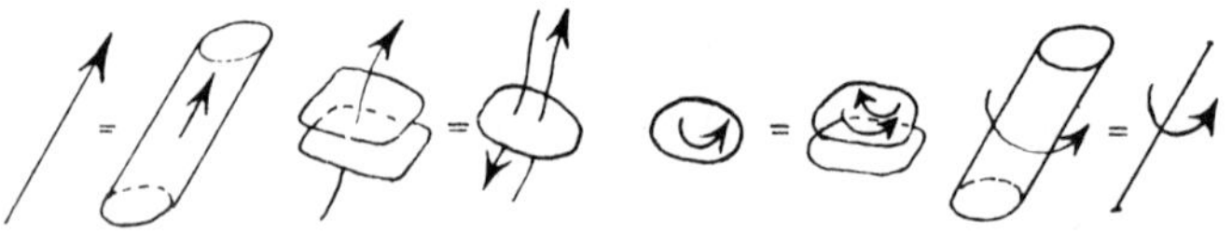

$\overset{+}{\underset{+}{}}v^1 = \overset{+}{\underset{+}{}}\mathfrak{h}^1 =$	$\overset{+}{\underset{+}{}}w_1 = \overset{+}{\underset{+}{}}\mathfrak{f}_1 =$	$\overset{+}{\underset{+}{}}f^{23} = \overset{+}{\underset{+}{}}\mathfrak{w}^{23} =$	$\overset{+}{\underset{+}{}}h_{23} = \overset{+}{\underset{+}{}}\mathfrak{v}_{23} =$
$= \overset{+}{\underset{+}{}}\tilde{\mathfrak{v}}_{23} = \overset{+}{\underset{+}{}}\tilde{h}_{23}$	$= \overset{-}{\underset{+}{}}\tilde{\mathfrak{w}}^{23} = \overset{-}{\underset{+}{}}\tilde{f}^{23}$	$= \overset{-}{\underset{+}{}}\tilde{\mathfrak{f}}_1 = \overset{-}{\underset{+}{}}\tilde{w}_1$	$= \overset{+}{\underset{+}{}}\tilde{\mathfrak{h}}^1 = \overset{+}{\underset{+}{}}\tilde{v}^1.$

FIG. 9.

The four quantities each have four different notations, and two different geometrical representations. In agreement with their simplest representation they are often called *polar vector*, *polar bivector*, *axial bivector*, and *axial vector*.

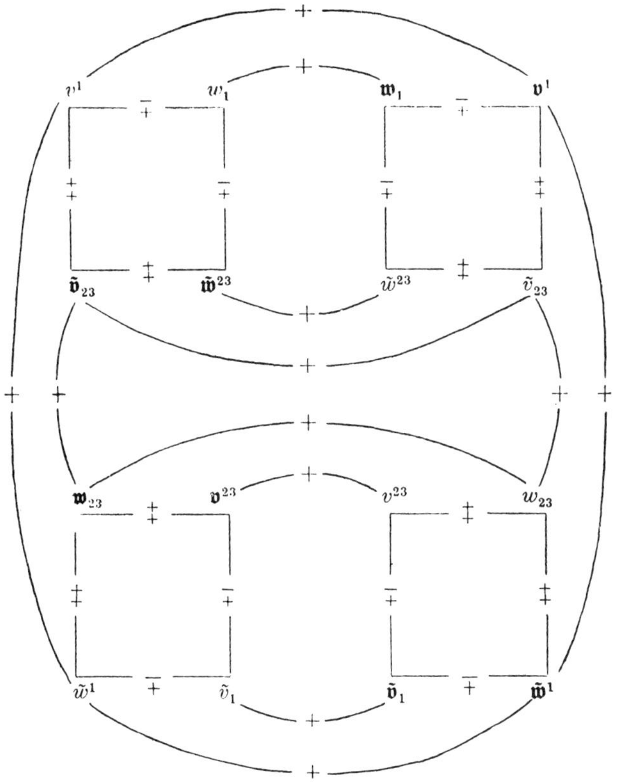

(4.21)

After introduction of the fundamental tensor $g_{\lambda\kappa}$ we get (group G_{or})

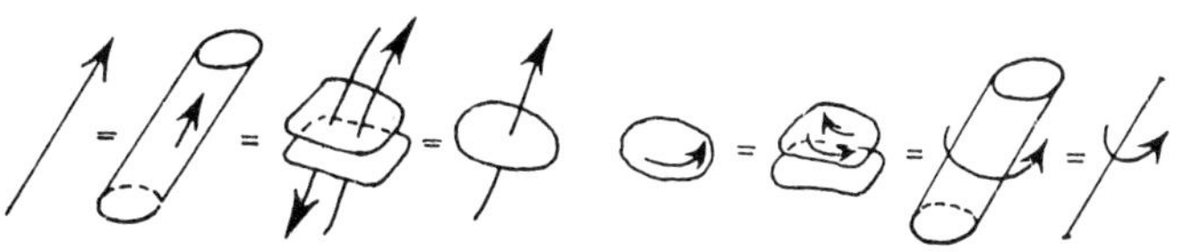

$${}^+_+v^1 = {}^+_+\mathfrak{h}^1 = {}^-_+w_1 = {}^-_+\mathfrak{f}_1 = {}^+_+\tilde{\mathfrak{v}}_{23} = {}^+_+\tilde{h}_{23} = {}^+_+\tilde{\mathfrak{w}}^{23} = {}^+_+\tilde{f}^{23}$$

$${}^+_+f^{23} = {}^+_+\mathfrak{w}^{23} = {}^+_+h_{23} = {}^+_+\mathfrak{v}_{23} = {}^-_+\tilde{\mathfrak{f}}_1 = {}^-_+\tilde{w}_1 = {}^+_+\tilde{\mathfrak{h}}^1 = {}^+_+\tilde{v}^1.$$

FIG. 10.

We now have two quantities each with eight different notations and four different geometrical representations. In agreement with their simplest representation they are often called *polar vector* and *axial vector*.

If a unit volume and a screw-sense are introduced instead of a fundamental tensor we get (group G_{sa})

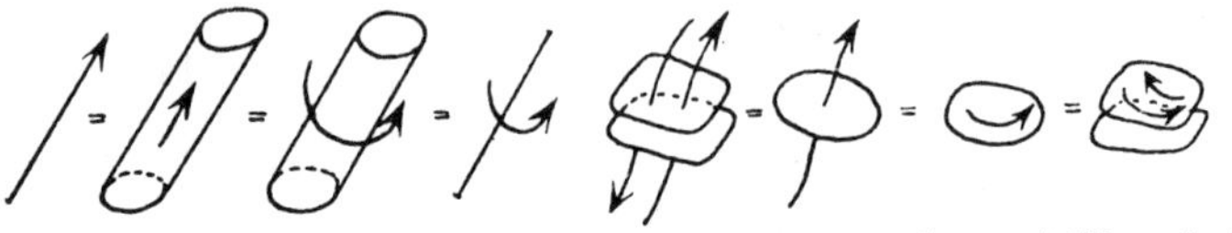

$$ {}^{+}_{+}v^1 = {}^{+}_{+}\mathfrak{h}^1 = {}^{+}_{+}h_{23} = {}^{+}_{+}\mathfrak{v}_{23} = = {}^{+}_{+}\tilde{\mathfrak{v}}_{23} = {}^{+}_{+}\tilde{h}_{23} = {}^{+}_{+}\tilde{\mathfrak{h}}^1 = {}^{+}_{+}\tilde{v}^1 \qquad \Big| \qquad {}^{-}_{+}w_1 = {}^{-}_{+}\mathfrak{f}_1 = {}^{+}_{+}f^{23} = {}^{+}_{+}\mathfrak{w}^{23} = = {}^{+}_{+}\tilde{\mathfrak{w}}^{23} = {}^{+}_{+}\tilde{f}^{23} = {}^{-}_{+}\tilde{\mathfrak{f}}_1 = {}^{-}_{+}\tilde{w}_1. $$

FIG. 11.

We now have two quantities each with eight different notations and four different geometrical representations. In agreement with their simplest representation they are often called *vector* and *bivector*.

After introduction of a fundamental tensor and a screw-sense (group G_{ro}) there remains only one quantity, the *vector* with sixteen different notations and eight different geometric representations:

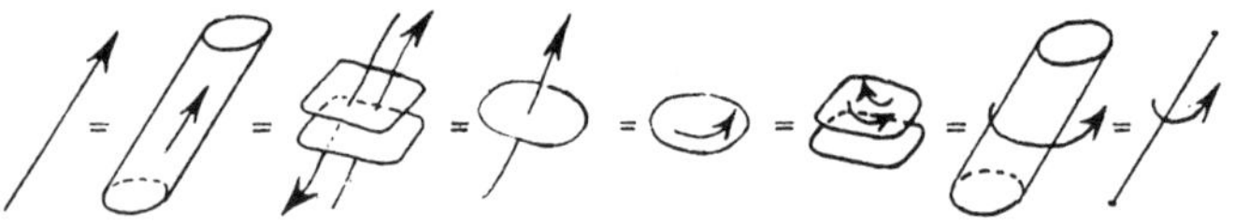

$$ {}^{+}_{+}v^1 = {}^{+}_{+}\mathfrak{h}^1 = {}^{-}_{+}w_1 = {}^{-}_{+}\mathfrak{f}_1 = {}^{+}_{+}f^{23} = {}^{+}_{+}\mathfrak{w}^{23} = {}^{+}_{+}h_{23} = {}^{+}_{+}\mathfrak{v}_{23} = = {}^{+}_{+}\tilde{\mathfrak{v}}_{23} = {}^{+}_{+}\tilde{h}_{23} = {}^{+}_{+}\tilde{\mathfrak{w}}^{23} = {}^{+}_{+}\tilde{f}^{23} = {}^{-}_{+}\tilde{\mathfrak{f}}_1 = {}^{-}_{+}\tilde{w}_1 = {}^{+}_{+}\tilde{\mathfrak{h}}^1 = {}^{+}_{+}\tilde{v}^1. $$

FIG. 12.

Of course, the terms polar vector, polar bivector, etc., originally only have a sense with respect to the transformation group used in their definition. But we may also, when using only cartesian right-handed systems, employ these terms to indicate the special geometrical representation we prefer. In this case we always use the same letter with or without $\sim$ or the corresponding gothic letter for the kernel:

$$ \begin{array}{llllllllr} v^h, & v_i, & v^{hi}, & v_{ih}, & \tilde{v}^h, & \tilde{v}_i, & \tilde{v}^{hi}, & \tilde{v}_{ih}, & \\ \tilde{\mathfrak{v}}_{ih}, & \tilde{\mathfrak{v}}^{hi}, & \tilde{\mathfrak{v}}_i, & \tilde{\mathfrak{v}}^h, & \mathfrak{v}_{ih}, & \mathfrak{v}^{hi}, & \mathfrak{v}_i, & \mathfrak{v}^h & (h, i = 1, 2, 3). \end{array} \tag{4.22} $$

This is also convenient if we wish to introduce more general coordinate systems during the investigation.

Fig. 13 illustrates the different kinds of identification in E_3:

In the theory of relativity we have $n = 4$ and the odd signature $- - - +$. After introducing the fundamental tensor we get for G_{or} (cf. (4.13))

$$ \begin{aligned} &\alpha_1 = \beta_1 = \gamma_1 = \delta_1 = 1, \\ &\alpha_2 = \delta_2 = 1, \qquad \beta_2 = \gamma_2 = 0, \\ &\alpha_3 = \beta_3 = \gamma_3 = \delta_3 = 1, \end{aligned} \tag{4.23} $$

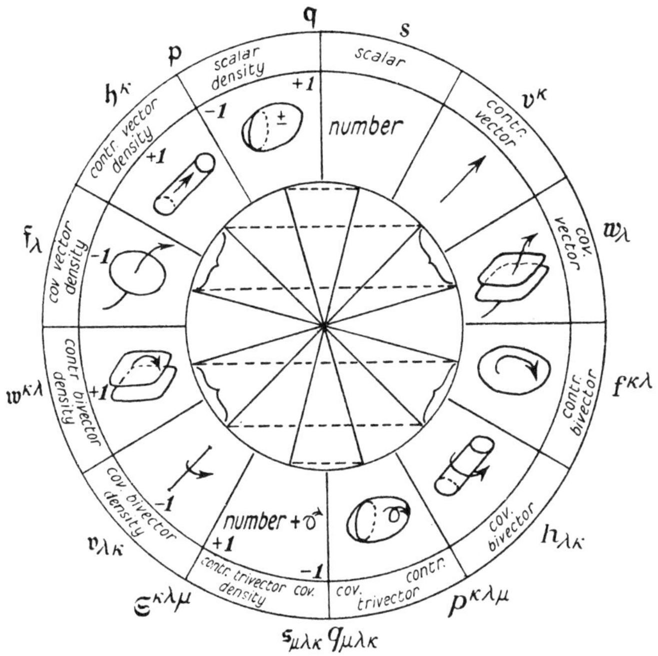

FIG. 13.

and the following table:

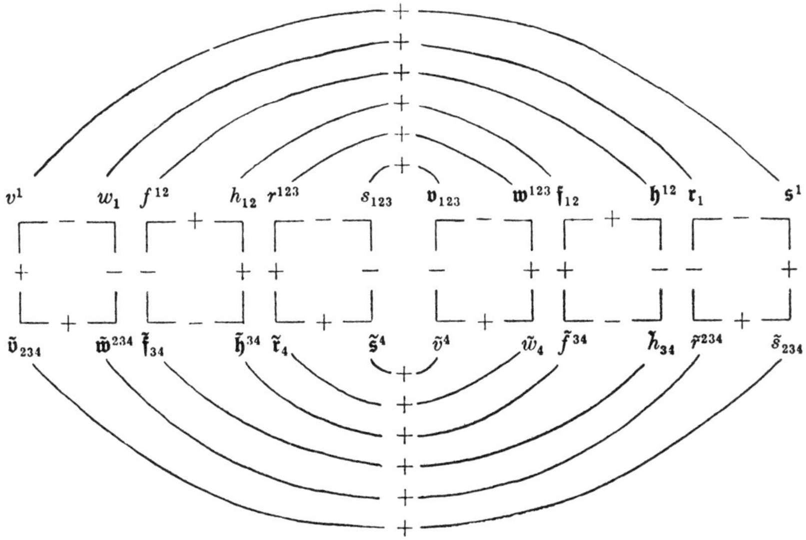

(4.24)

Writing down the right-hand side of this table for $\tilde{v}^1$, $\tilde{w}_1$, $\tilde{f}^{12}$, $\tilde{h}_{12}$, $\tilde{r}^{123}$, and $\tilde{s}_{123}$ we get

$$\begin{array}{ccc|ccc|ccc}
\tilde{v}^1 & & \tilde{w}_1 & \tilde{f}^{12} & & \tilde{h}_{12} & \tilde{r}^{123} & & \tilde{s}_{123} \\
\ulcorner & - & \urcorner & \ulcorner & + & \urcorner & \ulcorner & - & \urcorner \\
+ & & - & + & & - & + & & - \\
\llcorner & + & \lrcorner & \llcorner & - & \lrcorner & \llcorner & + & \lrcorner \\
\mathfrak{v}_{234} & & \mathfrak{w}^{234} & \mathfrak{f}_{34} & & \mathfrak{h}^{34} & \mathfrak{r}_4 & & \mathfrak{s}^4
\end{array} \tag{4.25}$$

and comparing this with the left-hand side of (4.24) we see that by introducing a screw-sense (group G_{ro}) further identifications can only be obtained for vectors and trivectors but not for bivectors.

Hence in an R_4 with signature $---+$ we have, for G_{or}, three different quantities:

(1) Vectors with four different geometrical representations and eight different kinds of components:

$$\left.\begin{aligned} v^1 &= \tilde{\mathfrak{v}}_{234} = -v_1 = \tilde{\mathfrak{v}}^{234} = -\mathfrak{v}_1 = \tilde{v}^{234} = \mathfrak{v}^1 = \tilde{v}_{234} \\ v^4 &= -\tilde{\mathfrak{v}}_{123} = v_4 = \tilde{\mathfrak{v}}^{123} = \mathfrak{v}_4 = \tilde{v}^{123} = \mathfrak{v}^4 = -\tilde{v}_{123} \end{aligned}\right\} \text{ cycl. } 1, 2, 3. \tag{4.26}$$

(2) Bivectors with four different geometrical representations and eight different kinds of components:

$$\left.\begin{aligned} f^{12} &= -\tilde{\mathfrak{f}}_{34} = f_{12} = \tilde{\mathfrak{f}}^{34} = \mathfrak{f}_{12} = \tilde{f}^{34} = \mathfrak{f}^{12} = -\tilde{f}_{34} \\ f^{14} &= -\tilde{\mathfrak{f}}_{23} = -f_{14} = -\tilde{\mathfrak{f}}^{23} = -\mathfrak{f}_{14} = -\tilde{f}^{23} = \mathfrak{f}^{14} = -\tilde{f}_{23} \end{aligned}\right\} \text{cycl. } 1, 2, 3. \tag{4.27}$$

(3) Trivectors with four different geometrical representations and eight different kinds of components:

$$\left.\begin{aligned} \tilde{r}^1 &= \mathfrak{r}_{234} = -\tilde{r}_1 = \mathfrak{r}^{234} = -\tilde{\mathfrak{r}}_1 = r^{234} = \tilde{\mathfrak{r}}^1 = r_{234} \\ \tilde{r}^4 &= -\mathfrak{r}_{123} = \tilde{r}_4 = \mathfrak{r}^{123} = \tilde{\mathfrak{r}}_4 = r^{123} = \tilde{\mathfrak{r}}^4 = -r_{123} \end{aligned}\right\} \text{ cycl. } 1, 2, 3. \tag{4.28}$$

The simplest geometrical representations are

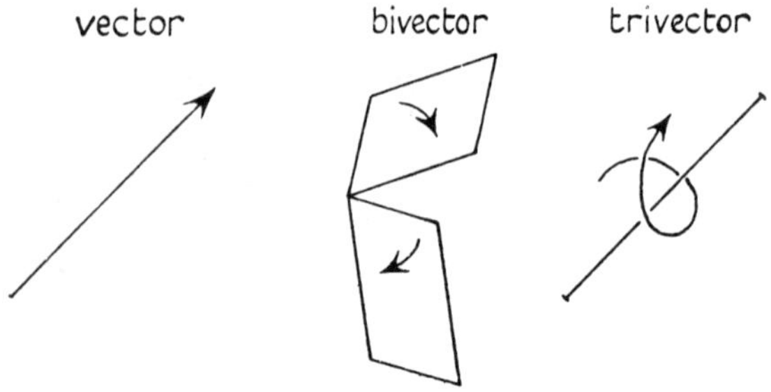

FIG. 14.

If a screw-sense is introduced (group G_{ro}) a trivector can also be represented by an arrow and we have only two quantities, the vector and the bivector. Every bivector can be split up into one special bivector of the first kind and one of the second kind.

5. Ordinary vector algebra in R_3

In ordinary vector algebra in R_3 all possible identifications are carried out. If we leave aside densities and W-quantities we have only four different notations for the components of a vector:

$$v^1 = \mp v_1 = v^{23} = v_{23}. \tag{5.1}$$

Using these we can go back to the different affine invariant forms of the scalar product $\bar{v}.\bar{w}$ and the vector product $\bar{u} = \bar{v}\times\bar{w}$ of two vectors. For the scalar product we get $\binom{5}{2}$ expressions:

$$\begin{aligned}
\bar{v}.\bar{w} &= v_1 w^1 + v_2 w^2 + v_3 w^3 = v_i w^i = v_\lambda w^\lambda \\
&= \mp v_{23} w^{23} \mp v_{31} w^{31} \mp v_{12} w^{12} = \mp \tfrac{1}{2} v_{ij} w^{ij} = \mp \tfrac{1}{2} v_{\lambda\kappa} w^{\lambda\kappa} \\
&= \mp v^{23} w^1 \mp v^{31} w^2 \mp v^{12} w^3 = \mp \tfrac{1}{2} i_{hij} v^{hi} w^j = \mp \tfrac{1}{2} i_{\kappa\lambda\mu} v^{\kappa\lambda} w^\mu \\
&= v_{23} w_1 + v_{31} w_2 + v_{12} w_3 = \tfrac{1}{2} I^{hij} v_{hi} w_j = \tfrac{1}{2} I^{\kappa\lambda\mu} v_{\kappa\lambda} w_\mu \\
&= \mp v^1 w^1 \mp v^2 w^2 \mp v^3 w^3 = g_{ih} v^i w^h = g_{\lambda\kappa} v^\lambda w^\kappa \\
&= \mp v_1 w_1 \mp v_2 w_2 \mp v_3 w_3 = g^{hi} v_h w_i = g^{\kappa\lambda} v_\kappa w_\lambda \qquad (5.2) \\
&= \mp v^{23} w^{23} \mp v^{31} w^{31} \mp v^{12} w^{12} = \mp \tfrac{1}{2} g_{hj} g_{ik} v^{hi} w^{jk} = \mp \tfrac{1}{2} g_{\kappa\mu} g_{\lambda\nu} v^{\kappa\lambda} w^{\mu\nu} \\
&= \mp v_{23} w_{23} \mp v_{31} w_{31} \mp v_{12} w_{12} = \mp \tfrac{1}{2} g^{hj} g^{ik} v_{hi} w_{jk} = \mp \tfrac{1}{2} g^{\kappa\mu} g^{\lambda\nu} v_{\kappa\lambda} w_{\mu\nu} \\
&= \mp v_{23} w^1 \mp v_{31} w^2 \mp v_{12} w^3 = \tfrac{1}{2} I^{hij} g_{jk} v_{hi} w^k = \tfrac{1}{2} I^{\kappa\lambda\mu} g_{\mu\nu} v_{\kappa\lambda} w^\nu \\
&= v^{23} w_1 + v^{31} w_2 + v^{12} w_3 = \mp \tfrac{1}{2} i_{hij} g^{jk} v^{hi} w_k = \mp \tfrac{1}{2} i_{\kappa\lambda\mu} g^{\mu\nu} v^{\kappa\lambda} w_\nu
\end{aligned}$$

$$I^{\kappa\lambda\mu} = 6 \underset{1}{i}^{[\kappa} \underset{2}{i}^{\lambda} \underset{3}{i}^{\mu]}, \qquad i_{\kappa\lambda\mu} = 6 \overset{1}{i}_{[\kappa} \overset{2}{i}_{\lambda} \overset{3}{i}_{\mu]}, \qquad I_{\kappa\lambda\mu} = \mp i_{\kappa\lambda\mu}$$

$$(h, i, j, k = 1, 2, 3).$$

The upper sign holds for the signature $- - -$, the lower for $+ + +$. We get forty expressions for the vector product. These can all be derived from the following six by raising and lowering of indices:

$$\begin{aligned}
u^{23} &= v^2 w^3 - v^3 w^2 \\
u^{23} &= v_{31} w^3 - v_{12} w^2 = -I^{23i} v_{ij} w^j \\
u^{23} &= v^{31} w^{12} - v^{12} w^{31} = \mp g_{ij} v^{[2|i|} w^{3]j} \\
u^1 &= -v^{31} w_3 + v^{12} w_2 = v^{1i} w_i \\
u^1 &= v_2 w_3 - v_3 w_2 = I^{1ij} v_i w_j \\
u^1 &= v^{31} w^{12} - v^{12} w^{31} = -\tfrac{1}{2} i_{ijk} v^{ij} w^{k1} \quad (i, j, k = 1, 2, 3),
\end{aligned} \tag{5.3}$$

or in general coordinates

$$
\begin{aligned}
u^{\kappa\lambda} &= 2v^{[\kappa}w^{\lambda]} \\
u^{\kappa\lambda} &= -I^{\kappa\lambda\mu}v_{\mu\nu}w^{\nu} \\
u^{\kappa\lambda} &= \mp 2g_{\mu\nu}v^{[\kappa|\mu|}w^{\lambda]\nu} \\
u^{\kappa} &= v^{\kappa\lambda}w_{\lambda} \\
u^{\kappa} &= I^{\kappa\lambda\mu}v_{\lambda}w_{\mu} \\
u^{\kappa} &= -\tfrac{1}{2}i_{\mu\lambda\nu}v^{\mu\lambda}w^{\nu\kappa}.
\end{aligned}
\tag{5.4}
$$

EXERCISES

III.1. What is the geometric meaning of the following equations in E_3?—

$$
(a)\quad \begin{cases} \mathfrak{h}^{\kappa}\mathfrak{f}_{\kappa} = 1 \\ \tilde{h}_{\lambda\kappa}\tilde{f}^{\lambda\kappa} = \mp 2, \end{cases} \qquad (b)\quad \begin{cases} h_{\lambda\kappa}f^{\lambda\kappa} = \mp 2 \\ \tilde{\mathfrak{h}}^{\kappa}\tilde{\mathfrak{f}}_{\kappa} = 1. \end{cases} \tag{1 α}
$$

III.2. Suppose that a contravariant vector be given by an arrow in an E_3. How can the other geometrical representations be constructed after the introduction of:

(1) a unit volume (one other representation),
(2) a fundamental tensor (three other representations),
(3) a unit volume and a screw-sense (three other representations),
(4) a fundamental tensor and a screw-sense (seven other representations).

III.3. In the E_6 of all bivectors in R_4 the simple bivectors fill a quadratic cone. What is the equation of this cone if the six expressions (4.16 a) are used as coordinates in E_6?

III.4.† Let $v^{\kappa_1\ldots\kappa_p}$ be a simple p-vector and $'v_{\lambda_1\ldots\lambda_{n-p}}$ a simple $(n-p)$-vector both with the p-direction of a given E_p; let $w^{\kappa_1\ldots\kappa_q}$ be a simple q-vector and $'w_{\lambda_1\ldots\lambda_{n-q}}$ a simple $(n-q)$-vector both with the q-direction of a given E_q; all these quantities being determined to within a scalar factor.

Form the transvections

$$v^{\kappa_1\ldots\kappa_z\mu_{z+1}\ldots\mu_p}\,'w_{\mu_{z+1}\ldots\mu_p\lambda_{p+q-z+1}\ldots\lambda_n}$$

for all possible values of z. If s is the minimum value of z (including 0!) for which this transvection is not zero, E_p and E_q intersect in an E_s and

$$v^{\kappa_1\ldots\kappa_s\mu_{s+1}\ldots\mu_p}\,'w_{\mu_{s+1}\ldots\mu_p\lambda_{p+q-s+1}\ldots\lambda_n}$$

is the product of a contravariant s-vector with the s-direction of E_s and a covariant $(n-p-q+s)$-vector with the $(p+q-s)$-direction of the join of E_p and E_q.

III.5. If

$$u^{\kappa_1\ldots\kappa_p} = \underset{1}{e}^{[\kappa_1}\ldots\underset{p}{e}^{\kappa_p]} \tag{5 α}$$

and if $v^{\kappa_1\ldots\kappa_q}$ is a simple q-vector $(p+q-n = r)$ such that

$$v^{1\ldots r(p+1)\ldots n} \neq 0, \tag{5 β}$$

the E_p of $u^{\kappa_1\ldots\kappa_p}$ and the E_q of $v^{\kappa_1\ldots\kappa_q}$ intersect exactly in an E_r.‡

† Givens 1937. 1, p. 360.
‡ P.P. 1949. 1, p. 27.

IV

GEOMETRIC OBJECTS IN X_n†

1. The X_n‡

We consider all sets of n values of n variables ξ^κ ($\kappa = 1,..., n$) and call each set a *point*. Instead of the ξ^κ another set of n variables $\xi^{\kappa'}$ ($\kappa' = 1',..., n'$) can be introduced by the equations

$$\xi^{\kappa'} = \xi^{\kappa'}(\xi^\kappa) \tag{1.1}$$

together with the conditions that the functions $\xi^{\kappa'}$ are analytic§ in some region and that the matrix of

$$A_\kappa^{\kappa'} \stackrel{\text{def}}{=} \partial \xi^{\kappa'}/\partial \xi^\kappa \tag{1.2}$$

has rank n in that region. Then the inverse transformation

$$\xi^\kappa = \xi^\kappa(\xi^{\kappa'}) \tag{1.3}$$

exists in some region and the functions ξ^κ are analytic in that region. Now the manifold of all the points considered is provided with the original coordinates ξ^κ and with all coordinate systems that can be derived from the ξ^κ in the way described. This manifold is called an X_n. The difference between an X_n and an E_n is that in an E_n the only coordinate systems that are allowable are those that can be derived from each other by linear transformations, but in an X_n all invertible analytic transformations are allowed. Naturally the notions straight line, plane, etc., do not exist in an X_n. Accordingly the coordinates in an X_n are called *curvilinear*.

2. Definition of geometric objects in X_n‖

If at a definite point ξ^κ of X_n there is a correspondence between ordered sets of N numbers Φ_Λ ($\Lambda = 1,..., N$) and the allowable coordinate systems (κ) in a neighbourhood of ξ^κ such that

(1) *to every (κ) there corresponds one and only one set Φ_Λ;*

(2) *the set $\Phi_{\Lambda'}$ corresponding to (κ') can be expressed in terms of Φ_Λ and the values of $A_\kappa^{\kappa'}$, $\partial_\lambda A_\kappa^{\kappa'}$, $\partial_\mu \partial_\lambda A_\kappa^{\kappa'}$,..., in ξ^κ only*

† General references: R.K. 1924. 1; Eisenhart 1926. 1; Levi Civita 1927. 2; Veblen and Whitehead 1932. 1; E I. 1935. 1; Brillouin 1938. 1; Lichnerowicz 1947. 1; Brandt 1947. 2; Michal 1947. 3; P.P. 1949. 1; Schouten 1951. 1; R.C. Ch. II.

‡ For a more elaborate exposition cf. P.P. Ch. II, 1951. 1 § 1, 2.

§ Instead it is possible to assume only that the functions have continuous derivatives up to a certain order.

‖ Cf. II, § 1 and, also for literature, Veblen and Whitehead 1932. 1; E I. 1935. 1; R.C. II § 3.

then the Φ_Λ *are said to be the components with respect to* (κ) *of a geometric object at the point* ξ^κ.

Geometric objects in X_n are classified according to the laws of transformation of their components. If in the condition (2) the expression for $\Phi_{\Lambda'}$ is linear and homogeneous in the Φ_Λ, is algebraically homogeneous in the $A_\kappa^{\kappa'}$, and does not involve the derivatives of the $A_\kappa^{\kappa'}$, then the Φ_Λ are the components of a *geometric quantity at* ξ^κ.

If the object is defined for each point of a definite region of X_n we have a *field* of objects in that region.

A point ξ^κ is an object with the transformation

$$\xi^{\kappa'} = \xi^{\kappa'}(\xi^\kappa), \qquad \mathrm{Det}(A_\kappa^{\kappa'}) \neq 0, \tag{2.1}$$

but this object is not a quantity because the transformation is not linear homogeneous in ξ^κ.

By differentiation of (2.1) we get

$$d\xi^{\kappa'} = A_\kappa^{\kappa'} d\xi^\kappa. \tag{2.2}$$

From this equation we see that the components of the linear element $d\xi^\kappa$ always undergo a linear homogeneous transformation. Hence the transformation $(\kappa) \to (\kappa')$ in X_n induces a linear homogeneous transformation in every point of the region considered, and this means that to every point of this region there belongs a centred E_n. We call this E_n the *local* E_n *of the point* ξ^κ and identify the centre with the point ξ^κ. The X_n and a local E_n have no other point in common. But in ξ^κ the directions in X_n are in one-to-one correspondence with the directions in the local E_n of ξ^κ. Two local E_n's belonging to different points of X_n are entirely independent. *Note particularly that they have no points in common.* If an X_n happens to be embedded in an E_ν $(\nu > n)$, it may sometimes be useful to identify the local E_n's with the tangent E_n's of X_n in E_ν. But if this is done the sections of two tangent E_n's must be left out of consideration. Often the local E_n in ξ^κ is identified with an 'infinitesimal neighbourhood' of ξ^κ in X_n. Though not correct, this identification may sometimes have some heuristic value. Of course consideration of common points of 'neighbouring infinitesimal neighbourhoods' is not permitted.

From the definition of a geometric quantity at a point ξ^κ we see that such a quantity may be considered as a geometric quantity in the local E_n of ξ^κ as defined in Chapter II, § 1. Hence we have vectors, bivectors, densities, etc., in each local E_n and consequently vector fields, bivector fields, etc., in X_n. Two quantities in two different local E_n's cannot be handled as quantities in the same E_n. They cannot be added, multiplied, or transvected.

If for some reason or other a set of coordinate systems in an X_n is privileged, transforming into each other for all transformations of G_a, the X_n may be considered as an E_n. In this case all quantities at different points of this E_n can be added, multiplied, transvected, etc., in the way described in Chapter II, provided that no components are used other than those with respect to these privileged coordinate systems.

In order to find the measuring vectors at each point of the X_n we consider the n scalars $\overset{\kappa}{\xi}$ defined by the equations

$$\overset{\kappa}{\xi} \overset{*}{=} \xi^{\kappa}. \tag{2.3}$$

Then by differentiation we get the following quantities:

(1) The contravariant and covariant measuring vectors belonging to (κ)

$$\underset{\lambda}{e^{\kappa}} \overset{\text{def}}{=} \partial\xi^{\kappa}/\partial\overset{\lambda}{\xi}, \qquad \overset{\kappa}{e}_{\lambda} \overset{\text{def}}{=} \partial\overset{\kappa}{\xi}/\partial\xi^{\lambda}. \tag{2.4}$$

These vectors have the components *1* and *0* with respect to (κ).

(2) The unity affinor

$$A^{\kappa}_{\lambda} \overset{\text{def}}{=} \partial\xi^{\kappa}/\partial\xi^{\lambda}. \tag{2.5}$$

(3) The n^2 scalars of the Kronecker symbol

$$\delta^{\kappa}_{\lambda} = \partial\overset{\kappa}{\xi}/\partial\overset{\lambda}{\xi}. \tag{2.6}$$

We now give an example of curvilinear coordinates in an ordinary R_2. The equation

$$\frac{x^2}{\frac{1}{2}c^2+\lambda}+\frac{y^2}{-\frac{1}{2}c^2+\lambda} = 1 \quad (c = \text{constant}) \tag{2.7}$$

represents a system of ∞^1 ellipses and hyperbolas with the foci at the points $x = \pm c$; $y = 0$. For different values of λ we have

$$\begin{array}{rll} \lambda > \frac{1}{2}c^2 & & \text{ellipse} \\ \lambda = \frac{1}{2}c^2\colon & y = 0 & \text{degenerate ellipse or hyperbola} \\ -\frac{1}{2}c^2 < \lambda < \frac{1}{2}c^2 & & \text{hyperbola} \\ \lambda = -\frac{1}{2}c^2\colon & x = 0 & \text{degenerate hyperbola} \\ \lambda < -\frac{1}{2}c^2 & & \text{imaginary curve.} \end{array} \tag{2.8}$$

(2.7) is an equation of degree *2* in λ. Its roots are

$$\left.\begin{matrix}\lambda_1\\ \lambda_2\end{matrix}\right\} = \tfrac{1}{2}(x^2+y^2) \pm \sqrt{\left\{\frac{(x^2+y^2)^2}{4} - \tfrac{1}{2}c^2(x^2-y^2-\tfrac{1}{2}c^2)\right\}}, \tag{2.9}$$

and x and y can be expressed in terms of λ_1 and λ_2:

$$x^2 = \frac{(\lambda_1+\frac{1}{2}c^2)(\frac{1}{2}c^2+\lambda_2)}{c^2},$$
$$y^2 = \frac{(\lambda_1-\frac{1}{2}c^2)(\frac{1}{2}c^2-\lambda_2)}{c^2}. \tag{2.10}$$

We now introduce λ_1 and λ_2 as new coordinates in R_2. Then to every

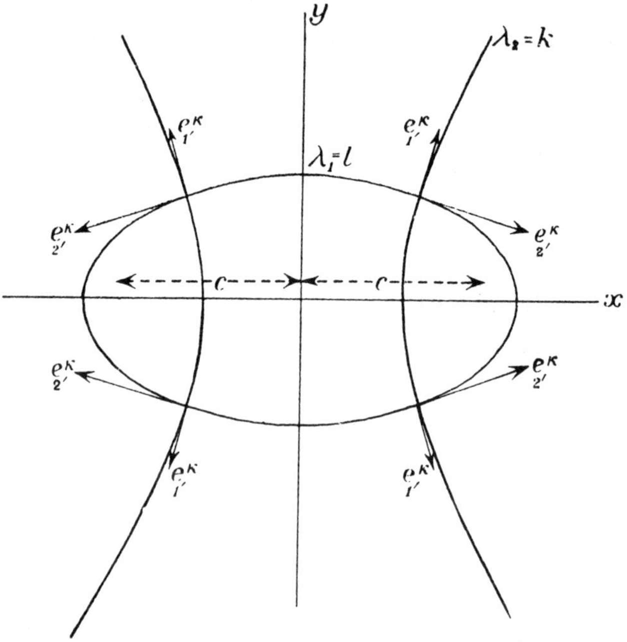

FIG. 15.

general set of values λ_1 and λ_2 there belong four points, one in each quadrant. The points on the x-axis have $\lambda_1 = \frac{1}{2}c^2$ if they lie between the foci and $\lambda_2 = \frac{1}{2}c^2$ if they belong to the outer segments. At the foci $\lambda_1 = \lambda_2 = \frac{1}{2}c^2$. The points on the y-axis have $\lambda_2 = -\frac{1}{2}c^2$. λ_1 takes the values from $+\infty$ to $+\frac{1}{2}c^2$ and λ_2 the values from $+\frac{1}{2}c^2$ to $-\frac{1}{2}c^2$. The parametric curves of λ_1 are the hyperbolas

$$\lambda_2 = k = \text{const.} \quad \text{or} \quad \frac{x^2}{\frac{1}{2}c^2+k}+\frac{y^2}{k-\frac{1}{2}c^2} = 1, \qquad -\tfrac{1}{2}c^2 \leqslant k \leqslant +\tfrac{1}{2}c^2, \tag{2.11}$$

and the parametric curves of λ_2 are the ellipses

$$\lambda_1 = l = \text{const.} \quad \text{or} \quad \frac{x^2}{\frac{1}{2}c^2+l}+\frac{y^2}{l-\frac{1}{2}c^2} = 1, \qquad l \geqslant \tfrac{1}{2}c^2. \tag{2.12}$$

The positive sense on these parametric curves is indicated in Fig. 15.

We need this sense to determine the right sign of $\partial x/\partial\lambda_1$, etc., in the functional matrix derived from (2.10):

$$\left.\begin{aligned}\frac{\partial x}{\partial\lambda_1}&=\substack{+\\-\\-\\+}\frac{1}{2c}\sqrt{\left(\frac{\lambda_2+\frac{1}{2}c^2}{\lambda_1+\frac{1}{2}c^2}\right)}; &\quad \frac{\partial x}{\partial\lambda_2}&=\substack{+\\+\\-\\-}\frac{1}{2c}\sqrt{\left(\frac{\lambda_1+\frac{1}{2}c^2}{\lambda_2+\frac{1}{2}c^2}\right)}\\ \frac{\partial y}{\partial\lambda_1}&=\substack{+\\-\\-\\+}\frac{1}{2c}\sqrt{\left(\frac{\frac{1}{2}c^2-\lambda_2}{\lambda_1-\frac{1}{2}c^2}\right)}; &\quad \frac{\partial y}{\partial\lambda_2}&=\substack{-\\-\\+\\+}\frac{1}{2c}\sqrt{\left(\frac{\lambda_1-\frac{1}{2}c^2}{\frac{1}{2}c^2-\lambda_2}\right)}\end{aligned}\right\}\text{in the}\left\{\begin{array}{l}\text{first,}\\ \text{second,}\\ \text{third,}\\ \text{fourth}\\ \text{quadrant.}\end{array}\right. \tag{2.13}$$

From (2.13) we get the functional determinant

$$\Delta^{-1}=\begin{vmatrix}\frac{\partial x}{\partial\lambda_1} & \frac{\partial x}{\partial\lambda_2}\\ \frac{\partial y}{\partial\lambda_1} & \frac{\partial y}{\partial\lambda_2}\end{vmatrix}=-\frac{\lambda_1-\lambda_2}{4c^2xy}. \tag{2.14}$$

Inversely we get from (2.9)

$$\begin{aligned}\frac{\partial\lambda_1}{\partial x}&=2x\frac{\lambda_1-\frac{1}{2}c^2}{\lambda_1-\lambda_2}; &\quad \frac{\partial\lambda_1}{\partial y}&=2y\frac{\frac{1}{2}c^2+\lambda_1}{\lambda_1-\lambda_2},\\ \frac{\partial\lambda_2}{\partial x}&=2x\frac{\frac{1}{2}c^2-\lambda_2}{\lambda_1-\lambda_2}; &\quad \frac{\partial\lambda_2}{\partial y}&=-2y\frac{\frac{1}{2}c^2+\lambda_2}{\lambda_1-\lambda_2},\end{aligned} \tag{2.15}$$

and consequently
$$\Delta=\begin{vmatrix}\frac{\partial\lambda_1}{\partial x} & \frac{\partial\lambda_1}{\partial y}\\ \frac{\partial\lambda_2}{\partial x} & \frac{\partial\lambda_2}{\partial y}\end{vmatrix}=-\frac{4c^2xy}{\lambda_1-\lambda_2}, \tag{2.16}$$

which is in agreement with (2.14).

The parametric curves of λ_1 and λ_2 form in the second and fourth quadrants a system with the same sense as x, y and in the first and third quadrant a system with the opposite sense. The x-axis and the y-axis are singular lines. On the y-axis we have $\Delta = 0$ and on the x-axis $\Delta = 0$ for $x \neq \pm c$ and $\Delta = 0/0$ for $x = \pm c$.

If we write $x = \xi^1$, $y = \xi^2$, $\lambda_1 = \xi^{1'}$, $\lambda_2 = \xi^{2'}$ from (2.13, 15) we get for the measuring vectors

$$\begin{aligned}\overset{1}{e}_{1'}&=\underset{1'}{e^1}=A^1_{1'}=\substack{+\\-\\-\\+}\frac{1}{2c}\sqrt{\left(\frac{\lambda_2+\frac{1}{2}c^2}{\lambda_1+\frac{1}{2}c^2}\right)}; &\quad \overset{1}{e}_{2'}&=\underset{2'}{e^1}=A^1_{2'}=\substack{+\\+\\-\\-}\frac{1}{2c}\sqrt{\left(\frac{\lambda_1+\frac{1}{2}c^2}{\lambda_2+\frac{1}{2}c^2}\right)},\\ \overset{2}{e}_{1'}&=\underset{1'}{e^2}=A^2_{1'}=\substack{+\\-\\-\\+}\frac{1}{2c}\sqrt{\left(\frac{\frac{1}{2}c^2-\lambda_2}{\lambda_1-\frac{1}{2}c^2}\right)}; &\quad \overset{2}{e}_{2'}&=\underset{2'}{e^2}=A^2_{2'}=\substack{-\\-\\+\\+}\frac{1}{2c}\sqrt{\left(\frac{\lambda_1-\frac{1}{2}c^2}{\frac{1}{2}c^2-\lambda_2}\right)},\\ \underset{1}{e^{1'}}&=\overset{1'}{e}_1=A^{1'}_1=2x\frac{\lambda_1-\frac{1}{2}c^2}{\lambda_1-\lambda_2}; &\quad \underset{1}{e^{2'}}&=\overset{2'}{e}_1=A^{2'}_1=2y\frac{\frac{1}{2}c^2-\lambda_1}{\lambda_1-\lambda_2},\\ \underset{2}{e^{1'}}&=\overset{1'}{e}_2=A^{1'}_2=2x\frac{\frac{1}{2}c^2+\lambda_2}{\lambda_1-\lambda_2}; &\quad \underset{2}{e^{2'}}&=\overset{2'}{e}_2=A^{2'}_2=-2y\frac{\frac{1}{2}c^2+\lambda_2}{\lambda_1-\lambda_2}.\end{aligned} \tag{2.17}$$

From (2.17) we may easily deduce that $\underset{1'}{e^\kappa}$ and $\underset{2'}{e^\kappa}$ are perpendicular to each other and that the length of these measuring vectors is

$$\begin{aligned} \sqrt{(\underset{1'}{e^1}\underset{1'}{e^1}+\underset{1'}{e^2}\underset{1'}{e^2})} &= \frac{1}{2}\sqrt{\left(\frac{\lambda_1-\lambda_2}{\lambda_1^2-\frac{1}{2}c^4}\right)}, \\ \sqrt{(\underset{2'}{e^1}\underset{2'}{e^1}+\underset{2'}{e^2}\underset{2'}{e^2})} &= \frac{1}{2}\sqrt{\left(\frac{\lambda_1-\lambda_2}{\frac{1}{2}c^4-\lambda_2^2}\right)}. \end{aligned} \tag{2.18}$$

The points where the rectangle of $\underset{1'}{e^\kappa}$ and $\underset{2'}{e^\kappa}$ is a square lie on the curve of degree 4

$$(x^2+y^2)^2-c^2(x^2-y^2-\tfrac{1}{2}c^2) = \tfrac{1}{2}c^4 \tag{2.19}$$

through the foci.

3. Invariant differential operators, I: Grad, Div, and Rot§

The derivatives of an affinor or affinor density with a valence different from zero do not form a geometrical object. For instance

$$\begin{aligned} \partial_{\mu'} P^{\kappa'}_{\cdot\lambda'} &= A^{\mu}_{\mu'}\partial_{\mu} A^{\lambda\kappa'}_{\lambda'\kappa} P^{\kappa}_{\cdot\lambda} \\ &= A^{\mu\lambda\kappa'}_{\mu'\lambda'\kappa}\partial_{\mu} P^{\kappa}_{\cdot\lambda}+A^{\mu\kappa'}_{\mu'\kappa} P^{\kappa}_{\cdot\lambda}\partial_{\mu} A^{\lambda}_{\lambda'}+A^{\mu\lambda}_{\mu'\lambda'} P^{\kappa}_{\cdot\lambda}\partial_{\mu} A^{\kappa'}_{\kappa}.\dagger \end{aligned} \tag{3.1}$$

This is due to the fact that the $A^{\kappa'}_{\kappa}$ are not in general constants. Only in an E_n are the $A^{\kappa'}_{\kappa}$ constants and is $\partial_\mu P^{\kappa}_{\cdot\lambda}$ an affinor. Now in an X_n there are invariant differential operators for scalars, W-scalars, covariant p-vectors and W-p-vectors, and contravariant p-vector densities and p-vector-Δ-densities of weight $+1$.

The derivatives of a scalar p or a W-scalar $\tilde{p}$ are the components of a covariant vector or W-vector respectively:

$$\partial_{\mu'} p = A^{\mu}_{\mu'}\partial_\mu p, \qquad \partial_{\mu'}\tilde{p} = (\Delta/|\Delta|)A^{\mu}_{\mu'}\partial_\mu \tilde{p}. \tag{3.2}$$

This vector is called the *natural derivative*‡ or *gradient* and we write for brevity

$$Dp = \operatorname{Grad} p, \qquad D\tilde{p} = \operatorname{Grad}\tilde{p}. \tag{3.3}$$

If the $(n-1)$-direction of a field w_λ is at every point tangent to an X_{n-1} of a system of ∞^1 X_{n-1}'s with the equation q = const. we call w_λ, X_{n-1}-*building*. In this case an equation of the form

$$w_\lambda = \alpha\partial_\lambda q \tag{3.4}$$

exists. Hence every gradient vector is X_{n-1}-building and every vector that is X_{n-1}-building is the product of a gradient vector and a scalar.

† The $\partial_\mu P^{\kappa}_{\cdot\lambda}$ together with the $P^{\kappa}_{\cdot\lambda}$ form a geometric object but not a geometric quantity.

‡ We use this term in order to have a general expression for the three invariant differential operations.

§ Cf. R.C. II § 6.

If $w_\lambda = \partial_\lambda q$ the X_{n-1}'s with the equations $q = \text{const.}$ are called the *equiscalar* X_{n-1}'s *of the field* q. Consider two of these X_{n-1}'s

$$q = c, \qquad q = c+d\epsilon \tag{3.5}$$

and a point $\underset{0}{\xi}{}^\kappa$ on the first of them. In this point we have

$$dq = w_\lambda d\xi^\lambda, \tag{3.6}$$

and the equation

$$d\epsilon = w_\lambda d\xi^\lambda \tag{3.7}$$

expresses the fact that $\xi^\kappa + d\xi^\kappa$ lies on the second X_{n-1}. Hence the vector $w_\lambda/d\epsilon$ is represented by the two tangent E_{n-1}'s at the points ξ^κ and $\xi^\kappa + d\xi^\kappa$. Now let k be an arbitrary constant and consider the equiscalar X_{n-1}'s $q = c$, $q = c+k$, $q = c+2k$, etc. Then we see that in any point of the field on one of these X_{n-1}'s the tangent E_{n-1} at this point and the tangent E_{n-1} at the neighbouring point on the next X_{n-1} together represent approximately the vector w_λ/k and this approximation tends to an exact representation if k tends to zero. Hence a gradient field is not only X_{n-1}-building but on an 'infinitesimal scale' its double-E_{n-1}'s fit together and build double-X_{n-1}'s filling the whole X_n without gaps.

The alternated derivatives of a covariant q-vector are components of a covariant $(q+1)$-vector:

$$\begin{aligned} \partial_{[\mu'} w_{\lambda'_1 \ldots \lambda'_q]} &= A^{\mu\lambda_1 \ldots \lambda_q}_{\mu'\lambda'_1 \ldots \lambda'_q} \partial_{[\mu} w_{\lambda_1 \ldots \lambda_q]} + \partial_{[\mu'} (A^{\lambda_1}_{\lambda'_1} \ldots A^{\lambda_q}_{\lambda'_q]}) \, w_{\lambda_1 \ldots \lambda_q} \\ &= A^{\mu\lambda_1 \ldots \lambda_q}_{\mu'\lambda'_1 \ldots \lambda'_q} \partial_{[\mu} w_{\lambda_1 \ldots \lambda_q]}. \end{aligned} \tag{3.8}$$

This is due to the fact that

$$\partial_{[\mu'} A^{\lambda}_{\lambda']} = \partial_{[\mu'} \partial_{\lambda']} \xi^\lambda = 0. \tag{3.9}$$

The $(q+1)$-vector $(q+1)\partial_{[\mu} w_{\lambda_1 \ldots \lambda_q]}$† is called the *natural derivative* or *rotation* of $w_{\lambda_1 \ldots \lambda_q}$ and we write for brevity

$$(q+1)Dw = \operatorname{Rot} w. \tag{3.10}$$

The same holds *mutatis mutandis* for a covariant W-q-vector.

From these definitions it follows immediately that

$$\begin{aligned} \operatorname{Rot} \operatorname{Grad} p &= 0, & \operatorname{Rot} \operatorname{Grad} \tilde{p} &= 0, \\ \operatorname{Rot} \operatorname{Rot} w &= 0, & \operatorname{Rot} \operatorname{Rot} \tilde{w} &= 0. \end{aligned} \tag{3.11}$$

If $\mathfrak{w}^{\kappa_1 \ldots \kappa_p}$ is a contravariant p-vector density of weight $+1$ it can be proved that $\partial_\mu \mathfrak{w}^{\mu\kappa_2 \ldots \kappa_p}$ is a contravariant $(p-1)$-vector density of weight $+1$. The same holds *mutatis mutandis* for a contravariant p-vector-Δ-

† Cf. Ch. V, § 1, where Rot w is written with the covariant operator ∇_μ.

density of weight $+1$. In order to prove this we take $\delta_p = 0$ for all values of p and $q = n-p$ in (II.8.13):

$$\tilde{w}_{\lambda_1\ldots\lambda_p} = \frac{1}{q!}\,\tilde{\mathfrak{e}}_{\lambda_1\ldots\lambda_p\kappa_1\ldots\kappa_q}\,\mathfrak{w}^{\kappa_1\ldots\kappa_q}. \tag{3.12}$$

Then

$$\begin{aligned}(p+1)\,\partial_{[\mu}\tilde{w}_{\lambda_1\ldots\lambda_p]} &= \frac{p+1}{q!}\,\tilde{\mathfrak{e}}_{[\lambda_1\ldots\lambda_p|\kappa_1\ldots\kappa_q|}\,\partial_{\mu]}\,\mathfrak{w}^{\kappa_1\ldots\kappa_q}\\ &= \frac{1}{q!}\,\tilde{\mathfrak{e}}_{\lambda_1\ldots\lambda_p\kappa_1\ldots\kappa_q}\,\partial_\mu\,\mathfrak{w}^{\kappa_1\ldots\kappa_q} -\\ &\quad -\frac{1}{q!}\sum_s^{1\ldots p}\tilde{\mathfrak{e}}_{\lambda_1\ldots\lambda_{s-1}\mu\lambda_{s+1}\ldots\lambda_p\kappa_1\ldots\kappa_q}\,\partial_{\lambda_s}\,\mathfrak{w}^{\kappa_1\ldots\kappa_q}\end{aligned} \tag{3.13}$$

and because an alternation over $n+1$ indices vanishes identically

$$\begin{aligned}(p+1)\,\partial_{[\mu}\tilde{w}_{\lambda\ldots\lambda_p]} &= \frac{1}{q!}\sum_s^{1\ldots p}\tilde{\mathfrak{e}}_{\lambda_1\ldots\lambda_{s-1}\mu\lambda_{s+1}\ldots\lambda_p\kappa_1\ldots\kappa_q}\,\partial_{\lambda_s}\,\mathfrak{w}^{\kappa_1\ldots\kappa_q}+\\ &\quad +\frac{1}{q!}\sum_t^{1\ldots q}\tilde{\mathfrak{e}}_{\lambda_1\ldots\lambda_p\kappa_1\ldots\kappa_{t-1}\mu\kappa_{t+1}\ldots\kappa_q}\,\partial_{\kappa_t}\,\mathfrak{w}^{\kappa_1\ldots\kappa_q}-\\ &\quad -\frac{1}{q!}\sum_s^{1\ldots p}\tilde{\mathfrak{e}}_{\lambda_1\ldots\lambda_{s-1}\mu\lambda_{s+1}\ldots\lambda_p\kappa_1\ldots\kappa_q}\,\partial_{\lambda_s}\,\mathfrak{w}^{\kappa_1\ldots\kappa_q}\\ &= \frac{1}{q!}\sum_t^{1\ldots q}\tilde{\mathfrak{e}}_{\lambda_1\ldots\lambda_p\kappa_1\ldots\kappa_{t-1}\mu\kappa_{t+1}\ldots\kappa_q}\,\partial_{\kappa_t}\,\mathfrak{w}^{\kappa_1\ldots\kappa_q}\\ &= (-1)^p\frac{1}{(q-1)!}\,\tilde{\mathfrak{e}}_{\mu\lambda_1\ldots\lambda_p\kappa_2\ldots\kappa_q}\,\partial_{\kappa_1}\,\mathfrak{w}^{\kappa_1\ldots\kappa_q}.\dagger\end{aligned} \tag{3.14}$$

From this we see that $\partial_\lambda\,\mathfrak{w}^{\lambda\kappa_{p+2}\ldots\kappa_n}$ is a $(q-1)$-vector density of weight $+1$ adjoint to $(-1)^p$ Rot $\tilde{w}$.‡ This quantity is called the *natural derivative* or *divergence* of $\mathfrak{w}^{\kappa_1\ldots\kappa_p}$ and we write for brevity Div $\mathfrak{w}$.§ The same holds *mutatis mutandis* for p-vector-Δ-densities of weight $+1$. From the definitions it follows immediately that

$$\text{Div Div}\,\mathfrak{w} = 0, \qquad \text{Div Div}\,\tilde{\mathfrak{w}} = 0 \tag{3.15}$$

for $q \geqslant 2$.

It can be proved that no other invariant differential operators exist in X_n if no auxiliary fields are given.

† For $\delta_p \neq 0$ we get a factor $(-1)^{p-\delta_p}$ instead of $(-1)^p$ in (3.14).
‡ For $\delta_p \neq 0$ we get a factor $(-1)^{p-\delta_p+\delta_{p+1}}$ instead of $(-1)^p$.
§ Cf. Ch. V, § 1, where Rot $\tilde{w}$ and Div $\tilde{\mathfrak{w}}$ are written with the covariant operator ∇_μ.

We will prove directly that $\partial_\mu \mathfrak{w}^\mu$ is a scalar Δ-density of weight $+1$ if $\mathfrak{w}^\mu$ is a vector Δ-density of weight $+1$. For $\partial_\mu \mathfrak{w}^{\mu'}$ we get

$$\begin{aligned}\partial_{\mu'}\mathfrak{w}^{\mu'} &= A^\lambda_{\mu'}\partial_\lambda \Delta^{-1} A^{\mu'}_\mu \mathfrak{w}^\mu = \Delta^{-1}\partial_\mu \mathfrak{w}^\mu + \mathfrak{w}^\mu \partial_\mu \Delta^{-1} + \mathfrak{w}^\mu \Delta^{-1} A^\lambda_{\mu'}\partial_\lambda A^{\mu'}_\mu \\ &= \Delta^{-1}\partial_\mu \mathfrak{w}^\mu + \mathfrak{w}^\mu (\partial_\mu \Delta^{-1} + \Delta^{-1} A^\lambda_{\mu'}\partial_\lambda A^{\mu'}_\mu).\end{aligned} \tag{3.16}$$

Now from the definition of $A^\lambda_{\mu'}$ it follows that (cf. I.1.5)

$$A^\lambda_{\mu'} = \frac{1}{\Delta}\frac{\partial \Delta}{\partial A^{\mu'}_\lambda} = \frac{\partial \log \Delta}{\partial A^{\mu'}_\lambda}, \tag{3.17}$$

or

$$\partial_\mu \Delta^{-1} = -\Delta^{-1} A^\lambda_{\mu'}\partial_\mu A^{\mu'}_\lambda. \tag{3.18}$$

Hence

$$\partial_{\mu'} \mathfrak{w}^{\mu'} = \Delta^{-1}\partial_\mu \mathfrak{w}^\mu. \tag{3.19}$$

The operator div in ordinary vector analysis corresponds to the operator Div if $\bar{v}$ is considered as a contravariant vector density

$$\operatorname{div}\bar{v} = \partial_1 v^1 + \partial_2 v^2 + \partial_3 v^3 = \partial_h v^h \quad (h = 1, 2, 3), \tag{3.20}$$

and to the operator Rot if $\bar{v}$ is considered as a covariant bivector

$$\operatorname{div}\bar{v} = \partial_1 v_{23} + \partial_2 v_{31} + \partial_3 v_{12} = 3\partial_{[1} v_{23]}. \tag{3.21}$$

The operator rot in ordinary vector analysis corresponds to the operator $\mp$Rot if $\bar{v}$ is considered as a covariant vector

$$w_{23} = \mp\partial_2 v_3 \pm \partial_3 v_2 = \mp 2\partial_{[2} v_{3]}; \quad \text{cycl. } 1,2,3 \tag{3.22}$$

and to the operator $-$Div if $\bar{v}$ is considered as a contravariant bivector density

$$w^1 = \partial_2 v^{12} - \partial_3 v^{31} = -\partial_h v^{h1}; \text{ cycl. } 1,2,3 \quad (h = 1,2,3).$$

4. Invariant differential operators, II: The theorem of Stokes§

In X_n we consider a simply connected part τ_{q+1} of an X_{q+1}† and its boundary X_q, τ_q. Let the q-dimensional element of X_q with some *inner* orientation, fixed for the whole X_q, be $df^{\kappa_1 \ldots \kappa_q}$ and the $(q+1)$-dimensional element of X_{q+1} with some *inner* orientation, fixed for the whole X_{q+1}, be $df^{\kappa_1 \ldots \kappa_{q+1}}$. Let the orientations be chosen in such a way that the direction from a point of τ_{q+1} towards the boundary followed by the orientation of $df^{\kappa_1 \ldots \kappa_q}$ gives the orientation of $df^{\kappa_1 \ldots \kappa_{q+1}}$. Now let $v_{\lambda_1 \ldots \lambda_q}$ be an m-vector field in τ_{q+1}. Then we have

$$\int_{\tau_{q+1}} \partial_{[\mu} v_{\lambda_1 \ldots \lambda_q]} df^{\mu\lambda_1 \ldots \lambda_q} = \int_{\tau_q} v_{\lambda_1 \ldots \lambda_q} df^{\lambda_1 \ldots \lambda_q}‡ \tag{4.1}$$

† A simply connected part of an X_{q+1} is a part of X_n that can be transformed by a bicontinuous point transformation into the part fixed by the relations

$$\begin{matrix} 0 \leqslant \xi^1 \leqslant 1; & \xi^{q+2} = 0 \\ \vdots & \vdots \\ 0 \leqslant \xi^{q+1} \leqslant 1; & \xi^n = 0 \end{matrix}$$

with respect to some allowable coordinate system.

‡ Cf. also for literature R.K. 1924. 1, pp. 95 ff.; P.P. 1949, 1, pp. 67 ff.

§ R.C. II § 8.

if the following conditions are satisfied:

(1) $v_{\lambda_1 \ldots \lambda_q}$ is continuous in τ_{q+1} and on τ_q;

(2) the derivatives of $v_{\lambda_1 \ldots \lambda_q}$ which occur in $\partial_{[\mu} v_{\lambda_1 \ldots \lambda_q]}$ exist at all points of τ_{q+1};

(3) these derivatives are continuous at all points of τ_{q+1} with the exception of the points of a finite number of X_q's.

We prove this for the case when the form of τ_q satisfies certain conditions specified below.

Let the coordinate system be chosen in such a way that the equations of τ_{q+1} are

$$\xi^{q+2} = 0, \quad \ldots, \quad \xi^n = 0, \tag{4.2}$$

and let the form of τ_q be such that the curves on τ_{q+1}

$$\xi^2 = \text{constant}, \quad \ldots, \quad \xi^{q+1} = \text{constant} \tag{4.3}$$

each intersect τ_q in at most two points. Let the inner orientation on τ_{q+1} be fixed by $\underset{1}{e^\kappa}, \ldots, \underset{q+1}{e^\kappa}$ in this order. Then $df^{\kappa_1 \ldots \kappa_{q+1}}$ may be written in the form

$$df^{\kappa_1 \ldots \kappa_{q+1}} = (q+1)!\, \underset{1}{d}\xi^{[\kappa_1} \ldots \underset{q+1}{d}\, \xi^{\kappa_{q+1}]}. \tag{4.4}$$

If the vectors $\underset{1}{d}\xi^\kappa, \ldots, \underset{q+1}{d}\, \xi^\kappa$ are chosen in such a way that

$$\begin{aligned}
&\underset{1}{d}\xi^1 = d\xi^1, \quad \underset{1}{d}\xi^2 = 0, \quad \ldots, \quad \underset{1}{d}\xi^n = 0,\\
&\underset{2}{d}\xi^1 = 0, \quad \underset{2}{d}\xi^2 = d\xi^2, \quad \underset{2}{d}\xi^3 = 0, \quad \ldots, \quad \underset{2}{d}\xi^n = 0,\\
&\vdots\\
&\underset{q+1}{d}\, \xi^1 = 0, \quad \ldots, \quad \underset{q+1}{d}\, \xi^q = 0, \quad \underset{q+1}{d}\, \xi^{q+1} = d\xi^{q+1}, \quad \underset{q+1}{d}\, \xi^{q+2} = 0, \quad \ldots, \quad \underset{q+1}{d}\, \xi^n = 0,
\end{aligned} \tag{4.5}$$

we have

$$df^{1 \ldots q+1} = d\xi^1 \ldots d\xi^{q+1}, \tag{4.6}$$

and all components of $df^{\kappa_1 \ldots \kappa_{q+1}}$ which have an index $q+2, \ldots, n$ vanish. Now we choose one of the curves (4.3) which intersects τ_q in the points P_1 and P_2:

$$\begin{aligned}
&P_1\colon\ \xi^1 = \underset{1}{\xi^1}, \quad \xi^2 = \underset{0}{\xi^2}, \quad \ldots, \quad \xi^{q+1} = \underset{0}{\xi^{q+1}},\\
&P_2\colon\ \xi^1 = \underset{2}{\xi^1}, \quad \xi^2 = \underset{0}{\xi^2}, \quad \ldots, \quad \xi^{q+1} = \underset{0}{\xi^{q+1}},
\end{aligned} \qquad \underset{2}{\xi^1} > \underset{1}{\xi^1}. \tag{4.7}$$

The points of τ_{q+1} which satisfy the equations

$$\begin{gathered}
\underset{1}{\xi^1} \leqslant \xi^1 \leqslant \underset{2}{\xi^1}, \qquad \underset{0}{\xi^\mathfrak{a}} \leqslant \xi^\mathfrak{a} \leqslant \underset{0}{\xi^\mathfrak{a}} + d\xi^\mathfrak{a} \quad (\mathfrak{a} = 2, \ldots, q+1),\\
\xi^\mathfrak{x} = 0 \quad (\mathfrak{x} = q+2, \ldots, n)
\end{gathered} \tag{4.8}$$

constitute a $(q+1)$-dimensional tube in τ_{q+1} and this tube cuts a q-dimensional element from τ_q in each of the points P_1 and P_2. Fig. 16 illustrates the case $n = 3, q = 2$. If in accordance with our presumptions we fix the orientation on τ_q in P_2 by $\underset{2}{e^\kappa}, \ldots, \underset{q+1}{e^\kappa}$ in this order, then the orientation in P_1 is opposite to the orientation fixed by $\underset{2}{e^\kappa}, \ldots, \underset{q+1}{e^\kappa}$ in P_1. Accordingly we have for the q-dimensional element of τ_q in P_2

$$\overset{2}{df}{}^{2\ldots q+1} = d\xi^2 \ldots d\xi^{q+1}, \tag{4.9}$$

and in P_1

$$\overset{1}{df}{}^{2\ldots q+1} = -d\xi^2 \ldots d\xi^{q+1}. \tag{4.10}$$

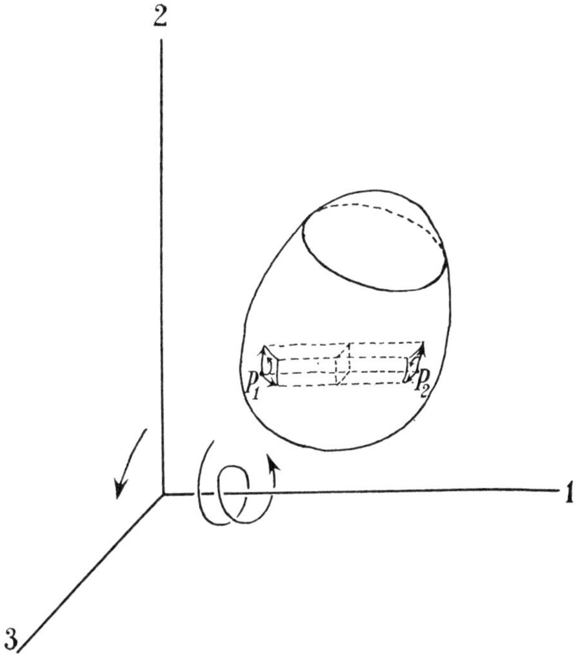

Fig. 16.

Now we consider the integral over τ_{q+1}

$$\int_{\tau_{q+1}} \partial_1 v_{2\ldots q+1} df^{1\ldots q+1} = \int_{\tau_{q+1}} \partial_1 v_{2\ldots q+1} d\xi^1 \ldots d\xi^{q+1}. \tag{4.11}$$

The part of this integral over the tube is

$$(\overset{2}{v}_{2\ldots q+1} - \overset{1}{v}_{2\ldots q+1}) d\xi^2 \ldots d\xi^{q+1} = \overset{2}{v}_{2\ldots q+1} \overset{2}{df}{}^{2\ldots q+1} + \overset{1}{v}_{2\ldots q+1} \overset{1}{df}{}^{2\ldots q+1}, \tag{4.12}$$

where $\overset{2}{v}_{2\ldots q+1}$ and $\overset{1}{v}_{1\ldots q+1}$ are the values of $v_{2\ldots q+1}$ in P_2 and P_1 respectively. Hence

$$\int_{\tau_{q+1}} \partial_1 v_{2\ldots q+1} df^{1\ldots q+1} = \int_{\tau_q} v_{2\ldots q+1} df^{2\ldots q+1}. \tag{4.13}$$

By using other sets of curves instead of (4.3) we can get in the same way equations with $\partial_2, \ldots, \partial_{q+1}$ instead of ∂_1. Adding them together and multiplying by a suitable factor we get (4.1). Now we still have to get rid of the condition of intersection imposed on the form of τ_q. Let it be possible to split up τ_{q+1} into a finite number of τ_{q+1}'s such that the condition of intersection holds for the boundary τ_q of each of them. Then (4.1) holds for each τ_{q+1} and by summation of the integrals on the left-hand side all integrals over boundaries between two adjacent τ_{q+1}'s cancel each other out because the orientations in the common boundary are opposite. This finishes the proof for the case when τ_{q+1} can be split up into a finite number of parts, each of which satisfies the condition of intersection.

(4.1) is a form of the generalized theorem of Stokes (in Germany also named after Gauss and in France after Ostrogradski). Many other forms can be deduced from it. First $\tilde{\mathfrak{v}}^{\kappa_1 \ldots \kappa_p}$, $\tilde{\mathfrak{f}}_{\lambda_1 \ldots \lambda_{p-1}}$, $\tilde{\mathfrak{f}}_{\lambda_1 \ldots \lambda_p}$ ($p = n-q$) can be introduced instead of $v_{\lambda_1 \ldots \lambda_q}$, $f^{\mu \lambda_1 \ldots \lambda_q}$, $f^{\lambda_1 \ldots \lambda_q}$. Then we get the four formulae†

$$p \int\limits_{\tau_{q+1}} \partial_{[\mu} v_{\lambda_1 \ldots \lambda_q} \, d\tilde{\mathfrak{f}}_{\kappa_1 \ldots \kappa_{p-1}]} = \int\limits_{\tau_q} v_{[\lambda_1 \ldots \lambda_q} \, d\tilde{\mathfrak{f}}_{\kappa_1 \ldots \kappa_{p-1} \mu]}, \tag{4.14}$$

$$p \int\limits_{\tau_{q+1}} \partial_\mu \tilde{\mathfrak{v}}^{\lambda_1 \ldots \lambda_{p-1} \mu} \, d\tilde{\mathfrak{f}}_{\lambda_1 \ldots \lambda_{p-1}} = \int\limits_{\tau_q} \tilde{\mathfrak{v}}^{\lambda_1 \ldots \lambda_p} \, d\tilde{\mathfrak{f}}_{\lambda_1 \ldots \lambda_p}, \tag{4.15}$$

$$\frac{n-q}{q+1} \int\limits_{\tau_{q+1}} \partial_\mu \tilde{\mathfrak{v}}^{[\lambda_1 \ldots \lambda_{p-1} |\mu|} \, df^{\lambda_p \kappa_1 \ldots \kappa_q]} = \int\limits_{\tau_{q+1}} \partial_\mu \tilde{\mathfrak{v}}^{[\lambda_1 \ldots \lambda_p} \, df^{|\mu| \kappa_1 \ldots \kappa_q]}$$

$$= \int\limits_{\tau_q} \tilde{\mathfrak{v}}^{[\lambda_1 \ldots \lambda_p} \, df^{\kappa_1 \ldots \kappa_q]}, \tag{4.16}$$

which all have the same geometrical significance.

If the elements of τ_q and τ_{q+1} have outer orientation (this is the case that occurs most frequently in physics), and if these orientations are chosen in such a way that the orientation of τ_{q+1} followed by the direction from a point of τ_{q+1} towards the boundary gives the orientation of τ_q, we get the following five formulae:

$$\int\limits_{\tau_{q+1}} \partial_{[\mu} \tilde{v}_{\lambda_1 \ldots \lambda_q]} \, d\tilde{f}^{\mu \lambda_1 \ldots \lambda_q} = \int\limits_{\tau_q} \tilde{v}_{\lambda_1 \ldots \lambda_q} \, d\tilde{f}^{\lambda_1 \ldots \lambda_q}, \tag{4.17}$$

$$p \int\limits_{\tau_{q+1}} \partial_{[\mu} \tilde{v}_{\lambda_1 \ldots \lambda_q} \, d\mathfrak{f}_{\kappa_1 \ldots \kappa_{p-1}]} = \int\limits_{\tau_q} \tilde{v}_{[\lambda_1 \ldots \lambda_q} \, d\mathfrak{f}_{\kappa_1 \ldots \kappa_{p-1} \mu]}, \tag{4.18}$$

$$p \int\limits_{\tau_{q+1}} \partial_\mu \mathfrak{v}^{\lambda_1 \ldots \lambda_{p-1} \mu} \, d\mathfrak{f}_{\lambda_1 \ldots \lambda_{p-1}} = \int\limits_{\tau_q} \mathfrak{v}^{\lambda_1 \ldots \lambda_p} \, d\mathfrak{f}_{\lambda_1 \ldots \lambda_p}, \tag{4.19}$$

† Cf. R.K. 1924. 1, p. 97; Weyssenhoff 1937. 2; Schouten and v. Dantzig 1940. 1, p. 471. In this latter paper the sign ~ is used in a different and less efficient way.

$$\frac{n-q}{q+1}\int\limits_{\tau_{q+}} \partial_\mu \mathfrak{v}^{[\lambda_1\ldots\lambda_{p-1}|\mu|}\, d\tilde{f}^{\lambda_p\kappa_1\ldots\kappa_q]} = \int\limits_{\tau_{q+1}} \partial_\mu \mathfrak{v}^{[\lambda_1\ldots\lambda_p}\, d\tilde{f}^{|\mu|\kappa_1\ldots\kappa_q]}$$

$$= \int\limits_{\tau_q} \mathfrak{v}^{[\lambda_1\ldots\lambda_p}\, d\tilde{f}^{\kappa_1\ldots\kappa_q]}, \tag{4.20}$$

which all have the same geometrical significance.† If one of the ten formulae (4.1, 14–20) is proved the other nine can be derived from it.

In X_3 we have the following cases:

(1) $q = 2$.

(*a*) Integral of a covariant bivector over a surface with inner orientation:

$$\boxed{\int\limits_{\tau_3} \partial_{[\mu} v_{\lambda\kappa]}\, df^{\mu\lambda\kappa} = \int\limits_{\tau_2} v_{\lambda\kappa}\, df^{\lambda\kappa},} \tag{4.21}$$

$$\int\limits_{\tau_3} \partial_{[\mu} v_{\lambda\kappa]}\, d\tilde{\mathfrak{f}} = \int\limits_{\tau_2} v_{[\lambda\kappa}\, d\tilde{\mathfrak{f}}_{\mu]}, \tag{4.22}$$

$$\int\limits_{\tau_3} \partial_\mu \tilde{\mathfrak{v}}^\mu\, d\tilde{\mathfrak{f}} = \int\limits_{\tau_2} \tilde{\mathfrak{v}}^\mu\, d\tilde{\mathfrak{f}}_\mu, \tag{4 23}$$

$$\tfrac{1}{3}\int\limits_{\tau} \partial_\mu \tilde{\mathfrak{v}}^\mu\, df^{\nu\kappa_1\kappa_2} = \int\limits_{\tau_3} \partial_\mu \tilde{\mathfrak{v}}^{[\nu}\, df^{|\mu|\kappa_1\kappa_2]} = \int\limits_{\tau_2} \tilde{\mathfrak{v}}^{[\nu}\, df^{\kappa_1\kappa_2]}. \tag{4.24}$$

(*b*) Integral of a contravariant vector density of weight $+1$ over a surface with outer orientation:

$$\int\limits_{\tau_3} \partial_{[\mu} \tilde{v}_{\lambda\kappa]}\, d\tilde{f}^{\mu\lambda\kappa} = \int\limits_{\tau_2} \tilde{v}_{\lambda\kappa}\, d\tilde{f}^{\lambda\kappa}, \tag{4.25}$$

$$\int\limits_{\tau_3} \partial_{[\mu} \tilde{v}_{\lambda\kappa]}\, d\mathfrak{f} = \int\limits_{\tau_2} \tilde{v}_{[\lambda\kappa}\, d\mathfrak{f}_{\mu]}, \tag{4.26}$$

$$\boxed{\int\limits_{\tau_3} \partial_\mu \mathfrak{v}^\mu\, d\mathfrak{f} = \int\limits_{\tau_2} \mathfrak{v}^\mu\, d\mathfrak{f}_\mu,} \tag{4.27}$$

$$\tfrac{1}{3}\int\limits_{\tau_3} \partial_\mu \mathfrak{v}^\mu\, d\tilde{f}^{\nu\kappa_1\kappa_2} = \int\limits_{\tau_3} \partial_\mu \mathfrak{v}^{[\nu}\, d\tilde{f}^{|\mu|\kappa_1\kappa_2]} = \int\limits_{\tau_2} \mathfrak{v}^{[\nu}\, d\tilde{f}^{\kappa_1\kappa_2]}. \tag{4.28}$$

(2) $q = 1$.

(*a*) Integral of a covariant vector over a curve with inner orientation:

$$\boxed{\int\limits_{\tau_2} \partial_{[\mu} v_{\lambda]}\, df^{\mu\lambda} = \int\limits_{\tau_1} v_\lambda\, df^\lambda,} \tag{4.29}$$

$$2\int\limits_{\tau_2} \partial_{[\mu} v_\lambda\, d\tilde{\mathfrak{f}}_{\kappa]} = \int\limits_{\tau_1} v_{[\lambda}\, d\tilde{\mathfrak{f}}_{\kappa\mu]}, \tag{4.30}$$

† Cf. Weyssenhoff 1937. 2; Schouten and v. Dantzig 1940. 1, p. 472.

$$2\int\limits_{\tau_2} \partial_\mu \tilde{\mathfrak{v}}^{\lambda\mu}\, d\tilde{\mathfrak{f}}_\lambda = \int\limits_{\tau_1} \tilde{\mathfrak{v}}^{\lambda\mu}\, d\tilde{\mathfrak{f}}_{\lambda\mu}, \tag{4.31}$$

$$\int\limits_{\tau_3} \partial_\mu \tilde{\mathfrak{v}}^{[\lambda|\mu|}\, df^{\nu\kappa]} = \int\limits_{\tau_2} \partial_\mu \tilde{\mathfrak{v}}^{[\lambda\nu}\, df^{|\mu|\kappa]} = \int\limits_{\tau_1} \tilde{\mathfrak{v}}^{[\lambda\nu}\, df^{\kappa]}. \tag{4.32}$$

(*b*) Integral of a contravariant bivector density of weight $+1$ over a curve with outer orientation:

$$\int\limits_{\tau_2} \partial_{[\mu} \tilde{v}_{\lambda]}\, d\tilde{f}^{\mu\lambda} = \int\limits_{\tau_1} \tilde{v}_\lambda\, d\tilde{f}^{\lambda}, \tag{4.33}$$

$$2\int\limits_{\tau_2} \partial_{[\mu} \tilde{v}_{\lambda}\, d\mathfrak{f}_{\kappa]} = \int\limits_{\tau_1} \tilde{v}_{[\lambda}\, d\mathfrak{f}_{\kappa\mu]}, \tag{4.34}$$

$$\boxed{2\int\limits_{\tau_2} \partial_\mu \mathfrak{v}^{\lambda\mu}\, d\mathfrak{f}_\lambda = \int\limits_{\tau_1} \mathfrak{v}^{\lambda\mu}\, d\mathfrak{f}_{\lambda\mu}} \tag{4.35}$$

$$\int\limits_{\tau_3} \partial_\mu \mathfrak{v}^{[\lambda|\mu|}\, df^{\nu\kappa]} = \int\limits_{\tau_2} \partial_\mu \mathfrak{v}^{[\lambda\nu}\, df^{|\mu|\kappa]} = \int\limits_{\tau_1} \mathfrak{v}^{[\lambda\nu}\, df^{\kappa]}. \tag{4.36}$$

The most important formulae are (4.21), (4.27), (4.29), and (4.35). The integrals occurring in (4.27) can be interpreted hydrodynamically. Let v^κ be the velocity of a motion of a fluid and μ its mass density, i.e. the mass per parallelepiped of the measuring vectors. Then $\mathfrak{v}^\kappa = \mu v^\kappa$ is the vector density of the current and $\mathfrak{v}^\mu\, d\mathfrak{f}_\mu$ is the total mass flowing through the element $d\mathfrak{f}_\mu$ in unit time in the direction of the orientation of $d\mathfrak{f}_\mu$. If this orientation is directed towards the outside of τ_2, the integral $\int\limits_{\tau_2} \mathfrak{v}^\mu\, d\mathfrak{f}_\mu$ is the total mass flowing through τ_2 in unit time to the outside. The equation of continuity†

$$\partial_\lambda \mathfrak{v}^\lambda + \frac{d\mu}{dt} = 0, \tag{4.37}$$

expresses the fact that the divergence of the current density equals the decrease of mass density in unit time. Hence

$$\int\limits_{\tau_3} \partial_\lambda \mathfrak{v}^\lambda\, d\mathfrak{f} = -\int\limits_{\tau_3} \frac{d\mu}{dt}\, d\mathfrak{f} \tag{4.38}$$

is the total loss of mass in τ_3 in unit time. Obviously this loss has to be equal to the total mass flowing through τ_2 in unit time $\int\limits_{\tau_2} \mathfrak{v}^\mu\, d\mathfrak{f}_\mu$.

If $\partial_\mu \mathfrak{v}^\mu = 0$ the total mass in every τ_3 is constant. If k is a small constant the whole space can be split up into tubes with impenetrable walls such that the mass k flows through each section of each tube in

† In an X_3 no unit of volume exists, and two infinitesimal volumes at different points cannot be compared. Hence in an X_3 it is nonsense to speak of an 'incompressible' fluid.

unit time. The tube at a point ξ^κ represents approximately the vector density $\mathfrak{v}^\kappa/k$ and this approximation tends to an exact representation if k tends to zero. Hence the equation $\partial_\mu \mathfrak{v}^\mu = 0$ expresses the fact that the tubes of $\mathfrak{v}^\kappa$ on an 'infinitesimal scale' fill the whole X_3 without gaps.

We will prove that w_λ is always a gradient if $\operatorname{Rot} w = 0$. To do this we take a fixed point $\underset{0}{\xi}{}^\kappa$ and a variable point ξ^κ in the region considered. If these points are connected by two different curves s and s' we have, according to the theorem of Stokes,

$$\int_{s,\ \underset{0}{\xi}{}^\kappa}^{\xi^\kappa} w_\lambda \, d\xi^\lambda + \int_{s',\ \xi^\kappa}^{\underset{0}{\xi}{}^\kappa} w_\lambda \, d\xi^\lambda = \int_{\tau_2} \partial_{[\mu} w_{\lambda]} \, df^{\mu\lambda} = 0, \tag{4.39}$$

provided that the surface τ_2 bounded by s and s' lies wholly in the region where $\operatorname{Rot} w = 0$. Hence

$$p = \int_{s,\ \underset{0}{\xi}{}^\kappa}^{\xi^\kappa} w_\lambda \, d\xi^\lambda = \int_{s',\ \underset{0}{\xi}{}^\kappa}^{\xi^\kappa} w_\lambda \, d\xi^\lambda \tag{4.40}$$

is a function of ξ^κ which is independent of the choice of s and

$$w_\lambda = \partial_\lambda p. \tag{4.41}$$

Starting from this theorem the following more general theorem can be proved by induction:

THEOREM. *If the rotation of a covariant q-vector (W-q-vector) vanishes in some region of X_n, there always exists a region where the q-vector (W-q-vector) can be expressed as the rotation of a covariant (q—1)-vector (W-(q—1)-vector).*

This theorem can of course also be expressed in terms of contravariant p-vector densities (p-vector-Δ-densities) of weight $+1$. Then we get:

THEOREM. *If the divergence of a contravariant p-vector density (p-vector-Δ-density) of weight $+1$ vanishes in some region of X_n, there always exists a region where it can be expressed as the divergence of a contravariant $(p+1)$-vector density ($(p+1)$-vector-Δ-density) of weight $+1$ $(p < n)$.*

All integrals that have occurred in this section up till now have been scalars or n-vector-Δ-densities. This is due to the fact that in an X_n these are the only quantities that can be added if they belong to different points. But if we restrict ourselves to more special transformations of coordinates, other forms of the theorem of Stokes arise. For instance, if we only allow coordinate transformations with $\Delta > 0$, W-scalars can be added and besides (4.1) we have the formula

$$\int_{\tau_{q+1}} \partial_{[\mu} \tilde{v}_{\lambda_1 \ldots \lambda_q]} df^{\mu\lambda_1\ldots\lambda_q} = \int_{\tau_q} \tilde{v}_{\lambda_1\ldots\lambda_q} df^{\lambda_1\ldots\lambda_q}. \tag{4.42}$$

If only transformations of G_a are allowed, that is, if we are in an $\overset{\circ}{E}_n$, all quantities that belong to different points with the same manner of transformation can be added. Accordingly the following formulae hold in E_n:

$$\int_{\tau_{q+1}} df^{\mu\lambda_1\ldots\lambda_q}\partial_\mu \multimap \Omega = \int_{\tau_q} df^{\lambda_1\ldots\lambda_q} \multimap \Omega, \tag{4.43}$$

$$p\int_{\tau_{q+1}} d\tilde{\mathfrak{f}}_{[\lambda_1\ldots\lambda_{p-1}}\partial_{\lambda_p]} \multimap \Omega = \int_{\tau_q} d\tilde{\mathfrak{f}}_{\lambda_1\ldots\lambda_p} \multimap \Omega, \tag{4.44}$$

$$\int_{\tau_{q+1}} d\tilde{f}^{\mu\lambda_1\ldots\lambda_q}\partial_\mu \multimap \Omega = \int_{\tau_q} d\tilde{f}^{\lambda_1\ldots\lambda_q} \multimap \Omega, \tag{4.45}$$

$$p\int_{\tau_{q+1}} d\mathfrak{f}_{[\lambda_1\ldots\lambda_{p-1}}\partial_{\lambda_p]} \multimap \Omega = \int_{\tau_q} d\mathfrak{f}_{\lambda_1\ldots\lambda_p} \multimap \Omega, \tag{4.46}$$

where Ω is a symbol with suppressed indices standing for any quantity and where $\multimap$ symbolizes any operation performed on the indices $\lambda_1\ldots\lambda_q$ or $\lambda_1\ldots\lambda_p$ and the indices of Ω, which is composed of alternation, mixing, or transvection.

For example, if

$$\mathfrak{v}_\lambda = \partial_\mu \mathfrak{P}^{\mu}_{\cdot\lambda}, \tag{4.47}$$

we have in E_3

$$\int_{\tau_3} \mathfrak{v}_\lambda d\mathfrak{f} = \int_{\tau_2} \mathfrak{P}^{\mu}_{\cdot\lambda} d\mathfrak{f}_\mu, \tag{4.48}$$

but this equation does not hold in an X_3.

5. Invariant differential operators, III: The Lie derivative†

Let the points of a region R of X_n be subject to the point transformation

$$\eta^\kappa \overset{*}{=} f^\kappa(\xi^\lambda). \tag{5.1}$$

The functions f^κ are supposed to be analytic in R with a non-vanishing functional determinant

$$\left|\frac{\partial f^\kappa}{\partial \xi^\lambda}\right| \neq 0, \tag{5.2}$$

and to be chosen in such a way that they determine a one-to-one correspondence between the points of R and the points of another region R'. Now we introduce another coordinate system (κ') such that each point

† Cf. Slebodzinski 1931. 2; Schouten and v. Kampen 1933. 1; E I 1935. 1; P.P. 1949. 1; R.C. II § 10 also for further literature.

in R has the same coordinates with respect to (κ) as its image in R' has with respect to (κ'). Let ξ^κ and $\xi^{\kappa'}$ be the coordinates of a point of R' with respect to (κ) and (κ') respectively. The $\xi^{\kappa'}$ must be equal to the coordinates with respect to (κ) of the corresponding point of R. Hence, if

$$\xi^\kappa \overset{*}{=} \phi^\kappa(\eta^\lambda) \tag{5.3}$$

is the inversion of (5.1), the $\xi^{\kappa'}$ must be equal to the $\phi^\kappa(\xi^\lambda)$ and accordingly the transformation of (κ) into (κ') and vice versa is given by the equations

$$\xi^{\kappa'} = \delta^{\kappa'}_\kappa \phi^\kappa(\xi^\lambda), \qquad \xi^\kappa = f^\kappa(\xi^{\lambda'}). \tag{5.4}$$

This process we call the *dragging along*† *of the coordinate system* (κ) *by the point transformation* (5.1).

Now let some field, for example $\underset{1}{P}^{\kappa\lambda}{}_{\cdot\cdot\mu}$, be given in R. We form a second field in R', whose components $\underset{2}{P}^{\kappa'\lambda'}{}_{\cdot\cdot\mu'}$ with respect to (κ') in any point of R' are equal to the components $\underset{1}{P}^{\kappa\lambda}{}_{\cdot\cdot\mu}$ of the first field in the corresponding point of R. This process we call the *dragging along*† *of the field* $\underset{1}{P}^{\kappa\lambda}{}_{\cdot\cdot\mu}$ *by the point transformation* (5.1) and we call $\underset{2}{P}^{\kappa\lambda}{}_{\cdot\cdot\mu}$ the *field dragged along*. If R and R' have some region in common, the fields $\underset{1}{P}$ and $\underset{2}{P}$ can be compared. If then $\underset{1}{P} = \underset{2}{P}$, the field $\underset{1}{P}$ is called *invariant for the point transformation* (5.1).

We consider a system of n ordinary differential equations of the form

$$\frac{d\eta^\kappa}{dt} = \psi^\kappa(\eta^\nu), \tag{5.5}$$

with the independent variables ξ^κ and t and the unknown variables η^κ, with functions ψ^κ analytic in the region considered and with the initial conditions

$$\eta^\kappa = \xi^\kappa \quad \text{for} \quad t = 0. \tag{5.6}$$

If the solution of these equations is expanded into a power series in t we get

$$\begin{aligned} \eta^\kappa &= \xi^\kappa + \psi^\kappa(\xi^\nu)t + \dots, \\ \xi^\kappa &= \eta^\kappa - \psi^\kappa(\eta^\nu)t + \dots. \end{aligned} \tag{5.7}$$

These equations represent a point transformation. For this point transformation the coordinate system which is dragged along (κ') is given by the equations (cf. (5.4))

$$\begin{aligned} \xi^\kappa &= \delta^\kappa_{\kappa'}\xi^{\kappa'} + \psi^\kappa(\xi^{\nu'})t + \dots, \\ \xi^{\kappa'} &= \delta^{\kappa'}_\kappa \xi^\kappa - \delta^{\kappa'}_\kappa \psi^\kappa(\xi^\lambda)t + \dots. \end{aligned} \tag{5.8}$$

† 'Mitschleppen' in 1933. 1 and 1935. 1.

Differentiation gives, writing $v^\kappa \stackrel{\text{def}}{=} \psi^\kappa(\xi^\kappa)$:

$$\begin{aligned} A^{\kappa}_{\lambda'} &\stackrel{*}{=} \delta^{\lambda}_{\lambda'}(A^{\kappa}_{\lambda}+t\partial_\lambda v^\kappa+...),\\ A^{\kappa'}_{\lambda} &\stackrel{*}{=} \delta^{\kappa'}_{\kappa}(A^{\kappa}_{\lambda}-t\partial_\lambda v^\kappa+...).\end{aligned} \tag{5.9}$$

If we now neglect all higher powers of t and accordingly write dt instead of t, we get what is called an '*infinitesimal point transformation* $v^\kappa dt$'. The one-parameter group of transformations (5.7) is said to be *generated by the infinitesimal transformation* $v^\kappa dt$. All points of X_n undergo a *displacement* $v^\kappa dt$. If a coordinate system or a field is dragged along by *the infinitesimal transformation* $v^\kappa dt$ we call this process the *dragging along over* $v^\kappa dt$. The system which is dragged along (κ') is given by the equations

$$\begin{aligned} \xi^\kappa &\stackrel{*}{=} \delta^{\kappa}_{\kappa'}\xi^{\kappa'}+v^\kappa dt,\\ \xi^{\kappa'} &\stackrel{*}{=} \delta^{\kappa'}_{\kappa}\xi^{\kappa}-\delta^{\kappa'}_{\kappa}v^\kappa dt.\end{aligned} \tag{5.10}$$

Now since v^κ is given we may require the components with respect to (κ) of a field, for instance $P^{\kappa}_{\cdot\lambda}$, dragged along over $v^\kappa dt$. The components of the field dragged along at $\xi^\kappa+v^\kappa dt$ with respect to (κ') are equal to the components $P^{\kappa}_{\cdot\lambda}$ in ξ^κ. Hence the components of this field in $\xi^\kappa+v^\kappa dt$ with respect to (κ) are

$$\begin{aligned} A^{\rho'\kappa}_{\lambda\,\sigma'}\delta^{\rho}_{\rho'}\delta^{\sigma'}_{\sigma}P^{\sigma}_{\cdot\rho} &\stackrel{*}{=} (A^{\rho}_{\lambda}-\partial_\lambda v^\rho dt)(A^{\kappa}_{\sigma}+\partial_\sigma v^\kappa dt)P^{\sigma}_{\cdot\rho}\\ &\stackrel{*}{=} P^{\kappa}_{\cdot\lambda}-P^{\kappa}_{\cdot\rho}\partial_\lambda v^\rho dt+P^{\sigma}_{\cdot\lambda}\partial_\sigma v^\kappa dt.\end{aligned} \tag{5.11}$$

To get the components of the field dragged along in ξ^κ with respect to (κ) we have to subtract $v^\mu\partial_\mu P^{\kappa}_{\cdot\lambda}dt$; hence these components are

$$P^{\kappa}_{\cdot\lambda}-v^\mu\partial_\mu P^{\kappa}_{\cdot\lambda}dt-P^{\kappa}_{\cdot\rho}\partial_\lambda v^\rho dt+P^{\sigma}_{\cdot\lambda}\partial_\sigma v^\kappa dt. \tag{5.12}$$

The expression
$$(v^\mu\partial_\mu P^{\kappa}_{\cdot\lambda}+P^{\kappa}_{\cdot\rho}\partial_\lambda v^\rho-P^{\sigma}_{\cdot\lambda}\partial_\sigma v^\kappa)\,dt \tag{5.13}$$

is called the *Lie differential of* $P^{\kappa}_{\cdot\lambda}$ *with respect to* $v^\kappa dt$. Hence, *if a field is dragged along over* $v^\kappa dt$ *the increase of the field in the fixed point* ξ^κ *is equal to the negative Lie differential in* ξ^κ.

The expression†

$$\underset{v}{\pounds} P^{\kappa}_{\cdot\lambda} \stackrel{\text{def}}{=} v^\mu\partial_\mu P^{\kappa}_{\cdot\lambda}+P^{\kappa}_{\cdot\rho}\partial_\lambda v^\rho-P^{\sigma}_{\cdot\lambda}\partial_\sigma v^\kappa \tag{5.14}$$

is called the *Lie derivative of* $P^{\kappa}_{\cdot\lambda}$ *with respect to* v^κ. From its definition it is clear that it is an affinor with the same valences as $P^{\kappa}_{\cdot\lambda}$. This can be verified by a direct calculation of the manner of transformation. For an affinor field of higher valence, the Lie derivative is constructed in the same way. Each upper (lower) index gives rise to a term with a negative (positive) sign.

† Cf. Ch. V, § 1, where the Lie derivative is written with the covariant operator ∇_μ.

Obviously the following rules are valid:

(1) the Lie derivative of a sum is the sum of the Lie derivatives of the summands;

(2) the Lie derivative of a contraction is the contraction of the Lie derivative,

(3) the rule of Leibniz holds for the Lie derivative of products and transvections.

The Lie derivative of v^κ with respect to v^κ is zero. The Lie derivative of A^κ_λ is zero for every choice of v^κ.

For a scalar field we have

$$\underset{v}{\pounds}\, s = v^\mu \partial_\mu s, \tag{5.15}$$

for a vector field

$$\begin{aligned} &(a) \quad \underset{v}{\pounds}\, u^\kappa = v^\mu \partial_\mu u^\kappa - u^\mu \partial_\mu v^\kappa, \\ &(b) \quad \underset{v}{\pounds}\, w_\lambda = v^\mu \partial_\mu w_\lambda + w_\mu \partial_\lambda v^\mu, \end{aligned} \tag{5.16}$$

and for a p-vector field,

$$\begin{aligned} &(a) \quad \underset{v}{\pounds}\, u^{\kappa_1 \dots \kappa_p} = v^\mu \partial_\mu u^{\kappa_1 \dots \kappa_p} - p u^{\mu[\kappa_2 \dots \kappa_p} \partial_\mu v^{\kappa_1]}, \\ &(b) \quad \underset{v}{\pounds}\, w_{\lambda_1 \dots \lambda_p} = v^\mu \partial_\mu w_{\lambda_1 \dots \lambda_p} + p w_{\mu[\lambda_2 \dots \lambda_p} \partial_{\lambda_1]} v^\mu. \end{aligned} \tag{5.17}$$

To obtain the Lie derivative of a scalar density or Δ-density of weight $\mathfrak{k}$ we start from (5.17 b) for $p = n$:

$$\underset{v}{\pounds}\, w_{\lambda_1 \dots \lambda_n} = v^\mu \partial_\mu w_{\lambda_1 \dots \lambda_n} + n w_{\mu[\lambda_2 \dots \lambda_n} \partial_{\lambda_1]} v^\mu = v^\mu \partial_\mu w_{\lambda_1 \dots \lambda_n} + w_{\lambda_1 \dots \lambda_n} \partial_\mu v^\mu$$
$$= \partial_\mu v^\mu w_{\lambda_1 \dots \lambda_n}. \tag{5.18}$$

Now $w_{1 \dots n}$ may be considered as the component of a scalar Δ-density of weight $+1$. Hence for a scalar Δ-density of weight $+1$ we have

$$\underset{v}{\pounds}\, \tilde{\mathfrak{q}} = \partial_\mu \tilde{\mathfrak{q}} v^\mu = \mathrm{Div}(\tilde{\mathfrak{q}} v). \tag{5.19}$$

Since every scalar density $\mathfrak{q}$ of weight $+1$ is the product of $\tilde{\omega}$ (cf. III, § 3) with a scalar Δ-density of weight $+1$ we have

$$\underset{v}{\pounds}\, \mathfrak{q} = \partial_\mu \mathfrak{q} v^\mu = \mathrm{Div}(\mathfrak{q} v). \tag{5.20}$$

From this we get for a scalar density $\mathfrak{s}$ of weight $\mathfrak{k}$

$$\underset{v}{\pounds}\, \mathfrak{s} = v^\mu \partial_\mu \mathfrak{s} + \mathfrak{k} \mathfrak{s} \partial_\mu v^\mu, \tag{5.21}$$

and for an affinor density of weight $\mathfrak{k}$, for example $\mathfrak{P}^{\kappa\lambda}{}_{\cdot\cdot\nu}$,

$$\underset{v}{\pounds}\, \mathfrak{P}^{\kappa\lambda}{}_{\cdot\cdot\nu} = v^\mu \partial_\mu \mathfrak{P}^{\kappa\lambda}{}_{\cdot\cdot\nu} + \mathfrak{P}^{\kappa\lambda}{}_{\cdot\cdot\rho} \partial_\nu v^\rho - \mathfrak{P}^{\sigma\lambda}{}_{\cdot\cdot\nu} \partial_\sigma v^\kappa - \mathfrak{P}^{\kappa\sigma}{}_{\cdot\cdot\nu} \partial_\sigma v^\lambda + \mathfrak{k} \mathfrak{P}^{\kappa\lambda}{}_{\cdot\cdot\nu} \partial_\mu v^\mu. \tag{5.22}$$

Similar formulae hold for affinor-Δ-densities. It may easily be verified that the Lie derivatives of $\mathfrak{E}^{\kappa_1 \ldots \kappa_n}$ and $\mathfrak{e}_{\lambda_1 \ldots \lambda_n}$ vanish.

If the Lie derivative with respect to v^κ of some field vanishes at all points, the field is said to be *absolutely invariant with respect to the field* v^κ. An absolutely invariant field has the following properties:

(1) *the field value at any fixed point does not change if the field is dragged along by any one of the point transformations of the one-parameter group generated by* $v^\kappa\, dt$;

(2) *the components of the field at any point* ξ^κ *with respect to* (κ) *are equal to the components at the point* $\xi^\kappa + v^\kappa\, dt$ *with respect to the coordinate system* (κ') *that arises from* (κ) *by dragging along over* $v^\kappa\, dt$.

Obviously sums, products, and transvections of absolutely invariant fields are absolutely invariant.

The invariance of a field with respect to v^κ is generally lost if v^κ is replaced by σv^κ. This means that the field is not invariant for all point transformations which leave the streamlines of the field v^κ invariant, but only for those generated by $v^\kappa\, dt$. If the Lie derivative of a field vanishes for *all* values of σ, the field is said to be *absolutely invariant with respect to the streamlines of* v^κ. It follows from (5.17 *b*) that for a covariant p-vector $w_{\lambda_1 \ldots \lambda_p}$ the additional condition for this invariance is

$$v^\mu w_{\mu \lambda_2 \ldots \lambda_p} = 0. \tag{5.23}$$

The same holds for a covariant W-p-vector. In the same way we get the additional condition for a p-vector density (or p-vector-Δ-density) of weight $+1$:

$$\mathfrak{w}^{[\kappa_1 \ldots \kappa_p} v^{\kappa]} = 0. \tag{5.24}$$

From this we see that a scalar or an n-vector density of weight $+1$, which is absolutely invariant with respect to v^κ, is always absolutely invariant with respect to the streamlines of v^κ and that a covariant n-vector or a scalar density of weight $+1$ can never have this stronger invariance.†

6. Invariant differential operators, IV: The Lagrange derivative‡

Let Φ_Λ $(\Lambda = 1, \ldots, N)$ be a set of functions of the coordinates ξ^κ. The Φ_Λ may be scalars or components of a geometric object, for instance an affinor $P_{\kappa\lambda}$, or they may be quite arbitrary. At first we do not consider the manner of transformation of the Φ_Λ and accordingly the ξ^κ will

† Cf. for more particulars on absolutely and relatively invariant fields P.P. 1949. 1 pp. 73 ff.

‡ Cf. R.C. II § 11.

not be transformed. Let the index Λ be suppressed and the derivatives of the Φ with respect to ξ^κ be denoted by Φ_μ, $\Phi_{\nu\mu}$, etc.

Now let $\mathfrak{L}$ be a function of the Φ, Φ_μ, etc., up to a certain definite order. Then in a certain region of X_n $\mathfrak{L}$ is a function of the ξ^κ, but we do not know anything about its manner of transformation. We consider the integral

$$\int_{\tau_n} \mathfrak{L}\, d\xi^1 ... d\xi^n \tag{6.1}$$

over an arbitrary region τ_n of X_n where the Φ are analytic in the ξ^κ.

Let the field Φ be subjected to a variation $\overset{v}{d}\Phi$ such that the variations of Φ and of all its derivatives occurring in $\mathfrak{L}$ vanish at the boundary τ_{n-1} of τ_n. Then for the variation of $\mathfrak{L}$ we have

$$\overset{v}{d}\mathfrak{L} = \frac{\partial \mathfrak{L}}{\partial \Phi} \overset{v}{d}\Phi + \frac{\partial \mathfrak{L}}{\partial \Phi_\mu} \overset{v}{d}\Phi_\mu + ..., \tag{6.2}$$

and consequently

$$\overset{v}{d} \int_{\tau_n} \mathfrak{L}\, d\xi^1 ... d\xi^n = \int_{\tau_n} \left(\frac{\partial \mathfrak{L}}{\partial \Phi} \overset{v}{d}\Phi + \frac{\partial \mathfrak{L}}{\partial \Phi_\mu} \overset{v}{d}\Phi_\mu + ... \right) d\xi^1 ... d\xi^n. \tag{6.3}$$

By integration by parts we get, using Stokes's theorem (4.16) for $p = 1$,

$$\begin{aligned} \overset{v}{d} \int_{\tau_n} \mathfrak{L}\, d\xi^1 ... d\xi^n &= \int_{\tau_n} \left(\frac{\partial \mathfrak{L}}{\partial \Phi} \overset{v}{d}\Phi + \frac{\partial \mathfrak{L}}{\partial \Phi_\mu} \partial_\mu \overset{v}{d}\Phi + ... \right) d\xi^1 ... d\xi^n \\ &= \int_{\tau_n} \left(\frac{\partial \mathfrak{L}}{\partial \Phi} - \partial_\mu \frac{\partial \mathfrak{L}}{\partial \Phi_\mu} + \partial_\nu \partial_\mu \frac{\partial \mathfrak{L}}{\partial \Phi_{\mu\nu}} - ... \right) \overset{v}{d}\Phi d\xi^1 ... d\xi^n. \end{aligned} \tag{6.4}$$

Hence, if we write

$$[\mathfrak{L}] \overset{\text{def}}{=} \frac{\partial \mathfrak{L}}{\partial \Phi} - \partial_\mu \frac{\partial \mathfrak{L}}{\partial \Phi_\mu} + \partial_\nu \partial_\mu \frac{\partial \mathfrak{L}}{\partial \Phi_{\mu\nu}} - ..., \tag{6.5}$$

we have

$$\overset{v}{d} \int_{\tau_n} \mathfrak{L}\, d\xi^1 ... d\xi^n = \int_{\tau_n} [\mathfrak{L}] \overset{v}{d}\Phi d\xi^1 ... d\xi^n. \tag{6.6}$$

Now here is a difficulty. In (4.16) $\mathfrak{v}^\kappa$ was a vector density of weight $+1$. But we know nothing about the manner of transformation of $\frac{\partial \mathfrak{L}}{\partial \Phi_\mu} \overset{v}{d}\Phi$ for example. This difficulty clears itself when we notice that (4.16) holds for every coordinate system, and thus for the coordinate system (κ) used here, and also that we have made the condition that the coordinates should not be transformed. Hence the transformation of $\frac{\partial \mathfrak{L}}{\partial \Phi_\mu} \overset{v}{d}\Phi$ is not important.

$[\mathfrak{L}]$ is called the *Lagrange derivative* of $\mathfrak{L}$. If $[\mathfrak{L}] = 0$ the variation

(6.3) vanishes for every choice of τ_n, provided that the variation of Φ satisfies the boundary conditions.

The equation $[\mathfrak{L}] = 0$ is called the *Lagrange equation.* Equations of this form occur in classical dynamics. The ξ^κ reduce to *one* variable t and this variable *is not transformed.*

Now let us suppose that Φ is an affinor, e.g. $P_{\lambda\kappa}$, and that $\mathfrak{L}$ is a scalar Δ-density of weight $+1$, depending on $P_{\lambda\kappa}, \partial_\mu P_{\lambda\kappa}, \ldots$. Then the integral (6.1) is a scalar and the building of the Lagrange derivative is invariant for all coordinate transformations. The equation $[\mathfrak{L}] = 0$ is now also invariant. If Φ has the valences p, q, $[\mathfrak{L}]$ is an affinor-Δ-density of weight $+1$ with the valences q, p. If we only consider coordinate transformations with $\Delta > 0$ a scalar density of weight $+1$ can be taken for $\mathfrak{L}$, and $[\mathfrak{L}]$ is then an affinor density of weight $+1$ with the valences q, p.

If Φ is an affinor and $\tilde{\mathfrak{L}}$ a scalar Δ-density of weight $+1$ a very important relation exists between Φ, $[\tilde{\mathfrak{L}}]$ and their first derivatives. In order to obtain this relation we suppose that the variation $\overset{v}{d}\Phi$ is due to the dragging along of the field over $v^\kappa dt$, where v^κ is an arbitrary field such that $v^\kappa, \partial_\mu v^\kappa, \partial_\mu \partial_\lambda v^\mu, \ldots$ are zero on τ_{n-1}. We avoid complicated formulae by taking an affinor of valence two $\Phi = P_{\lambda\kappa}$. No extra difficulties arise for higher valences. We then have (cf. (5.12))

$$\overset{v}{d}P_{\lambda\kappa} = -v^\mu \partial_\mu P_{\lambda\kappa}\, dt - P_{\mu\kappa}\, \partial_\lambda v^\mu\, dt - P_{\lambda\mu}\, \partial_\kappa v^\mu\, dt, \tag{6.7}$$

and the field dragged along is

$$'P_{\lambda\kappa} = P_{\lambda\kappa} - v^\mu \partial_\mu P_{\lambda\kappa}\, dt - P_{\mu\kappa}\, \partial_\lambda v^\mu\, dt - P_{\lambda\mu}\, \partial_\kappa v^\mu\, dt. \tag{6.8}$$

Accordingly we have

$$\begin{aligned}\partial_\nu{}' P_{\lambda\kappa} = \partial_\nu P_{\lambda\kappa} &- (\partial_\nu v^\mu)\partial_\mu P_{\lambda\kappa}\, dt - v^\mu \partial_\nu \partial_\mu P_{\lambda\kappa}\, dt - (\partial_\nu P_{\mu\kappa})\partial_\lambda v^\mu\, dt - \\ &- P_{\mu\kappa}\, \partial_\nu \partial_\lambda v^\mu\, dt - (\partial_\nu P_{\lambda\mu})\partial_\kappa v^\mu\, dt - P_{\lambda\mu}\, \partial_\nu \partial_\kappa v^\mu\, dt,\end{aligned} \tag{6.9}$$

and from this we see that $\overset{v}{d}P_{\lambda\kappa}$ and $\overset{v}{d}\,\partial_\nu P_{\lambda\kappa}$ are zero on τ_{n-1}. For higher derivatives of $P_{\lambda\kappa}$ the same holds, provided that sufficiently high derivatives of v^κ vanish on τ_{n-1}.

If (6.7) is now substituted into (6.4) we get, on using (6.5) and Stokes's theorem,

$$\begin{aligned}&\overset{v}{d} \int_{\tau_n} \tilde{\mathfrak{L}}\, d\xi^1 \ldots d\xi^n \\ &\qquad = - \int_{\tau_n} [\tilde{\mathfrak{L}}]^{\lambda\kappa} (v^\mu \partial_\mu P_{\lambda\kappa} + P_{\mu\kappa}\, \partial_\lambda v^\mu + P_{\lambda\mu}\, \partial_\kappa v^\mu)\, dt d\xi^1 \ldots d\xi^n \\ &\qquad = - \int_{\tau_n} \{[\tilde{\mathfrak{L}}]^{\lambda\kappa} (\partial_\mu P_{\lambda\kappa} - \partial_\lambda P_{\mu\kappa} - \partial_\kappa P_{\lambda\mu}) - P_{\mu\kappa}\, \partial_\lambda [\tilde{\mathfrak{L}}]^{\lambda\kappa} - \\ &\qquad\qquad\qquad - P_{\lambda\mu}\, \partial_\kappa [\tilde{\mathfrak{L}}]^{\lambda\kappa}\} v^\mu\, dt d\xi^1 \ldots d\xi^n.\end{aligned} \tag{6.10}$$

But this variation has to be zero, because the field dragged along over $v^\kappa dt$ has exactly the same components with respect to (κ') as the original field with respect to (κ). Hence

$$[\tilde{\mathfrak{L}}]^{\lambda\kappa}(\partial_\mu P_{\lambda\kappa} - \partial_\lambda P_{\mu\kappa} - \partial_\kappa P_{\lambda\mu}) - P_{\mu\kappa}\partial_\lambda[\tilde{\mathfrak{L}}]^{\lambda\kappa} - P_{\lambda\mu}\partial_\kappa[\tilde{\mathfrak{L}}]^{\lambda\kappa} = 0.\dagger \quad (6.11)$$

As we shall see later, this identity is of special importance in the theory of relativity. If we only consider those coordinate transformations for which $\Delta > 0$, a scalar density of weight $+1$ can be taken for $\mathfrak{L}$. Then $[\mathfrak{L}]$ is an affinor density. (For further relations and literature, R.C. II § 11.)

7. Anholonomic coordinate systems in X_n‡

Starting from any system of allowable coordinates ξ^κ we always get covariant measuring vectors $\overset{\kappa}{e}_\lambda$ whose rotations vanish. It is often convenient to introduce systems of measuring vectors in every local E_n of X_n which are not connected with any system of allowable coordinates but which have some other desirable property.

If we introduce n arbitrary contravariant vector fields $\underset{i}{e}^\kappa$ $(i = 1,..., n)$ and the reciprocal system (cf. II, § 2) $\overset{h}{e}_\lambda$ $(h = 1,..., n)$, we have

$$\overset{h}{e}_\mu \underset{i}{e}^\mu = \delta_i^h; \qquad \overset{h}{e}_\lambda \underset{h}{e}^\kappa = A_\lambda^\kappa \quad (h, i = 1,..., n), \quad (7.1)$$

and in general the $\overset{h}{e}_\lambda$ are not gradient vectors.

We call such a system of measuring vectors an *anholonomic coordinate system* in X_n and the components of an object with respect to this system *anholonomic components*. If

$$A_\lambda^h \overset{\text{def}}{=} \underset{i}{e}^h \overset{i}{e}_\lambda; \qquad A_i^\kappa \overset{\text{def}}{=} \underset{h}{e}^\kappa \overset{h}{e}_i \quad (h, i = 1,..., n), \quad (7.2)$$

the anholonomic components of, for example, $P^{\kappa\lambda}{}_{..\mu}$ are

$$P^{hi}{}_{..j} = A^{hi\mu}_{\kappa\lambda j} P^{\kappa\lambda}{}_{..\mu} \quad (h, i, j = 1,..., n), \quad (7.3)$$

and the anholonomic components of $d\xi^\kappa$ are

$$(d\xi)^h = A_\kappa^h \, d\xi^\kappa \quad (h = 1,..., n). \quad (7.4)$$

We must not write $d\xi^h$ because in general $A_\kappa^h \, d\xi^\kappa$ is not a total differential and accordingly no variable ξ^h exists.

Using anholonomic coordinates the operator

$$\partial_j \overset{\text{def}}{=} A_j^\mu \partial_\mu \quad (j = 1,..., n) \quad (7.5)$$

† Cf. Ch. V, § 1, where this identity is written with the covariant operator ∇_μ.

‡ For references to literature cf. E I. 1935. 1, p. 94, R.C. II §§ 5, 9.

must be introduced instead of ∂_μ. Then the gradient of a scalar p can be written in anholonomic components

$$w_i = \partial_i p \quad (i = 1,..., n). \tag{7.6}$$

But the rotation of a vector w_λ and $2\partial_{[j} w_{i]}$ are connected in a more complicated way.

$$\begin{aligned}\partial_{[j} w_{i]} &= A^\mu_{[j} \partial_{|\mu|} A^\lambda_{i]} w_\lambda = A^{\mu\lambda}_{ji} \partial_{[\mu} w_{\lambda]} + (A^\mu_{[j} \partial_{|\mu|} A^\lambda_{i]}) w_\lambda \\ &= A^{\mu\lambda}_{ji} \partial_{[\mu} w_{\lambda]} - \Omega_{ji}^{\cdot\cdot h} w_h \quad (h, i, j = 1,..., n),\end{aligned} \tag{7.7}$$

where $\boxed{\Omega_{ji}^{\cdot\cdot h} \overset{\text{def}}{=} A^{\mu\lambda}_{ji} \partial_{[\mu} A^h_{\lambda]}} \quad (h, i, j = 1,..., n).$ (7.8)

We call $\Omega_{ji}^{\cdot\cdot h}$ the *object of anholonomity* of the system (h). The components $\Omega_{ji}^{\cdot\cdot h}$ vanish if and only if all vectors $\overset{h}{e}_\lambda$ are gradient vectors, that is if the system (h) is holonomic.

In the same way we can prove that for a contravariant vector density of weight $+1$

$$\partial_j \mathfrak{v}^j = \Delta^{-1} \partial_\mu \mathfrak{v}^\mu - 2\Omega_{ji}^{\cdot\cdot i} \mathfrak{v}^j, \qquad \Delta = \text{Det}(A^h_\lambda) \quad (i, j = 1,..., n).\dagger \tag{7.9}$$

EXERCISES

For additional exercises see R.C. Ch. II.

IV.1. Prove that in R_n in cartesian coordinates

$$\nabla^2 w_{i_1...i_p} \overset{\text{def}}{=} \partial^j \partial_j w_{i_1...i_p} = \partial^j (p+1) \partial_{[j} w_{i_1...i_p]} + p \partial_{[i_1} \partial^j w_{|j| i_2...i_p]}$$
$$(i_1,..., i_p, j = 1,..., n;\ p \geqslant 1), \tag{1 α}$$

or in short, $\nabla^2 w = \text{Div Rot}\, w + \text{Rot Div}\, w;\ p > 1.$ (1 β)

IV.2. If an X_m in X_n is given in the parametric form

$$\xi^\kappa = f^\kappa(\eta^a) \quad (a = 1,..., m), \tag{2 α}$$

with the condition that the rank of the connecting quantity

$$B^\kappa_b \overset{\text{def}}{=} \partial_b \xi^\kappa; \quad \partial_b = \partial/\partial\eta^b \tag{2 β}$$

in the region considered is m, and if a vector field w_λ is given in X_n, the vector field in X_m defined by

$$'w_b = B^\lambda_b w_\lambda \tag{2 γ}$$

is called the section of the field w_λ with the X_m. Prove that the rotation of the field w_b in X_m is equal to the section

$$2B^\mu_c B^\lambda_b \partial_{[\mu} w_{\lambda]} \tag{2 δ}$$

of the rotation of w_λ with the X_m.‡

IV.3. Prove that $\quad D \underset{v}{\pounds} w_{\lambda_1...\lambda_p} = \underset{v}{\pounds} D w_{\lambda_1...\lambda_p}.$ (3 α)

† A very short proof of (7.7) and (7.9) will be found in Ch. V, § 7.
‡ Cf. E I. 1923. 1, pp. 18, 92 ff.; P.P. 1949. 1, pp. 3, 52.

IV.4. If w stands for $w_{\lambda_1 \dots \lambda_q}$, $q \geqslant 1$, and the operator T for the transvection of v^κ over the first index, prove that†

$$TDw+DTw = \underset{v}{\pounds} w. \tag{4 α}$$

IV.5. Prove the identity (6.11) for $P_{\lambda\kappa} = g_{\lambda\kappa}$; $\mathfrak{L} = \mathfrak{g}^{\frac{1}{2}}$; $\Delta > 0$. Cf. Ch. V, § 5.

IV.6. If the components of an affinor field, for instance, $P^{\kappa\lambda}_{\cdot\cdot\mu}$, are expressed as functions of the ξ^κ and the components of another affinor field, for instance $Q_{\nu\cdot\sigma}^{\cdot\rho}$

$$P^{\kappa\lambda}_{\cdot\cdot\mu} = f^{\kappa\lambda}_{\cdot\cdot\mu}(\xi^\tau, Q_{\nu\cdot\sigma}^{\cdot\rho}) \tag{6 α}$$

we have

$$dP^{\kappa\lambda}_{\cdot\cdot\mu} = (\bar{\partial}_\omega P^{\kappa\lambda}_{\cdot\cdot\mu})d\xi^\omega + \frac{\partial P^{\kappa\lambda}_{\cdot\cdot\mu}}{\partial Q_{\nu\cdot\sigma}^{\cdot\rho}} dQ_{\nu\cdot\sigma}^{\cdot\rho}, \tag{6 β}$$

where $\bar{\partial}_\omega$ denotes the partial derivation with respect to ξ^ω ($\omega = 1, \dots, n$) considering the $Q_{\nu\cdot\sigma}^{\cdot\rho}$ as constants. Prove that

$$R^{\kappa\lambda\cdot\nu\cdot\sigma}_{\cdot\cdot\mu\cdot\rho\cdot} \overset{\text{def}}{=} \frac{\partial P^{\kappa\lambda}_{\cdot\cdot\mu}}{\partial Q_{\nu\cdot\sigma}^{\cdot\rho}} \tag{6 γ}$$

is an affinor.

IV.7. If Φ_Λ is an affinor $Q^{\kappa\lambda}$ and $\tilde{\mathfrak{L}}$ a scalar Δ-density of weight $+1$, the identity mentioned in IV, § 6, takes the form

$$[\tilde{\mathfrak{L}}]_{\kappa\lambda}\partial_\mu Q^{\kappa\lambda} + [\tilde{\mathfrak{L}}]_{\mu\lambda}\partial_\kappa Q^{\kappa\lambda} + [\tilde{\mathfrak{L}}]_{\kappa\mu}\partial_\lambda Q^{\kappa\lambda} + Q^{\kappa\lambda}\partial_\kappa[\tilde{\mathfrak{L}}]_{\mu\lambda} + Q^{\kappa\lambda}\partial_\lambda[\tilde{\mathfrak{L}}]_{\kappa\mu} = 0.$$

† Cf. P.P. 1949. 1, p. 77.

V

GEOMETRY OF MANIFOLDS WHICH HAVE A GIVEN DISPLACEMENT†

1. Displacements

The local E_n's of different points of an X_n are entirely independent. So far the quantities in neighbouring local E_n's have not been connected in any way. It is, however, possible to define a process for the transplantation of the quantities from the local E_n of a point ξ^κ to the local E_n of a point $\xi^\kappa+d\xi^\kappa$. This process, valid for all points in the region considered and for all directions in these points, is called a *displacement*. A quantity and the displaced quantity are called *pseudo-parallel* (see Note D, p. 272). In an E_n a displacement is given *a priori*, viz. the ordinary parallel displacement. But in an X_n no ordinary parallelism exists.

If a contravariant vector field v^κ be given, the value of the field in $\xi^\kappa+d\xi^\kappa$ is $v^\kappa+dv^\kappa = v^\kappa+d\xi^\mu\partial_\mu v^\kappa$. Let $v^\kappa+\overset{*}{d}v^\kappa$ be the displaced vector in $\xi^\kappa+d\xi^\kappa$. Then we have

$$\begin{aligned} v^{\kappa'}+\overset{*}{d}v^{\kappa'} &= (A^{\kappa'}_\kappa+dA^{\kappa'}_\kappa)(v^\kappa+\overset{*}{d}v^\kappa) \\ &= A^{\kappa'}_\kappa v^\kappa+A^{\kappa'}_\kappa \overset{*}{d}v^\kappa+v^\kappa\, dA^{\kappa'}_\kappa; \end{aligned} \tag{1.1}$$

hence

$$\overset{*}{d}v^{\kappa'} = A^{\kappa'}_\kappa \overset{*}{d}v^\kappa+v^\kappa\, dA^{\kappa'}_\kappa. \tag{1.2}$$

Now we know that

$$dv^{\kappa'} = d(A^{\kappa'}_\kappa v^\kappa) = A^{\kappa'}_\kappa\, dv^\kappa+v^\kappa\, dA^{\kappa'}_\kappa, \tag{1.3}$$

and accordingly

$$dv^{\kappa'}-\overset{*}{d}v^{\kappa'} = A^{\kappa'}_\kappa(dv^\kappa-\overset{*}{d}v^\kappa). \tag{1.4}$$

From this we see that, although neither dv^κ nor $\overset{*}{d}v^\kappa$ transforms like a vector, the difference

$$\delta v^\kappa \overset{\text{def}}{=} dv^\kappa-\overset{*}{d}v^\kappa \tag{1.5}$$

is a vector. We call δv^κ the *covariant differential of the field v^κ with respect to the given displacement.*

If the coordinate system in that local E_n of ξ^κ which belongs to (κ) is displaced pseudo-parallelly to $\xi^\kappa+d\xi^\kappa$, δv^κ is the differential of v^κ with respect to this displaced coordinate system. This can be expressed in another way by saying that δv^κ *is the differential of the field from the*

† General references: Levi Civita 1917. 2; Schouten 1918. 1; R.K. 1924. 1; Eisenhart 1926. 1, 1927. 3; Levi Civita 1927. 2; E I. 1935. 1; R.C. Ch. III.

point of view of an observer whose local system of reference is subjected to a pseudo-parallel displacement.

Now it would be possible to define independently of each other displacements for all kinds of quantities, but this would not be a sensible thing to do. Hence we introduce a set of axioms in order to connect the displacements of different kinds of quantities. The displacements satisfying these axioms will be called *linear displacements*:

(1) if Φ is a quantity (indices suppressed), $\delta\Phi$ is a quantity of the same kind;
(2) $\delta(\Phi+\Psi) = \delta\Phi+\delta\Psi$;
(3) $\delta(\Phi\Psi) = (\delta\Phi)\Psi+\Phi\,\delta\Psi$ (Leibniz's rule);
(4) $\delta\Phi$ is linear and homogeneous in the $d\xi^\kappa$;
(5) Leibniz's rule holds for transvections;
(6) the covariant differential of a scalar is identical with the ordinary differential.

An X_n with a general linear displacement is called an L_n.

From (1), (2), and (4) it follows that the covariant differential of a contravariant vector has the form

$$\begin{aligned}\delta v^\kappa &= dv^\kappa+\Gamma^\kappa_{\mu\lambda}\,v^\lambda\,d\xi^\mu\\ &= (\partial_\mu\,v^\kappa+\Gamma^\kappa_{\mu\kappa}\,v^\lambda)\,d\xi^\mu,\end{aligned} \tag{1.6}$$

in which the n^3 parameters $\Gamma^\kappa_{\mu\lambda}$ are arbitrary functions of the ξ^κ. If $\Gamma^{\kappa'}_{\mu'\lambda'}$ are the parameters with respect to (κ') we have

$$\delta v^{\kappa'} = dv^{\kappa'}+\Gamma^{\kappa'}_{\mu'\lambda'}\,v^{\lambda'}\,d\xi^{\mu'}. \tag{1.7}$$

From (1.6) and (1.7) it follows that

$$A^{\kappa'}_\kappa\,dv^\kappa+A^{\kappa'}_\kappa\Gamma^\kappa_{\mu\lambda}\,v^\lambda\,d\xi^\mu = A^{\kappa'}_\kappa\,dv^\kappa+v^\kappa dA^{\kappa'}_\kappa+\Gamma^{\kappa'}_{\mu'\lambda'}\,v^{\lambda'}\,d\xi^{\mu'} \tag{1.8}$$

for every choice of v^κ and $d\xi^\kappa$. Hence

$$\begin{aligned}\Gamma^{\kappa'}_{\mu'\lambda'} &= A^{\kappa'\mu\lambda}_{\kappa\mu'\lambda'}\,\Gamma^\kappa_{\mu\lambda}-A^{\mu\lambda}_{\mu'\lambda'}\,\partial_\mu\,A^{\kappa'}_\lambda\\ &= A^{\kappa'\mu\lambda}_{\kappa\mu'\lambda'}\,\Gamma^\kappa_{\mu\lambda}+A^{\kappa'}_\lambda\,\partial_{\mu'}\,A^\lambda_{\lambda'},\end{aligned} \tag{1.9}$$

and from this we see that the $\Gamma^\kappa_{\mu\lambda}$ constitute a geometric object which is not a geometric quantity because the derivatives of the $A^{\kappa'}_\lambda$ occur in its transformation formula.

If w_λ is a covariant vector we have, according to (5) and (6),

$$\begin{aligned}d(v^\lambda w_\lambda) = \delta(v^\lambda w_\lambda) &= w_\lambda\,\delta v^\lambda+v^\lambda\delta w_\lambda\\ &= w_\lambda\,dv^\lambda+w_\lambda\,\Gamma^\lambda_{\mu\nu}\,v^\nu\,d\xi^\mu+v^\lambda\delta w_\lambda\end{aligned} \tag{1.10}$$

for every choice of v^λ. Hence

$$\delta w_\lambda = dw_\lambda-\Gamma^\kappa_{\mu\lambda}\,w_\kappa\,d\xi^\mu. \tag{1.11}$$

From (1.6), (1.11), and (2, 3) for an affinor, for instance $P^{\kappa\lambda}_{\cdot\cdot\nu}$, we get

$$\delta P^{\kappa\lambda}_{\cdot\cdot\nu} = dP^{\kappa\lambda}_{\cdot\cdot\nu} + \Gamma^{\kappa}_{\mu\rho} P^{\rho\lambda}_{\cdot\cdot\nu} d\xi^{\mu} + \Gamma^{\lambda}_{\mu\rho} P^{\kappa\rho}_{\cdot\cdot\nu} d\xi^{\mu} - \Gamma^{\sigma}_{\mu\nu} P^{\kappa\lambda}_{\cdot\cdot\sigma} d\xi^{\mu}. \tag{1.12}$$

For every upper (lower) index there is a term with a positive (negative) sign.

If we write $\Gamma_{\mu} \overset{\text{def}}{=} \Gamma^{\lambda}_{\mu\lambda}$, we have according to (1.9) and (I 1.5)

$$\begin{aligned} \Gamma_{\mu'} &= A^{\mu}_{\mu'} \Gamma_{\mu} - A^{\lambda}_{\kappa'} \partial_{\mu'} A^{\kappa'}_{\lambda} \\ &= A^{\mu}_{\mu'} \Gamma_{\mu} - \partial_{\mu'} \log \Delta. \end{aligned} \tag{1.13}$$

Hence Γ_{μ} is a geometric object but not a geometric quantity. Γ_{μ} plays an important role in the covariant differentiation of densities. If $w_{\lambda_1 \dots \lambda_n}$ is a covariant n-vector the covariant differential is

$$\begin{aligned} \delta w_{\lambda_1 \dots \lambda_n} &= dw_{\lambda_1 \dots \lambda_n} - \Gamma^{\nu}_{\mu\lambda_1} w_{\nu\lambda_2 \dots \lambda_n} d\xi^{\mu} - \dots - \Gamma^{\nu}_{\mu\lambda_n} w_{\lambda_1 \dots \lambda_{n-1}\nu} d\xi^{\mu} \\ &= dw_{\lambda_1 \dots \lambda_n} - n\Gamma^{\nu}_{\mu[\lambda_n} w_{\lambda_1 \dots \lambda_{n-1}]\nu} d\xi^{\mu}, \end{aligned} \tag{1.14}$$

and this implies that

$$\overset{*}{d} w_{1 \dots n} = n\Gamma^{\nu}_{\mu[n} w_{1 \dots n-1]\nu} d\xi^{\mu} = \Gamma_{\mu} w_{1 \dots n} d\xi^{\mu}. \tag{1.15}$$

Now $w_{1 \dots n}$ can be considered as the component of a scalar-Δ-density of weight $+1$, hence, if $\tilde{\mathfrak{q}}$ is such a density we have

$$\delta \tilde{\mathfrak{q}} = d\tilde{\mathfrak{q}} - \Gamma_{\mu} \tilde{\mathfrak{q}} \, d\xi^{\mu}. \tag{1.16}$$

If only those transformations with $\Delta > 0$ are allowed, ordinary densities and Δ-densities transform in the same way. Since the displacement is independent of the choice of the coordinate system, for an ordinary scalar density we have the formula

$$\delta \mathfrak{q} = d\mathfrak{q} - \Gamma_{\mu} \mathfrak{q} \, d\xi^{\mu}. \tag{1.17}$$

From this it follows for the covariant differential of an affinor-density of weight $\mathfrak{k}$, for instance $\mathfrak{P}^{\kappa}_{\cdot\lambda}$, that

$$\delta \mathfrak{P}^{\kappa}_{\cdot\lambda} = d\mathfrak{P}^{\kappa}_{\cdot\lambda} + \Gamma^{\kappa}_{\mu\nu} \mathfrak{P}^{\nu}_{\cdot\lambda} d\xi^{\mu} - \Gamma^{\nu}_{\mu\lambda} \mathfrak{P}^{\kappa}_{\cdot\nu} d\xi^{\mu} - \mathfrak{k} \mathfrak{P}^{\kappa}_{\cdot\lambda} \Gamma_{\mu} d\xi^{\mu}. \tag{1.18}$$

If there are more indices, there is a term with a positive (negative) sign for every upper (lower) index and an extra term containing $\mathfrak{k}$ and Γ_{μ}.

The same formula holds for affinor-Δ-densities, and consequently the covariant differential of a W-scalar is identical with the ordinary differential.

Every covariant differential is the transvection of $d\xi^{\mu}$ with another quantity. This latter quantity is called the *covariant derivative* and is symbolized by ∇_{μ} in connexion with the quantity differentiated:

$$\delta \Phi = d\xi^{\mu} \nabla_{\mu} \Phi \quad \text{(indices of } \Phi \text{ suppressed).} \tag{1.19}$$

So we have, for instance,

$$\begin{aligned} \nabla_\mu v^\kappa &= \partial_\mu v^\kappa + \Gamma^\kappa_{\mu\lambda} v^\lambda, \\ \nabla_\mu w_\lambda &= \partial_\mu w_\lambda - \Gamma^\kappa_{\mu\lambda} w_\kappa. \end{aligned} \tag{1.20}$$

Obviously the covariant derivatives of A^κ_λ, $\tilde{\mathfrak{E}}^{\kappa_1 \ldots \kappa_n}$ and $\tilde{\mathfrak{e}}_{\lambda_1 \ldots \lambda_n}$ are zero.

From (1.9) it follows that

$$\Gamma^{\kappa'}_{[\mu'\lambda']} = A^{\mu\lambda\kappa'}_{\mu'\lambda'\kappa} \Gamma^\kappa_{[\mu\lambda]}; \tag{1.21}$$

hence $\Gamma^\kappa_{[\mu\lambda]}$ is an affinor. We write

$$S_{\mu\lambda}^{\cdot\cdot\kappa} \overset{\text{def}}{=} \Gamma^\kappa_{[\mu\lambda]}. \tag{1.22}$$

If $S_{\mu\lambda}^{\cdot\cdot\kappa} = 0$, the displacement is called *symmetrical*. An X_n with a symmetrical displacement is called an A_n. If $S_{\mu\lambda}^{\cdot\cdot\kappa}$ has the form

$$S_{\mu\lambda}^{\cdot\cdot\kappa} = S_{[\mu} A^\kappa_{\lambda]}, \tag{1.23}$$

the displacement is called *semi-symmetrical*.

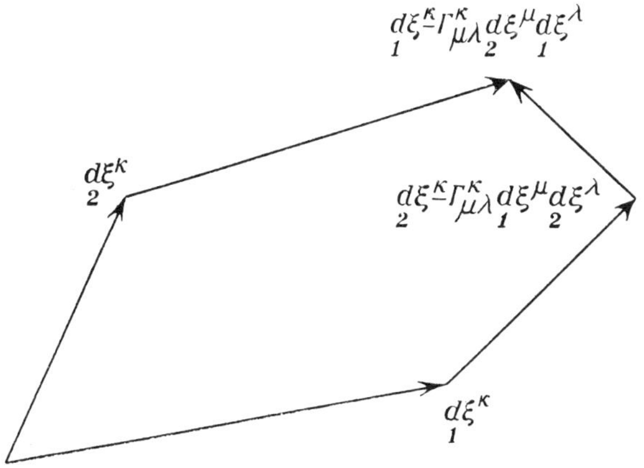

FIG. 17.

If two linear elements $\underset{1}{d\xi^\kappa}$ and $\underset{2}{d\xi^\kappa}$ in the same point are displaced pseudo-parallelly along each other we get a pentagon. The closing vector is

$$2\underset{1}{d\xi^\mu}\underset{2}{d\xi^\lambda} S_{\mu\lambda}^{\cdot\cdot\kappa}. \tag{1.24}$$

This implies that only in an A_n do infinitesimal parallelograms exist in all 2-directions. If the displacement in an X_n is semi-symmetrical the closing vector is

$$2S_\mu \underset{1}{d\xi^{[\mu}} \underset{2}{d\xi^{\kappa]}}. \tag{1.25}$$

Hence infinitesimal parallelograms only exist in the $(n-1)$-direction of the vector S_λ.

In an A_n the symbol ∂_μ can be replaced by ∇_μ in all invariant differential operators defined in IV, § 3. This can be proved by writing out the expressions with ∇. All terms with $\Gamma^\kappa_{\mu\lambda}$ vanish because $S_{\dot\mu\dot\lambda}{}^{\kappa} = 0$. For instance, if w_λ is a covariant vector and $\mathfrak{v}^\kappa$ a contravariant vector-density of weight $+1$, we have

$$\begin{aligned} &(a) \qquad \nabla_{[\mu} w_{\lambda]} = \partial_{[\mu} w_{\lambda]} - \Gamma^\kappa_{[\mu\lambda]} w_\kappa = \partial_{[\mu} w_{\lambda]}, \\ &(b) \qquad \nabla_\mu \mathfrak{v}^\mu = \partial_\mu \mathfrak{v}^\mu + \Gamma^\mu_{\mu\lambda} \mathfrak{v}^\lambda - \Gamma^\mu_{\lambda\mu} \mathfrak{v}^\lambda = \partial_\mu \mathfrak{v}^\mu. \end{aligned} \tag{1.26}$$

The same holds for the Lie derivative defined in IV, § 5; for instance

$$\underset{v}{\pounds} \mathfrak{P}^{\kappa}_{.\lambda} = v^\mu \nabla_\mu \mathfrak{P}^{\kappa}_{.\lambda} + \mathfrak{P}^{\kappa}_{.\rho} \nabla_\lambda v^\rho - \mathfrak{P}^{\sigma}_{.\lambda} \nabla_\sigma v^\kappa + \mathfrak{k}\mathfrak{P}^{\kappa}_{.\lambda} \nabla_\mu v^\mu, \tag{1.27}$$

and also for identities of the form (IV. (6.11)), for instance

$$[\tilde{\mathfrak{L}}]^{\lambda\kappa}(\nabla_\mu P_{\lambda\kappa} - \nabla_\lambda P_{\mu\kappa} - \nabla_\kappa P_{\lambda\mu}) - P_{\mu\kappa} \nabla_\lambda [\tilde{\mathfrak{L}}]^{\lambda\kappa} - P_{\lambda\mu} \nabla_\kappa [\tilde{\mathfrak{L}}]^{\lambda\kappa} = 0. \tag{1.28}$$

If for $P_{\lambda\kappa}$ we take the fundamental tensor $g_{\lambda\kappa}$ (1.28) takes the form

$$\nabla_\mu [\tilde{\mathfrak{L}}]^{\mu\kappa} = 0. \tag{1.29}$$

2. Geodesics

Let

$$\xi^\kappa = \xi^\kappa(z) \tag{2.1}$$

be the parametric form of a curve in an L_n with the parameter z. $d\xi^\kappa/dz$ is a contravariant vector tangent to the curve. A curve in L_n is called a *geodesic* if it is 'as straight as possible', that is, if the tangent vector remains tangent under psuedo-parallel displacement along the curve. The necessary and sufficient condition is

$$\frac{\delta}{dz}\frac{d\xi^\kappa}{dz} = \alpha(z)\frac{d\xi^\kappa}{dz}, \tag{2.2}$$

where $\alpha(z)$ is some function of z. If another parameter $t = t(z)$ is introduced instead of z we have

$$\frac{\delta}{dz}\left(\frac{d\xi^\kappa}{dt}\frac{dt}{dz}\right) = \alpha\frac{d\xi^\kappa}{dt}\frac{dt}{dz}, \tag{2.3}$$

or

$$\left(\frac{dt}{dz}\right)^2 \frac{\delta}{dt}\frac{d\xi^\kappa}{dt} + \frac{d^2t}{dz^2}\frac{d\xi^\kappa}{dt} = \alpha\frac{d\xi^\kappa}{dt}\frac{dt}{dz}, \tag{2.4}$$

or

$$\frac{\delta}{dt}\frac{d\xi^\kappa}{dt} = \frac{\alpha\dfrac{dt}{dz} - \dfrac{d^2t}{dz^2}}{\left(\dfrac{dt}{dz}\right)^2}\frac{d\xi^\kappa}{dt}. \tag{2.5}$$

The right-hand side of this equation will only vanish if t is a solution of

$$\frac{d^2t}{dz^2} - \alpha(z)\frac{dt}{dz} = 0. \tag{2.6}$$

The general solution of this equation is

$$t = C_1 \int e^{\int \alpha(z)dz}\,dz + C_2. \tag{2.7}$$

From this we see that on every geodesic in L_n there is a parameter t such that the equation (2.2) takes the form

$$\frac{\delta}{dt}\frac{d\xi^\kappa}{dt} = \frac{d^2\xi^\kappa}{dt^2} + \Gamma^\kappa_{\mu\lambda}\frac{d\xi^\mu}{dt}\frac{d\xi^\lambda}{dt} = 0. \tag{2.8}$$

Such a parameter is called a *natural parameter* of the geodesic. If $\underset{0}{t}$ is a natural parameter of a geodesic, every other natural parameter on the same geodesic can be written in the form

$$t = C_1 \underset{0}{t} + C_2, \tag{2.9}$$

where C_1 ($\neq 0$) and C_2 are arbitrary constants. Hence the natural parameter on a geodesic is determined to within affine transformations with *constant* coefficients. Changing C_2 displaces the null point and changing C_1 means changing the 'affine measure' on the geodesic. Segments of a geodesic in L_n have no 'length' in the ordinary sense, but two different segments on the *same* geodesic have a definite ratio that can be determined by using an arbitrary natural parameter. According to (2.8) geodesics in L_n depend on $\Gamma^\kappa_{(\mu\lambda)}$ only and are independent of $S_{\mu\lambda}{}^{\cdot\cdot\kappa}$.

3. Normal coordinates†

We will prove that in an A_n there is always a coordinate system (κ') such that $\Gamma^{\kappa'}_{\mu'\lambda'} = 0$ at some previously given point $\underset{a}{\xi}{}^\kappa$. Let

$$\underset{0}{t}{}^\kappa = f^\kappa(\alpha_1,\ldots,\alpha_{n-1}) \tag{3.1}$$

be a parametric representation of ∞^{n-1} contravariant vectors in the local E_n of $\underset{0}{\xi}{}^\kappa$, one and only one in each direction. On each geodesic through $\underset{0}{\xi}{}^\kappa$ we choose a natural parameter z such that $z = 0$ at $\underset{0}{\xi}{}^\kappa$ and

$$\underset{0}{t}{}^\kappa = \frac{d\xi^\kappa}{dz} \tag{3.2}$$

at $\underset{0}{\xi}{}^\kappa$ on every geodesic. Now on each geodesic the equation

$$\frac{d^2\xi^\kappa}{dz^2} + \Gamma^\kappa_{\mu\lambda}\frac{d\xi^\mu}{dz}\frac{d\xi^\lambda}{dz} = 0 \tag{3.3}$$

holds and by differentiation of this equation we get an infinite series of equations from which the values of all derivatives of ξ^κ with respect

† Cf. E I. 1935. 1, pp. 100 ff., 122 f.; R.C. III §§ 7, 8.

to z for $z = 0$ can be solved. When this has been done ξ^κ can be expanded into a series

$$\begin{aligned}\xi^\kappa &= \underset{0}{\xi^\kappa} + \left(\frac{d\xi^\kappa}{dz}\right)_{z=0} z + \frac{1}{2}\left(\frac{d^2\xi^\kappa}{dz^2}\right)_{z=0} z^2 + \dots \\ &= \underset{0}{\xi^\kappa} + \underset{0}{t^\kappa} z - \tfrac{1}{2}\Gamma^\kappa_{\mu\lambda}(\underset{0}{\xi^\nu})\underset{0}{t^\mu}\underset{0}{t^\lambda} z^2 + \dots,\end{aligned} \tag{3.4}$$

of which only the first three terms are interesting for our present purpose.

If now we introduce the $\underset{0}{t^\kappa} z$ as new coordinates ξ^h,

$$\xi^h \overset{\text{def}}{=} \delta^h_\kappa \underset{0}{t^\kappa} z \quad (h = 1,\dots, \mathrm{n}), \tag{3.5}$$

we have

$$\xi^h = \delta^h_\kappa\{(\xi^\kappa - \underset{0}{\xi^\kappa}) + \tfrac{1}{2}\Gamma^\kappa_{\rho\sigma}(\underset{0}{\xi^\nu})(\xi^\rho - \underset{0}{\xi^\rho})(\xi^\sigma - \underset{0}{\xi^\sigma}) + \dots\} \quad (h = 1,\dots, \mathrm{n}), \tag{3.6}$$

and by differentiation it follows that

$$\begin{aligned} A^h_\lambda &= \delta^h_\lambda + \delta^h_\kappa \Gamma^\kappa_{\rho\lambda}(\underset{0}{\xi^\nu})(\xi^\rho - \underset{0}{\xi^\rho}) + \dots, \\ \partial_\mu A^h_\lambda &= \delta^h_\kappa \Gamma^\kappa_{\mu\lambda}(\underset{0}{\xi^\nu}) + \dots \quad (h = 1,\dots, \mathrm{n}),\end{aligned} \tag{3.7}$$

hence

$$\left.\begin{aligned} A^h_\lambda &= \delta^h_\lambda \\ \partial_\mu A^h_\lambda &= \delta^h_\kappa \Gamma^\kappa_{\mu\lambda}(\underset{0}{\xi^\nu})\end{aligned}\right\} \quad \text{for } \xi^\kappa = \underset{0}{\xi^\kappa} \quad (h = 1,\dots, \mathrm{n}). \tag{3.8}$$

Now, according to (1.9) and (3.8) it follows that, in $\underset{0}{\xi^\kappa}$,

$$\begin{aligned}\Gamma^h_{ji} &= A^{h\mu\lambda}_{\kappa ji}\Gamma^\kappa_{\mu\lambda} - A^{\mu\lambda}_{ji}\partial_\mu A^h_\lambda \\ &= \delta^{h\mu\lambda}_{\kappa ji}\Gamma^\kappa_{\mu\lambda} - \delta^{\mu\lambda h}_{ji\kappa}\Gamma^\kappa_{\mu\lambda} = 0 \quad (h, i, j = 1,\dots, \mathrm{n}).\end{aligned} \tag{3.9}$$

The coordinates derived in this way are called *normal coordinates* for the point $\underset{0}{\xi^\kappa}$ in A_n.† They depend only on the choice of $\underset{0}{\xi^\kappa}$, the choice of the ∞^{n-1} vectors $\underset{0}{t^\kappa}$ in $\underset{0}{\xi^\kappa}$ and the choice of (κ). If $\underset{0}{\xi^\kappa}$ and the $\underset{0}{t^\kappa}$ are left unchanged and if (κ) is transformed into an arbitrary new system (κ') we get new normal coordinates $\xi^{h'}$:

$$\begin{aligned}\xi^{h'} &= \delta^{h'}_{\kappa'}\underset{0}{t^{\kappa'}} z = \delta^{h'}_{\kappa'} A^{\kappa'}_\kappa(\underset{0}{\xi^\nu})\underset{0}{t^\kappa} z \\ &= \delta^{h'\kappa}_{\kappa' h} A^{\kappa'}_\kappa(\underset{0}{\xi^\nu})\xi^h,\end{aligned} \tag{3.10}$$

and this equation shows that the normal coordinates are subjected to a linear homogeneous transformation with constant coefficients. Hence,

† Normal coordinates were first introduced by Veblen 1922. 1. For references to literature cf. E I. 1935. 1, p. 100, R.C. III § 7.

if $\underset{0}{\xi}^{\kappa}$ and the $\underset{0}{t}^{\kappa}$ are fixed, the normal coordinates are determined to within linear homogeneous transformations.

It is also possible to define normal coordinates such that the Γ_{ji}^{h} vanish at every point of a given curve and in special cases also at all points of a given X_m† in A_n.

Normal coordinates are very convenient because by using them it is often possible to avoid long calculations.

4. The V_n

If a *real* tensor $g_{\lambda\kappa}$ of rank n is introduced in an X_n the X_n is called a V_n and $g_{\lambda\kappa}$ its fundamental tensor.‡ Every local E_n of a V_n contains a fundamental tensor $g_{\lambda\kappa}$ and is accordingly an R_n. In a V_n the raising and lowering of indices will be affected by means of $g_{\lambda\kappa}$ and its inverse $g^{\kappa\lambda}$ in the same way as in an R_n.

The expression
$$v \overset{\text{def}}{=} |\sqrt{(g_{\lambda\kappa} v^{\lambda} v^{\kappa})}| \tag{4.1}$$
is called the *length* of the vector v^{κ}. The term *duration* is used instead of length in the theory of relativity if $g_{\lambda\kappa} v^{\lambda} v^{\kappa} > 0$. If $v = 0$, v^{κ} is said to be a *null vector* and its direction is called a *null direction*. Null vectors can only be real if $g_{\lambda\kappa}$ is indefinite.

We write ds for the length of the linear element $d\xi^{\kappa}$ defined by
$$ds^2 = |g_{\lambda\kappa} d\xi^{\lambda} d\xi^{\kappa}|. \tag{4.2}$$
If $ds = 0$ in every point of a curve the curve is called a *null curve* of V_n.

In a V_n, every segment of a curve has a length $\int ds$. The most important property of a V_n is that its fundamental tensor determines uniquely a *symmetric* displacement that leaves the length of a vector invariant. It therefore satisfies the equations
$$\nabla_{\mu} g_{\lambda\kappa} = 0. \tag{4.3}$$
If this equation is written out we get
$$\partial_{\mu} g_{\lambda\kappa} - g_{\rho\kappa} \Gamma_{\mu\lambda}^{\rho} - g_{\lambda\sigma} \Gamma_{\mu\kappa}^{\sigma} = 0. \tag{4.4}$$
By permutation of the indices we get the equivalent equations
$$\begin{aligned} \partial_{\lambda} g_{\kappa\mu} - g_{\rho\mu} \Gamma_{\lambda\kappa}^{\rho} - g_{\kappa\sigma} \Gamma_{\lambda\mu}^{\sigma} &= 0, \\ -\partial_{\kappa} g_{\mu\lambda} + g_{\rho\lambda} \Gamma_{\kappa\mu}^{\rho} + g_{\mu\sigma} \Gamma_{\kappa\lambda}^{\sigma} &= 0; \end{aligned} \tag{4.5}$$

† E I. 1935. 1, pp. 106 ff., R.C. III § 8, also for references to literature.

‡ In a V_n orthogonal curvilinear coordinate systems in all directions are possible for $n \leqslant 3$. They are possible for $n > 3$ if and only if a scalar σ exists such that a V_n with the fundamental tensor $\sigma g_{\lambda\kappa}$ is an R_n. Cf. Schouten 1921. 5, p. 84, 1927. 4, p. 719; R.K. 1924. 1, p. 196; Eisenhart 1926. 1, p. 122; E II. 1938. 3, p. 204; R.C. VI § 5.

hence, by addition, since $\Gamma^{\kappa}_{\mu\lambda} = \Gamma^{\kappa}_{\lambda\mu}$,

$$\partial_\mu g_{\lambda\kappa} + \partial_\lambda g_{\kappa\mu} - \partial_\kappa g_{\mu\lambda} = 2g_{\rho\kappa}\Gamma^{\rho}_{\mu\lambda}, \tag{4.6}$$

or

$$\Gamma^{\kappa}_{\mu\lambda} = \{^{\kappa}_{\mu\lambda}\} \stackrel{\text{def}}{=} \tfrac{1}{2}g^{\kappa\rho}(\partial_\mu g_{\lambda\rho} + \partial_\lambda g_{\mu\rho} - \partial_\rho g_{\mu\lambda}). \tag{4.7}$$

The symbol $\{^{\kappa}_{\mu\lambda}\}$ is called the Christoffel symbol after its discoverer. It was originally written $\{^{\mu\lambda}_{\kappa}\}$, but this is not in agreement with modern practice for the position of indices.

From (4.7) we get

$$\Gamma^{\lambda}_{\mu\lambda} = \tfrac{1}{2}g^{\lambda\rho}\,\partial_\mu g_{\lambda\rho}. \tag{4.8}$$

Now from the definition of $g^{\kappa\lambda}$ it follows that (cf. (I.1.5))

$$g^{\kappa\lambda} = \frac{1}{\mathfrak{g}}\frac{\partial \mathfrak{g}}{\partial g_{\kappa\lambda}}; \tag{4.9}$$

hence

$$\Gamma^{\lambda}_{\mu\lambda} = \tfrac{1}{2}\partial_\mu \log \mathfrak{g}, \tag{4.10}$$

and consequently

$$\nabla_\mu \mathfrak{g} = \partial_\mu \mathfrak{g} - 2\mathfrak{g}\,\Gamma^{\lambda}_{\mu\lambda} = 0. \tag{4.11}$$

This formula can be used to write the contracted covariant derivative of a contravariant p-vector, $\nabla_\mu v^{\mu\kappa_2...\kappa_p}$, in a form containing ∂_μ only and not the parameters $\Gamma^{\kappa}_{\mu\lambda}$:

$$\nabla_\mu v^{\mu\kappa_2...\kappa_p} = \mathfrak{g}^{-\frac{1}{2}}\,\nabla_\mu \mathfrak{g}^{\frac{1}{2}} v^{\mu\kappa_2...\kappa_p} = \mathfrak{g}^{-\frac{1}{2}}\partial_\mu \mathfrak{g}^{\frac{1}{2}} v^{\mu\kappa_2...\kappa_p} \tag{4.12}$$

(cf. (1.26)).

The geodesics in a V_n are not only the straightest curves but are also the shortest or longest curves between two of their points. In order to prove this we consider a curve s connecting the two points $\underset{0}{\xi^{\kappa}}$ and $\underset{1}{\xi^{\kappa}}$ whose tangent does not have a null direction anywhere. Besides s we consider all curves that can be derived from s by an infinitesimal point transformation $v^{\kappa}\,dt$, v^{κ} being an arbitrary vector field defined in all points of a neighbourhood of s. Then we wish to establish the relations between the length of s and the length of such a neighbouring curve s' between $\underset{0}{\xi^{\kappa}} + \underset{0}{v^{\kappa}}\,dt$ and $\underset{1}{\xi^{\kappa}} + \underset{1}{v^{\kappa}}\,dt$, where $\underset{0}{v^{\kappa}}$ and $\underset{1}{v^{\kappa}}$ are the values of v^{κ} in $\underset{0}{\xi^{\kappa}}$ and $\underset{1}{\xi^{\kappa}}$. To do this we have to determine the length of the linear element of s' at some point $\xi^{\kappa} + v^{\kappa}\,dt$. At $\xi^{\kappa} + v^{\kappa}\,dt$ we have to use the value $g_{\lambda\kappa} + v^{\mu}\partial_\mu g_{\lambda\kappa}\,dt$ of the fundamental tensor at that point. Now this gets rather complicated and it can be done in a more elegant way.

If $\xi^{\kappa} \to \eta^{\kappa}(\xi^{\nu})$ is any point transformation in V_n and if the curve s and the field $g_{\lambda\kappa}$ are both dragged along by this transformation, the length of the curve dragged along measured by means of the fundamental tensor dragged along is obviously the same as the length of the original curve. This is due to the fact that the coordinates of a point of s dragged along and the components of the fundamental tensor dragged along, both with respect to (κ'), which is the coordinate-system dragged along, are just the same as the original coordinates and components with respect to (κ). Now we consider the point transformation $\xi^\kappa \to \xi^\kappa - v^\kappa dt$. If s' is dragged along by this transformation it becomes s and if $g_{\lambda\kappa}$ is dragged along, this field becomes $g_{\lambda\kappa} + \underset{v}{£} g_{\lambda\kappa} dt$. Hence, instead of measuring s' by means of $g_{\lambda\kappa} + v^\mu \partial_\mu g_{\lambda\kappa} dt$ we may just as well measure s by means of (see Note E, p. 272)

$$g_{\lambda\kappa} - \underset{v}{£} g_{\lambda\kappa} dt = g_{\lambda\kappa} + v^\mu \partial_\mu g_{\lambda\kappa} dt + g_{\rho\kappa} \partial_\lambda v^\rho dt + g_{\lambda\rho} \partial_\kappa v^\rho dt. \tag{4.13}$$

Hence, if τ is an arbitrary parameter on s, the variation of the length of the linear element $d\xi^\kappa$ of s is

$$\begin{aligned} \overset{v}{d}\, ds &= \overset{v}{d} \left(g_{\lambda\kappa} \frac{d\xi^\lambda}{d\tau} \frac{d\xi^\kappa}{d\tau} \right)^{\frac{1}{2}} d\tau \\ &= -\frac{1}{2} \left(g_{\lambda\kappa} \frac{d\xi^\lambda}{d\tau} \frac{d\xi^\kappa}{d\tau} \right)^{-\frac{1}{2}} \underset{v}{£} g_{\mu\nu} \frac{d\xi^\mu}{d\tau} \frac{d\xi^\nu}{d\tau} d\tau dt \\ &= \tfrac{1}{2} (v^\rho \partial_\rho g_{\mu\nu} + g_{\sigma\nu} \partial_\mu v^\sigma + g_{\mu\sigma} \partial_\nu v^\sigma) \frac{d\xi^\mu}{ds} \frac{d\xi^\nu}{ds} ds dt. \end{aligned} \tag{4.14}$$

Now we have

$$\begin{aligned} & v^\rho \partial_\rho g_{\mu\nu} + g_{\sigma\nu} \partial_\mu v^\sigma + g_{\mu\sigma} \partial_\nu v^\sigma \\ &\quad = v^\rho \nabla_\rho g_{\mu\nu} + v^\rho \Gamma^\sigma_{\rho\mu} g_{\sigma\nu} + v^\rho \Gamma^\sigma_{\rho\nu} g_{\mu\sigma} + g_{\sigma\nu} \partial_\mu v^\sigma + g_{\mu\sigma} \partial_\nu v^\sigma \\ &\quad = g_{\sigma\nu} \nabla_\mu v^\sigma + g_{\mu\sigma} \nabla_\nu v^\sigma = 2 \nabla_{(\mu} v_{\nu)}; \end{aligned} \tag{4.15}$$

hence

$$\overset{v}{d}\, ds = (\nabla_\mu v_\nu) \frac{d\xi^\mu}{ds} \frac{d\xi^\nu}{ds} ds dt = \frac{\delta v_\nu}{ds} \frac{d\xi^\nu}{ds} ds dt. \tag{4.16}$$

This gives

$$\overset{v}{d} \int_{\overset{0}{\xi^\kappa}}^{\overset{1}{\xi^\kappa}} ds = dt \int_{\overset{0}{\xi^\kappa}}^{\overset{1}{\xi^\kappa}} \frac{d\xi^\lambda}{ds} \delta v_\lambda \tag{4.17}$$

for the variation of the length of s between $\underset{0}{\xi^\kappa}$ and $\underset{1}{\xi^\kappa}$, and from this formula by means of integration by parts we get

$$\overset{v}{d} \int_{\overset{0}{\xi^\kappa}}^{\overset{1}{\xi^\kappa}} ds = dt \left\{ \left(v_\lambda \frac{d\xi^\lambda}{ds} \right)_{\xi^\kappa = \overset{1}{\xi^\kappa}} - \left(v_\lambda \frac{d\xi^\lambda}{ds} \right)_{\xi^\kappa = \overset{0}{\xi^\kappa}} \right\} - dt \int_{\overset{0}{\xi^\kappa}}^{\overset{1}{\xi^\kappa}} v_\lambda \delta \frac{d\xi^\lambda}{ds}. \tag{4.18}$$

If now v^κ is zero at $\underset{0}{\xi^\kappa}$ and at $\underset{1}{\xi^\kappa}$, the first term to the right drops out and we get

$$\overset{v}{d} \int_{\overset{0}{\xi^\kappa}}^{\overset{1}{\xi^\kappa}} ds = -dt \int_{\overset{0}{\xi^\kappa}}^{\overset{1}{\xi^\kappa}} v_\lambda \delta \frac{d\xi^\lambda}{ds}, \tag{4.19}$$

and this proves that the variation is zero for every choice of v^κ satisfying the boundary conditions if and only if

$$\delta \frac{d\xi^\lambda}{ds} = 0, \tag{4.20}$$

that is if s is a geodesic. Hence a geodesic between two points is a curve of extreme length and s is a natural parameter.

If, at one of its points, a geodesic s has a linear element with length zero it can be proved that s is a null curve. Let s^κ be a tangent vector of s satisfying the condition

$$\delta s^\kappa = 0. \tag{4.21}$$

Then the length of s^κ at the point considered is zero. Now the length of a vector does not change under pseudo-parallel displacements, and accordingly s^κ has length zero at every point of s. But the converse is not true. A curve with zero length is not necessarily a geodesic.

5. Curvature of the V_n and the A_n

We consider a sphere with radius 1 in ordinary space. In spherical coordinates its linear element is

$$ds^2 = \sin^2\theta\, d\phi^2 + d\theta^2. \tag{5.1}$$

Hence, if we write $\xi^1 = \phi$, $\xi^2 = \theta$, the components of the fundamental tensor are

$$\begin{aligned} g_{11} &= \sin^2\theta, & g_{12} &= 0, & g_{22} &= 1, \\ g^{11} &= \sin^{-2}\theta, & g^{12} &= 0, & g^{22} &= 1, \end{aligned} \tag{5.2}$$

and the $\Gamma^\kappa_{\mu\lambda}$ are

$$\begin{aligned} \Gamma^1_{11} &= 0, & \Gamma^1_{12} &= \cot\theta, & \Gamma^1_{22} &= 0, \\ \Gamma^2_{11} &= -\sin\theta\cos\theta, & \Gamma^2_{12} &= 0, & \Gamma^2_{22} &= 0. \end{aligned} \tag{5.3}$$

From this we get for the covariant differentiation

$$\begin{aligned} \delta v^1 &= dv^1 + v^1 \cot\theta\, d\theta + v^2 \cot\theta\, d\phi, \\ \delta v^2 &= dv^2 - v^1 \sin\theta\cos\theta\, d\phi; \\ \delta w_1 &= dw_1 - w_1 \cot\theta\, d\theta + w_2 \sin\theta\cos\theta\, d\phi, \\ \delta w_2 &= dw_2 - w_1 \cot\theta\, d\phi. \end{aligned} \tag{5.4}$$

Consequently the necessary and sufficient conditions for the pseudo-parallel displacement of v^κ and w_λ are

$$\begin{aligned} dv^1 &= -v^1\cot\theta\,d\theta - v^2\cot\theta\,d\phi, \\ dv^2 &= v^1\sin\theta\cos\theta\,d\phi; \\ dw_1 &= w_1\cot\theta\,d\theta - w_2\sin\theta\cos\theta\,d\phi, \\ dw_2 &= w_1\cot\theta\,d\phi. \end{aligned} \tag{5.5}$$

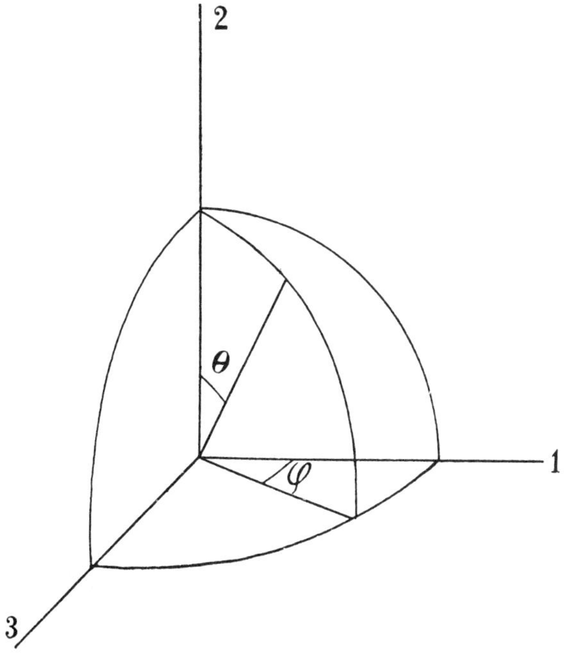

Fig. 18.

On a meridian we have $d\phi = 0$ and accordingly the pseudo-parallel displacement becomes

$$\begin{aligned} &(a)\quad dv^1 = -v^1\cot\theta\,d\theta, &&(c)\quad dw_1 = w_1\cot\theta\,d\theta, \\ &(b)\quad dv^2 = 0, &&(d)\quad dw_2 = 0. \end{aligned} \tag{5.6}$$

v^1 is the component in the direction of a parallel and v^2 the component in the direction of a meridian. Hence by (5.6 (a)) a meridian is a geodesic

On a parallel we have $d\theta = 0$ and accordingly the pseudo-parallel displacement becomes

$$\begin{aligned} &(a)\quad dv^1 = -v^2\cot\theta\,d\phi, &&(c)\quad dw_1 = -w_2\sin\theta\cos\theta\,d\phi, \\ &(b)\quad dv^2 = v^1\sin\theta\cos\theta\,d\phi, &&(d)\quad dw_2 = w_1\cot\theta\,d\phi. \end{aligned} \tag{5.7}$$

Hence a parallel is only a geodesic for $\theta = \pi/2$.

If a vector v^κ is given in the tangent at the point $\phi = 0$, $\theta = \theta_0$ of the parallel $\theta = \theta_0$ we have $v^2 = 0$ at that point. The length of this vector is $v^1 \sin\theta_0$. If this vector now undergoes a pseudo-parallel displacement of an amount $d\phi$ along the parallel, the components of the displaced vector at $d\phi$, θ_0 are v^1, $v^1 \sin\theta_0 \cos\theta_0\, d\phi$. The length of the vector 0, $v^1 \sin\theta_0 \cos\theta_0\, d\phi$ perpendicular to the parallel is

$$v^1 \sin\theta_0 \cos\theta_0\, d\phi.$$

Consequently the angle between the displaced vector and the parallel is $\cos\theta_0\, d\phi$. A pseudo-parallel displacement turns all vectors through the same angle, hence every vector at a point of the parallel $\theta = \theta_0$ turns through an angle $\cos\theta_0\, d\phi$ with respect to the parallel if it is displaced pseudo-parallelly along the parallel through $d\phi$. If the vector is displaced along the whole parallel till it returns to the starting-point the total angle is $2\pi\cos\theta_0$. This implies that the angle between the original position and the end position is $2\pi(1-\cos\theta_0)$. This is exactly the area of the segment of the sphere bounded by the parallel. It can be proved that this is also true for every closed curve on a sphere with radius 1.

If a vector is displaced parallel along a closed curve in a plane the angle between the original position and the end position is always zero. Hence the geometry on a sphere is not the same as the geometry in a plane. We shall see that this has something to do with the 'curvature' of the surface.

If we construct a cone which is tangent at the parallel $\theta = \theta_0$ to the sphere, this cone can be developed on a plane. Hence the geometry on a cone is the same as the geometry in a plane. An arc of a circle of radius $\tan\theta_0$ with length $2\pi\sin\theta_0$ (Fig. 19) corresponds to the parallel $\theta = \theta_0$. Hence, if a vector, tangent to the arc at one end point, is displaced in this plane parallel along the arc to the other end point, the angle between the displaced vector and the arc at this latter point is $2\pi\cos\theta_0$. This proves that the pseudo-parallel displacement on the sphere along a parallel is just the same as the parallel displacement on the cone which is tangent to the sphere in this parallel.

If an arbitrary curve is given on an arbitrary surface, the tangent planes in the points of this curve envelop a developable surface. In this developable surface we have an ordinary plane geometry, and it can be proved that the pseudo-parallel displacement on the surface along the curve is just the same as the parallel displacement on this developable surface along the same curve. This property can be used for the construction of the pseudo-parallel displacements on an arbitrary

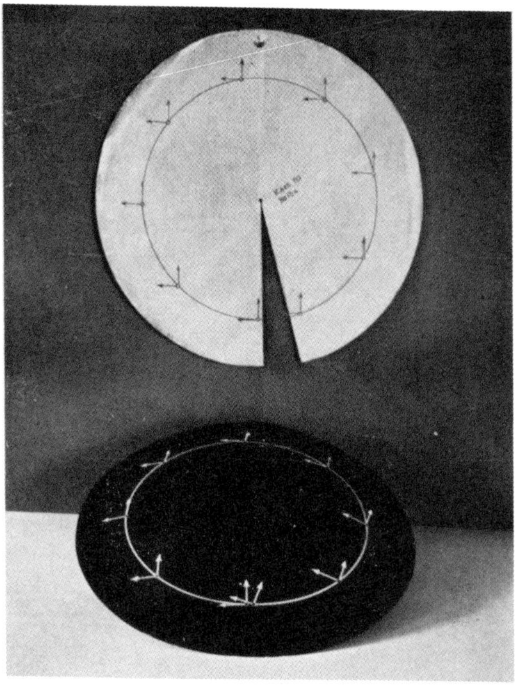

Fig. 20

Fig. 21

surface. Figs. 20 and 21 show the pseudo-parallel displacement on a sphere, a hyperbolic paraboloid, and a surface of constant negative curvature. (These are the original models from 1918.1.)

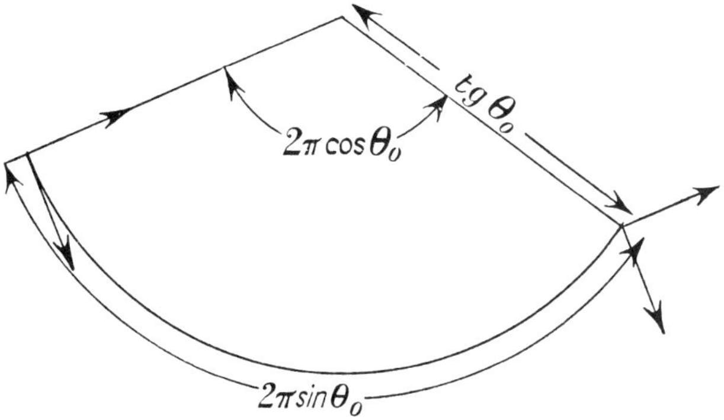

FIG. 19.

In order to investigate the behaviour of a vector in an A_n under pseudo-parallel displacement along a closed curve we consider an infinitesimal parallelogram at a point A. Let a vector v^κ in A be displaced pseudo-parallel first from A via B to D and then from A via C to D. The value of the displaced vector at B is

$$v^\kappa - \underset{2}{d}\xi^\mu \Gamma^\kappa_{\mu\lambda} v^\lambda. \tag{5.8}$$

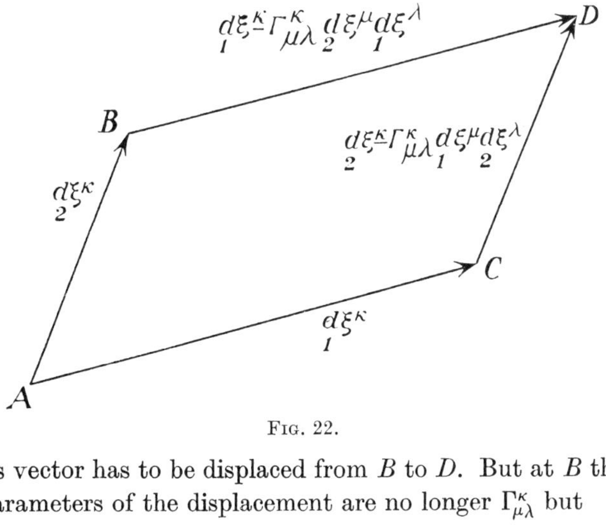

FIG. 22.

Now this vector has to be displaced from B to D. But at B the values of the parameters of the displacement are no longer $\Gamma^\kappa_{\mu\lambda}$ but

$$\Gamma^\kappa_{\mu\lambda} + \underset{2}{d}\xi^\nu \partial_\nu \Gamma^\kappa_{\mu\lambda}. \tag{5.9}$$

Hence the displaced vector at D which has come via B is

$$v^\kappa - \underset{2}{d}\xi^\mu \Gamma^\kappa_{\mu\lambda} v^\lambda - (\underset{1}{d}\xi^\rho - \Gamma^\rho_{\mu\lambda} \underset{2}{d}\xi^\mu \underset{1}{d}\xi^\lambda)(\Gamma^\kappa_{\rho\sigma} + \underset{2}{d}\xi^\nu \partial_\nu \Gamma^\kappa_{\rho\sigma})(v^\sigma - \underset{2}{d}\xi^\tau \Gamma^\sigma_{\tau\omega} v^\omega)$$

$$= v^\kappa - \underset{2}{d}\xi^\mu \Gamma^\kappa_{\mu\lambda} v^\lambda - \underset{1}{d}\xi^\mu \Gamma^\kappa_{\mu\lambda} v^\lambda + \underset{1}{d}\xi^\nu \underset{2}{d}\xi^\mu (\Gamma^\kappa_{\nu\rho} \Gamma^\rho_{\mu\lambda} - \partial_\mu \Gamma^\kappa_{\nu\lambda} + \Gamma^\rho_{\mu\nu} \Gamma^\kappa_{\rho\lambda}) v^\lambda \tag{5.10}$$

if all terms of third order are neglected. For the vector at D that has been displaced via C we find a similar expression that can be derived from (5.10) by interchanging $\underset{1}{d}\xi^\kappa$ and $\underset{2}{d}\xi^\kappa$. Hence the difference between the two displaced vectors at D is

$$\underset{1}{d}\xi^\nu \underset{2}{d}\xi^\mu \{2\partial_{[\nu} \Gamma^\kappa_{\mu]\lambda} + 2\Gamma^\kappa_{[\nu|\rho|} \Gamma^\rho_{\mu]\lambda}\} v^\lambda. \tag{5.11}$$

This expression must be a vector because it has a geometric meaning independent of the choice of the coordinate system, viz. the difference of two displaced vectors. This implies that the expression in curly brackets in (5.11) is an affinor. This affinor

$$R_{\nu\mu\lambda}^{\cdot\cdot\cdot\kappa} \stackrel{\text{def}}{=} 2\partial_{[\nu} \Gamma^\kappa_{\mu]\lambda} + 2\Gamma^\kappa_{[\nu|\rho|} \Gamma^\rho_{\mu]\lambda} \tag{5.12}$$

is called the *curvature affinor* of the A_n. The A_n is an E_n if and only if $R_{\nu\mu\lambda}^{\cdot\cdot\cdot\kappa}$ vanishes at every point.

$R_{\nu\mu\lambda}^{\cdot\cdot\cdot\kappa}$ also occurs if the process of covariant differentiation is applied twice in succession and if the result is alternated over the indices of differentiation:

$$\begin{aligned} \nabla_{[\nu} \nabla_{\mu]} v^\kappa &= \partial_{[\nu}(\nabla_{\mu]} v^\kappa) - \Gamma^\rho_{[\nu\mu]} \nabla_\rho v^\kappa + \Gamma^\kappa_{[\nu|\lambda|} \nabla_{\mu]} v^\lambda \\ &= \partial_{[\nu} \partial_{\mu]} v^\kappa + \partial_{[\nu} \Gamma^\kappa_{\mu]\lambda} v^\lambda + \Gamma^\kappa_{[\nu|\lambda|} \partial_{\mu]} v^\lambda + \Gamma^\kappa_{[\nu|\lambda|} \Gamma^\lambda_{\mu]\rho} v^\rho \\ &= \{\partial_{[\nu} \Gamma^\kappa_{\mu]\lambda} + \Gamma^\kappa_{[\nu|\rho|} \Gamma^\rho_{\mu]\lambda}\} v^\lambda = \tfrac{1}{2} R_{\nu\mu\lambda}^{\cdot\cdot\cdot\kappa} v^\lambda. \end{aligned} \tag{5.13}$$

In the same way we find

$$\nabla_{[\nu} \nabla_{\mu]} w_\lambda = -\tfrac{1}{2} R_{\nu\mu\lambda}^{\cdot\cdot\cdot\kappa} w_\kappa, \tag{5.14}$$

and for a general affinor, for instance, $P^\kappa_{\cdot\lambda\rho}$:

$$\nabla_{[\nu} \nabla_{\mu]} P^\kappa_{\cdot\lambda\rho} = \tfrac{1}{2} R_{\nu\mu\sigma}^{\cdot\cdot\cdot\kappa} P^\sigma_{\cdot\lambda\rho} - \tfrac{1}{2} R_{\nu\mu\lambda}^{\cdot\cdot\cdot\sigma} P^\kappa_{\cdot\sigma\rho} - \tfrac{1}{2} R_{\nu\mu\rho}^{\cdot\cdot\cdot\sigma} P^\kappa_{\cdot\lambda\sigma}. \tag{5.15}$$

For every upper (lower) index there is a term with a positive (negative) sign.

For a scalar density $\mathfrak{q}$ of weight $+1$ we get

$$\begin{aligned} \nabla_{[\mu} \nabla_{\lambda]} \mathfrak{q} &= \partial_{[\mu} \nabla_{\lambda]} \mathfrak{q} - \Gamma_{[\mu} \nabla_{\lambda]} \mathfrak{q} \\ &= \partial_{[\mu} \partial_{\lambda]} \mathfrak{q} - \partial_{[\mu} \Gamma_{\lambda]} \mathfrak{q} - \Gamma_{[\mu} \partial_{\lambda]} \mathfrak{q} + \Gamma_{[\mu} \Gamma_{\lambda]} \mathfrak{q} \\ &= -(\partial_{[\mu} \Gamma_{\lambda]}) \mathfrak{q} \\ &= -\tfrac{1}{2} R_{\mu\lambda\kappa}^{\cdot\cdot\cdot\kappa} \mathfrak{q}, \end{aligned} \tag{5.16}$$

and accordingly for a general affinor density of weight $\mathfrak{k}$, for instance $\mathfrak{P}^\kappa_{\cdot\lambda}$,

$$\nabla_{[\nu} \nabla_{\mu]} \mathfrak{P}^\kappa_{\cdot\lambda} = \tfrac{1}{2} R_{\nu\mu\rho}^{\cdot\cdot\cdot\kappa} \mathfrak{P}^\rho_{\cdot\lambda} - \tfrac{1}{2} R_{\nu\mu\lambda}^{\cdot\cdot\cdot\sigma} \mathfrak{P}^\kappa_{\cdot\sigma} - \tfrac{1}{2} \mathfrak{k} R_{\nu\mu\sigma}^{\cdot\cdot\cdot\sigma} \mathfrak{P}^\kappa_{\cdot\lambda}. \tag{5.17}$$

For every upper (lower) index there is a term with a positive (negative) sign and there is an additional term containing $R_{\nu\mu\lambda}^{\cdot\cdot\cdot\lambda}$ and a factor $\mathfrak{k}$.

The curvature affinor satisfies a number of identities. The *first identity*

$$\boxed{R_{(\nu\mu)\lambda}^{\cdot\cdot\cdot\kappa} = 0} \tag{5.18}$$

follows from the definition. The *second identity*

$$\boxed{R_{[\nu\mu\lambda]}^{\cdot\cdot\cdot\kappa} = 0} \tag{5.19}$$

follows immediately from (5.12).

In a V_n we write $K_{\nu\mu\lambda}^{\cdot\cdot\cdot\kappa}$ for the curvature affinor. For $K_{\nu\mu\lambda}^{\cdot\cdot\cdot\kappa}$ we have the *third identity*

$$\boxed{K_{\nu\mu(\lambda\kappa)} = 0.} \tag{5.20}$$

In fact

$$0 = \nabla_{[\omega}\nabla_{\mu]}g_{\lambda\kappa} = -\tfrac{1}{2}K_{\omega\mu\lambda}^{\cdot\cdot\cdot\rho}g_{\rho\kappa} - \tfrac{1}{2}K_{\omega\mu\kappa}^{\cdot\cdot\cdot\rho}g_{\lambda\rho} = -K_{\omega\mu(\lambda\kappa)}. \tag{5.21}$$

From the second and third identities the *fourth identity* follows:

$$\boxed{K_{\nu\mu\lambda\kappa} = K_{\lambda\kappa\nu\mu}.} \tag{5.22}$$

From the third identity we see that $K_{\nu\mu\sigma}^{\cdot\cdot\cdot\sigma} = 0$. Hence the displacement in a V_n is integrable for scalar densities. In fact a field $\mathfrak{g}$ exists whose covariant derivative vanishes.

The curvature affinor $K_{\nu\mu\lambda\kappa}$ of a V_n is alternating in $\nu\mu$ and $\lambda\kappa$ and symmetric in the sets $\nu\mu$ and $\lambda\kappa$. Hence this quantity plays the same role for bivectors as a tensor does for vectors. It may be called a *bivector-tensor*. A general bivector has $\frac{1}{2}n(n-1)$ independent components and consequently a general bivector tensor has

$$\tfrac{1}{2}\cdot\tfrac{1}{2}n(n-1)\{\tfrac{1}{2}n(n-1)+1\} = \tfrac{1}{8}(n^2-n)(n^2-n+2)$$

independent components. But we have only used the first, third, and fourth identity. It is easily shown that as a result of these three identities the second identity is equivalent to

$$K_{[\nu\mu\lambda\kappa]} = 0 \tag{5.23}$$

and these are exactly $\binom{n}{4}$ equations. Hence the number of independent components of $K_{\nu\mu\lambda\kappa}$ is

$$\tfrac{1}{8}(n^2-n)(n^2-n+2) - \binom{n}{4} = \tfrac{1}{12}n^2(n^2-1). \tag{5.24}$$

For $n = 2$ this number is equal to 1, for $n = 3$ equal to 6, and for $n = 4$ to 20.

There is another identity, containing the first derivatives of $R_{\nu\mu\lambda}^{\cdot\cdot\cdot\kappa}$:

$$\boxed{\nabla_{[\omega} R_{\nu\mu]\lambda}^{\cdot\cdot\cdot\kappa} = 0.} \tag{5.25}$$

This identity is named after Bianchi although it was earlier found by Ricci.†

To prove this identity for a certain point ξ^κ, for (κ) we take a system of normal coordinates belonging to that point (cf. V, § 3). Then we know that the $\Gamma_{\mu\lambda}^\kappa$ are zero at ξ^κ and consequently

$$\nabla_{[\omega} R_{\nu\mu]\lambda}^{\cdot\cdot\cdot\kappa} \overset{*}{=} \partial_{[\omega} R_{\nu\mu]\lambda}^{\cdot\cdot\cdot\kappa} = 2\partial_{[\omega}\{\partial_\nu \Gamma_{\mu]\lambda}^\kappa + \Gamma_{\nu|\rho|}^\kappa \Gamma_{\mu]\lambda}^\rho\} = 0. \tag{5.26}$$

From $R_{\nu\mu\lambda}^{\cdot\cdot\cdot\kappa}$, by transvection over $\nu\kappa$, we get the *affinor of Ricci*

$$R_{\mu\lambda} \overset{\text{def}}{=} R_{\nu\mu\lambda}^{\cdot\cdot\cdot\nu}. \tag{5.27}$$

In a V_n this affinor is a tensor, the *Ricci tensor*, and is written $K_{\mu\lambda}$:

$$K_{\mu\lambda} \overset{\text{def}}{=} K_{\nu\mu\lambda}^{\cdot\cdot\cdot\nu}. \tag*{‡(5.28)}$$

A scalar can be derived from $K_{\mu\lambda}$. This is

$$K \overset{\text{def}}{=} g^{\mu\lambda} K_{\mu\lambda}. \tag{5.29}$$

$\dfrac{1}{n(n-1)} K$ is called the *scalar curvature* of the V_n. For a sphere with radius one, $K_{\nu\mu\lambda\kappa}$ has only one independent component (cf. (5.3))

$$\begin{aligned} K_{1212} &= 2g_{2\kappa}\{\partial_{[1} \Gamma_{2]1}^\kappa + \Gamma_{[1|\rho|}^\kappa \Gamma_{2]1}^\rho\} \\ &= g_{22}\{-\partial_2 \Gamma_{11}^2 + \Gamma_{11}^2 \Gamma_{21}^1 - \Gamma_{2\rho}^2 \Gamma_{11}^\rho\} \\ &= \cos^2\theta - \sin^2\theta - \sin\theta\cos\theta\cot\theta = -\sin^2\theta. \end{aligned} \tag{5.30}$$

Hence $$K_{11} = \sin^2\theta, \qquad K_{12} = 0, \qquad K_{22} = 1, \tag{5.31}$$

and $$\tfrac{1}{2}K = \tfrac{1}{2}g^{\mu\lambda}K_{\mu\lambda} = 1. \tag{5.32}$$

Hence the scalar curvature of the sphere is 1, as it should be.

From Bianchi's identity (5.25) we get, by contraction over $\omega\kappa$,

$$\nabla_\omega R_{\nu\mu\lambda}^{\cdot\cdot\cdot\omega} + \nabla_\nu R_{\mu\omega\lambda}^{\cdot\cdot\cdot\omega} + \nabla_\mu R_{\omega\nu\lambda}^{\cdot\cdot\cdot\omega} = 0, \tag{5.33}$$

or $$\boxed{\nabla_\omega R_{\nu\mu\lambda}^{\cdot\cdot\cdot\omega} + 2\nabla_{[\mu} R_{\nu]\lambda} = 0.} \tag{5.34}$$

In a V_n this identity can be transvected with $g_{\mu\lambda}$ and this gives

$$2\nabla_\mu K_\nu^{\cdot\mu} - \nabla_\nu K = 0. \tag{5.35}$$

Hence, if we introduce the tensor

$$G_{\lambda\kappa} \overset{\text{def}}{=} K_{\lambda\kappa} - \tfrac{1}{2}Kg_{\lambda\kappa}, \tag{5.36}$$

† Cf. for literature R.K. 1924. 1, p. 91. According to Okaya, 1925. 1, the identity was originally found by Voss, 1880. 1.

‡ See Note F, p. 272.

the identity (5.35) takes the form

$$\boxed{\nabla_\mu G_\lambda^{\cdot\mu} = 0.} \tag{5.37}$$

The tensor $G_{\lambda\kappa}$ or rather the tensor density

$$\mathfrak{G}_{\lambda\kappa} \overset{\text{def}}{=} \mathfrak{g}^{\frac{1}{2}} G_{\lambda\kappa} \tag{5.38}$$

plays a very important role in the theory of relativity. This is due to the fact that $\mathfrak{G}^{\kappa\lambda}$ is the Lagrange derivative of $-\mathfrak{g}^{\frac{1}{2}}K$ as a function of $g_{\lambda\kappa}$, $\partial_\mu g_{\lambda\kappa}$, and $\partial_\omega \partial_\mu g_{\lambda\kappa}$ (cf. IV, § 6):

$$[\mathfrak{g}^{\frac{1}{2}}K]^{\kappa\lambda} = -\mathfrak{G}^{\kappa\lambda}. \tag{5.39}$$

A direct proof of (5.39) is rather long, but Palatini gave a very short and elegant proof by using a system of normal coordinates for (κ), which belonged to a definitely chosen point ξ^κ.† Then we know that the $\Gamma^\kappa_{\mu\lambda}$ vanish in ξ^κ. Consequently the $\partial_\mu g_{\lambda\kappa}$ also vanish at ξ^κ. But the derivatives at $\Gamma^\kappa_{\mu\lambda}$ do not vanish at that point. If $\overset{v}{d} g_{\lambda\kappa}$ is the variation of $g_{\lambda\kappa}$, the variations of $\Gamma^\kappa_{\mu\lambda}$ are not zero at $\underset{0}{\xi^\kappa}$ because the normal system is not a normal system for the new displacement. But the $\overset{v}{d}\Gamma^\kappa_{\mu\lambda}$ are the differences of the parameters of two different displacements and accordingly are *components of an affinor*. Now we have

$$K = 2g^{\mu\lambda}(\partial_{[\kappa}\Gamma^\kappa_{\mu]\lambda} + \Gamma^\kappa_{[\kappa|\rho|}\Gamma^\rho_{\mu]\lambda}) \tag{5.40}$$

and consequently, at ξ^κ,

$$\begin{aligned}
\overset{v}{d}(\mathfrak{g}^{\frac{1}{2}}K) &\overset{*}{=} 2\mathfrak{g}^{\frac{1}{2}}(\overset{v}{d}g^{\mu\lambda})\,\partial_{[\kappa}\Gamma^\kappa_{\mu]\lambda} + 2\mathfrak{g}^{\frac{1}{2}}g^{\mu\lambda}\,\partial_{[\kappa}\overset{v}{d}\Gamma^\kappa_{\mu]\lambda} + K\,\overset{v}{d}\mathfrak{g}^{\frac{1}{2}} \\
&\overset{*}{=} \mathfrak{g}^{\frac{1}{2}}K_{\mu\lambda}\overset{v}{d}g^{\mu\lambda} + \mathfrak{g}^{\frac{1}{2}}\partial_\kappa g^{\mu\lambda}\,\overset{v}{d}\Gamma^\kappa_{\mu\lambda} - \mathfrak{g}^{\frac{1}{2}}\partial_\mu g^{\mu\lambda}\,\overset{v}{d}\Gamma^\kappa_{\kappa\lambda} + \tfrac{1}{2}K\mathfrak{g}^{\frac{1}{2}}\overset{v}{d}\log\mathfrak{g} \\
&\overset{*}{=} \mathfrak{g}^{\frac{1}{2}}K_{\mu\lambda}\overset{v}{d}g^{\mu\lambda} + \mathfrak{g}^{\frac{1}{2}}\partial_\mu(g^{\nu\lambda}\,\overset{v}{d}\Gamma^\mu_{\nu\lambda} - g^{\mu\lambda}\,\overset{v}{d}\Gamma^\kappa_{\kappa\lambda}) - \tfrac{1}{2}\mathfrak{g}^{\frac{1}{2}}Kg_{\mu\lambda}\overset{v}{d}g^{\mu\lambda} \\
&\overset{*}{=} \mathfrak{g}^{\frac{1}{2}}G_{\mu\lambda}\overset{v}{d}g^{\mu\lambda} + \nabla_\mu\,\mathfrak{g}^{\frac{1}{2}}(g^{\nu\lambda}\,\overset{v}{d}\Gamma^\mu_{\nu\lambda} - g^{\mu\lambda}\,\overset{v}{d}\Gamma^\nu_{\nu\lambda}),
\end{aligned} \tag{5.41}$$

where the sign $*$ can at last be dropped because the equation is now in the invariant form.

Now if τ_n is a region of V_n on whose boundary the variations of $g_{\lambda\kappa}$, $\partial_\mu g_{\lambda\kappa}$, and $\partial_\nu \partial_\mu g_{\lambda\kappa}$ vanish, the $\overset{v}{d}\Gamma^\kappa_{\mu\lambda}$ must vanish there too and accordingly we have, using the theorem of Stokes for the last term of (5.41),

$$\overset{v}{d}\int_{\tau_n} \mathfrak{g}^{\frac{1}{2}}K\,d\xi^1 \ldots d\xi^n = -\int_{\tau_n} \mathfrak{g}^{\frac{1}{2}}G^{\lambda\kappa}\,\overset{v}{d}g_{\lambda\kappa}\,d\xi^1 \ldots d\xi^n, \tag{5.42}$$

† Cf. E I, 1935. 1, pp. 147 f., also for references to literature.

and this proves (5.39). If and only if $G_{\lambda\kappa}$ vanishes does the variation of the integral of $\mathfrak{g}^{\frac{1}{2}}K$ over τ_n vanish for every variation of $g_{\lambda\kappa}$ which satisfies the boundary conditions.

The identity (IV.6.11) now takes the form

$$\mathfrak{G}^{\lambda\kappa}(\partial_\mu g_{\lambda\kappa}-\partial_\lambda g_{\mu\kappa}-\partial_\kappa g_{\lambda\mu})-g_{\mu\kappa}\partial_\lambda \mathfrak{G}^{\lambda\kappa}-g_{\lambda\mu}\partial_\kappa \mathfrak{G}^{\lambda\kappa} = 0 \tag{5.43}$$

or

$$\nabla_\lambda \mathfrak{G}^{\lambda\kappa} = \partial_\lambda \mathfrak{G}^{\lambda\kappa}+\Gamma^{\kappa}_{\lambda\nu}\mathfrak{G}^{\lambda\nu} = 0, \tag{5.44}$$

in accordance with (5.37).

6. Curvature of a V_2 with a positive definite fundamental tensor

In an E_2 two bivectors can only differ by a scalar factor. Hence in a V_2 the curvature affinor $K_{\nu\mu\lambda\kappa}$ can only differ by a scalar factor from $g_{[\nu[\lambda}g_{\mu]\kappa]}$. It follows immediately from (5.28) and (5.29) that this factor is $-K$:

$$K_{\nu\mu\lambda\kappa} = -Kg_{[\nu[\lambda}g_{\mu]\kappa]}, \qquad K_{\mu\lambda} = \tfrac{1}{2}Kg_{\mu\lambda}, \qquad G_{\mu\lambda} = 0. \tag{6.1}$$

If $I^{\kappa\lambda}$ is a bivector which has the unit area and $I^{\kappa\lambda}\,d\sigma$ the bivector of a V_2-element with the area $d\sigma$, a vector v^κ after pseudo-parallel displacement along the boundary of $I^{\kappa\lambda}\,d\sigma$ shows the difference

$$dv^\kappa = I^{\nu\mu}K_{\nu\mu\lambda}^{\cdot\cdot\cdot\kappa}v^\lambda\,d\sigma \tag{6.2}$$

between the original position and the end position. Now taking cartesian local coordinates and $v^1 = 1$, $v^2 = 0$ in the original position, we have $I^{12} = -I^{21} = 1$ and $K_{1212} = -\frac{1}{2}K$, and accordingly

$$dv^1 = 0, \qquad dv^2 = \tfrac{1}{2}K\,d\sigma. \tag{6.3}$$

Hence the vector is turned through the angle $\frac{1}{2}K\,d\sigma$.

If v^κ is displaced pseudo-parallel along the boundary of a region τ_2 we get the integral $\int \frac{1}{2}K\,d\sigma$, that is the *integral of the scalar curvature over* τ_2 for the angle between the original position and the end position. In Figs. 20, 21 we see that a displacement in the clockwise direction gives a turn in the clockwise direction on a surface with positive curvature and in the counter clockwise direction if the curvature is negative.

7. Anholonomic coordinate systems†

For an anholonomic system (h) (cf. IV, § 7) the equations (1.9) remain valid:

$$\Gamma^h_{ji} = A^{h\mu\lambda}_{\kappa ji}\,\Gamma^\kappa_{\mu\lambda}-A^{\mu\lambda}_{ji}\,\partial_\mu A^h_\lambda \quad (h,i,j = 1,\ldots,n), \tag{7.1}$$

and from this it follows that we have (cf. IV.7.8)

$$\Gamma^h_{[ji]} = S_{ji}^{\cdot\cdot h}-\Omega_{ji}^{\cdot\cdot h} \quad (h,i,j = 1,\ldots,n) \tag{7.2}$$

† Cf. E I. 1935. 1, pp. 68, 94, R.C. III § 9, also for references to literature.

instead of (1.22). Consequently in an A_n we get

$$\Gamma^h_{[ji]} = -\Omega_{ji}^{\cdot\cdot h} \quad (h, i, j = 1,..., \mathrm{n}) \tag{7.3}$$

and (cf. (1.26))

$$\begin{aligned} &(a) \qquad \nabla_{[j} w_{i]} = \partial_{[j} w_{i]} + \Omega_{ji}^{\cdot\cdot h} w_h \\ &(b) \qquad \nabla_j \mathfrak{v}^j = \partial_j \mathfrak{v}^j + 2\Omega_{ji}^{\cdot\cdot i} \mathfrak{v}^j \end{aligned} \qquad (h, i, j = 1,..., \mathrm{n}). \tag{7.4}$$

From these equations the equations (IV.7.7) and (IV.7.9) can be derived immediately.

According to (7.2) get we in V_n

$$\begin{aligned} \Gamma^h_{ji} &= \{^{\,h}_{ji}\} - \Omega_{ji}^{\cdot\cdot h} + g_{jl} g^{hk} \Omega_{ik}^{\cdot\cdot l} + g_{il} g^{hk} \Omega_{jk}^{\cdot\cdot l} \\ \{^{\,h}_{ji}\} &\overset{\text{def}}{=} \tfrac{1}{2} g^{hk} (\partial_j g_{ik} + \partial_i g_{kj} - \partial_k g_{ji}) \end{aligned} \tag{7.5}$$

$$(h, i, j, k, l = 1,..., \mathrm{n})$$

instead of (4.7).

If the anholonomic system consists at every point of n mutually perpendicular unit vectors $\overset{h}{i}{}^{\kappa}$, $\underset{i}{i}{}_{\lambda}$; $h, i = 1,..., \mathrm{n}$, we have $\{^{\,h}_{ji}\} = 0$ and

$$\Gamma^h_{ji} = -\Omega_{ji}^{\cdot\cdot h} + g_{jl} g^{hk} \Omega_{ik}^{\cdot\cdot l} + g_{il} g^{hk} \Omega_{jk}^{\cdot\cdot l} \quad (h, i, j, k, l = 1,..., \mathrm{n}). \tag{7.6}$$

From (7.5) we get

$$\begin{aligned} \Gamma^i_{ji} &= \{^{\,i}_{ji}\} - \Omega_{ji}^{\cdot\cdot i} + g_{jl} g^{ik} \Omega_{ik}^{\cdot\cdot l} + \Omega_{ji}^{\cdot\cdot i} \\ &= \{^{\,i}_{ji}\} = \tfrac{1}{2} \partial_j \log \mathfrak{g} \qquad (i, j, k, l = 1,..., \mathrm{n}) \end{aligned} \tag{7.7}$$

in V_n (cf. (4.10, 11)). Hence (4.10) holds for anholonomic coordinate systems also.

In A_n we get from (5.12)

$$R_{kji}^{\cdot\cdot\cdot h} = 2\partial_{[k} \Gamma^h_{j]i} + 2\Gamma^h_{[k|l|} \Gamma^l_{j]i} + 2\Omega_{kj}^{\cdot\cdot l} \Gamma^h_{li} \quad (h, i, j, k, l = 1,..., \mathrm{n}) \tag{7.8}$$

for the curvature affinor in the anholonomic case.

Of course all equations which only contain quantities and signs ∇ are invariant under transformations into anholonomic coordinates. But equations which contain $\Gamma^\kappa_{\mu\lambda}$ or ∂_μ do not generally have this invariance. It is often convenient to use the sign $\overset{h}{=}$ instead of $=$ in equations that are only valid in holonomic coordinates.

8. Integral formulae in V_n and R_n†

In a V_n we apply (IV.4.15) for $p = 1$

$$\int_{\tau_{n-1}} \mathfrak{v}^\mu \, d\mathfrak{f}_\mu = \int_{\tau_n} \nabla_\mu \mathfrak{v}^\mu \, d\mathfrak{f} \tag{8.1}$$

† General references: Kronecker 1894. 1; Burkhardt and Meyer 1900. 2; Wilson 1913. 1.

(cf. V, § 1) to the special case when $\mathfrak{v}^\mu$ has the form

$$\mathfrak{v}^\mu = \phi \mathfrak{g}^{\frac{1}{2}} \nabla^\mu \psi, \qquad \nabla^\mu \psi \stackrel{\text{def}}{=} g^{\mu\lambda} \nabla_\lambda \psi. \tag{8.2}$$

Then we get *Green's first identity* for the V_n

$$\int_{\tau_{n-1}} \mathfrak{g}^{\frac{1}{2}} \phi \nabla^\mu \psi \, d\mathfrak{f}_\mu = \int_{\tau_n} \mathfrak{g}^{\frac{1}{2}} (\nabla_\mu \phi) \nabla^\mu \psi \, d\mathfrak{f} + \int_{\tau_n} \mathfrak{g}^{\frac{1}{2}} \phi \nabla_\mu \nabla^\mu \psi \, d\mathfrak{f}. \tag{8.3}$$

By interchanging ϕ and ψ and subtracting we get *Green's second identity* for the V_n

$$\int_{\tau_{n-1}} \mathfrak{g}^{\frac{1}{2}} (\phi \nabla^\mu \psi - \psi \nabla^\mu \phi) \, d\mathfrak{f}_\mu = \int_{\tau_n} \mathfrak{g}^{\frac{1}{2}} (\phi \nabla_\mu \nabla^\mu \psi - \psi \nabla_\mu \nabla^\mu \phi) \, d\mathfrak{f}. \tag{8.4}$$

ϕ and ψ have of course to be chosen in such a way that $\phi \nabla^\mu \psi$ and $\psi \nabla^\mu \phi$ satisfy the conditions of Ch. IV, § 4.

Now we consider an R_n, with a positive definite fundamental tensor and introduce cartesian coordinates x^h ($h = 1,\ldots, n$). If $n > 2$ we choose ψ so that

$$\psi = \frac{1}{n-2} a^{-n+2}, \qquad a \stackrel{\text{def}}{=} \sqrt{\{g_{ih}(x^i - {}'x^i)(x^h - {}'x^h)\}} \quad (h, i = 1,\ldots, n) \tag{8.5}$$

where ${}'x^h$ is a variable point. Then ψ is a function of x^h and ${}'x^h$ and

$$\partial_i \psi = -{}'\partial_i \psi = -a^{-n} a_i, \qquad {}'\partial_i \stackrel{\text{def}}{=} \partial/\partial\, {}'x^i, \qquad a_i \stackrel{\text{def}}{=} x_i - x_i' = a \partial_i a,$$
$$\nabla^2 \psi \stackrel{\text{def}}{=} g^{hi} \partial_h \partial_i \psi = 0. \tag{8.6}$$

Hence, if τ_n does not contain the point ${}'x^h$ we get, from (8.4),

$$\int_{\tau_{n-1}} \phi a^{-n} a^i \, d\mathfrak{f}_i + \frac{1}{n-2} \int_{\tau_{n-1}} a^{-n+2} \partial^i \phi \, d\mathfrak{f}_i = \frac{1}{n-2} \int_{\tau_n} a^{-n+2} \nabla^2 \phi \, d\mathfrak{f}. \tag{8.7}$$

If however τ_n contains the point ${}'x^h$ the integration on the right-hand side of (8.7) cannot be effected directly because $a^{-n+2} \to \infty$ for $x^h \to {}'x^h$. Therefore we consider this integral as the limit of an integral over the space $\tau_n - \epsilon$ between τ_{n-1} and a hypersphere ϵ with a small radius R around the point ${}'x^h$. Then the left-hand side of (8.7) becomes

$$\lim_{\epsilon \to 0} \left\{ \int_{\tau_{n-1}} \left(\phi a^{-n} a^i + \frac{1}{n-2} a^{-n+2} \partial^i \phi \right) d\mathfrak{f}_i - \right.$$
$$\left. - \int_{\epsilon} \left(\phi R^{-n+1} n^i + \frac{1}{n-2} R^{-n+2} \partial^i \phi \right) d\mathfrak{f}_i \right\} \quad (i = 1,\ldots, n), \tag{8.8}$$

where n^i is a unit vector and $d\mathfrak{f}_i = n_i \, d\tau_{n-1}$. Now

$$\int_{\epsilon} n^i \, d\mathfrak{f}_i = \underset{n}{\omega} R^{n-1} \quad (i = 1,\ldots, n), \tag{8.9}$$

with

$$\underset{n}{\omega} \stackrel{\text{def}}{=} \frac{2\pi^{n/2}}{\Gamma(\frac{1}{2}n)} = \begin{cases} \dfrac{n\,.\,\pi^{n/2}}{(\frac{1}{2}n)!} & \text{for } n \text{ even,} \\ \dfrac{2^{\frac{1}{2}(n+1)}\,.\,\pi^{\frac{1}{2}(n-1)}}{1.3.\,\ldots\,.n-2} & \text{for } n \text{ odd†} \end{cases} \tag{8.10}$$

is the $(n-1)$-dimensional area of the hypersphere ϵ. The last term in the second integral of (8.8) vanishes for $R \to 0$. Hence

$$\underset{n}{\omega}\phi = -\frac{1}{n-2}\int\limits_{\tau_n} a^{-n+2}\nabla^2\phi\, d\mathfrak{f}+\frac{1}{n-2}\int\limits_{\tau_{n-1}} a^{-n+2}\,\partial^i\phi\, d\mathfrak{f}_i+\int\limits_{\tau_{n-1}} a^{-n}\phi a^i\, d\mathfrak{f}_i. \tag{8.11}$$

In the integrals on the right-hand side of this equation the variables are the x^h and ϕ is considered as a function of the x^h, but on the left-hand side ϕ appears as a function of the $'x^h$. (8.11) is *Green's theorem* for an R_n with a positive definite fundamental tensor.

If in (8.11) we take the hypersphere $a = \rho$ for τ_n, it is possible that the last two integrals in (8.11) vanish when $\rho \to \infty$. In this case (8.11) takes the form

$$\underset{n}{\omega}\phi = -\frac{1}{n-2}\int\limits_{\infty} a^{-n+2}\nabla^2\phi\, d\mathfrak{f}, \tag{8.12}$$

or

$$\phi = \operatorname{Pot}\nabla^2\phi, \tag{8.13}$$

where Pot is an integral operator defined by

$$\operatorname{Pot} \stackrel{\text{def}}{=} -\frac{1}{(n-2)\underset{n}{\omega}}\int\limits_{\infty} \ldots a^{-n+2}\, d\mathfrak{f}. \tag{8.14}$$

A function ϕ satisfying (8.12) is called a *potential function*. Sufficient conditions are:‡

(1) ϕ and its first and second derivatives are finite everywhere;

(2) the average value of ϕ on a sphere with radius ρ tends to zero for $\rho \to \infty$;

(3)
$$\lim_{\rho\to\infty}\int\limits_{\tau_n} \rho^{-n+2}\nabla^2\phi\, d\mathfrak{f} = 0. \tag{8.15}$$

A function is always a potential function if the first condition is satisfied and if the function is zero at all points outside some finite sphere.

A function $f(x^h)$ is said to be *harmonic* in a region if its first and second derivatives are finite and continuous, except on a finite number of hypersurfaces where a finite discontinuity is allowed, and if $\nabla^2 f = 0$

† Cf. Kronecker 1894. 1, p. 266; Jahnke and Emde, 1933. 3, p. 87 f.

‡ Cf. Kronecker 1894. 1, p. 309; necessary and sufficient conditions on p. 307.

in that region. If we put $\psi = 1$ in (8.4) it follows that for every function ϕ which is harmonic in τ_n

$$\int_{\tau_{n-1}} \partial^i \phi \, d\mathfrak{f}_i = 0. \tag{8.16}$$

Now, if for τ_{n-1} we take the hypersphere $a = \rho$, it follows from (8.11) and (8.16) that

$$\underset{n}{\omega}\phi = \int_{\tau_{n-1}} \rho^{-n+1}\phi n^i \, d\mathfrak{f}_i, \tag{8.17}$$

or

$$\underset{n}{\omega}\rho^{n-1}\phi = \int_{\tau_{n-1}} \phi n^i \, d\mathfrak{f}_i, \tag{8.18}$$

and this equation expresses the fact that the value of ϕ at $'x^h$ is the mean of its values on any sphere with centre at $'x^h$, provided that ϕ is harmonic at all points inside the sphere and on its boundary.

ϕ at an inner point of τ_n can be expressed in terms of the values of $\nabla^2\phi$ in τ_n and the values of ϕ and $n^i\partial_i\phi$ on τ_{n-1} by means of (8.11). But obviously we have given too much here. If we take $\psi = \phi$ in (8.3) in R_n we get

$$\int_{\tau_{n-1}} \phi n^\mu \partial_\mu \phi \, d\tau_{n-1} = \int_{\tau_n} (\partial_\mu \phi)(\partial^\mu \phi) \, d\mathfrak{f} + \int_{\tau_n} \phi \nabla^2 \phi \, d\mathfrak{f}. \tag{8.19}$$

Hence, if $\nabla^2\phi = 0$ in τ_n and if either ϕ or $n^\mu\partial_\mu\phi$ vanish on τ_{n-1}, the first integral on the right-hand side must vanish, and, since $g_{\lambda\kappa}$ is positive definite, this is only possible if ϕ is constant in τ_n. Now let $\overset{1}{\phi}$ and $\overset{2}{\phi}$ be two fields such that $\nabla^2\overset{1}{\phi} = \nabla^2\overset{2}{\phi}$ in τ_n and either $\overset{1}{\phi} = \overset{2}{\phi}$ on τ_{n-1} or $n^\mu\partial_\mu\overset{1}{\phi} = n^\mu\partial_\mu\overset{2}{\phi}$ on τ_{n-1}. Then it follows that $\overset{2}{\phi} = \overset{1}{\phi}+$constant in τ_n. Note that this holds not only in an R_n but also in a V_n provided that $g_{\lambda\kappa}$ is positive definite.

As an example in R_3 we consider the flow of heat in an isotropic medium. If T is the temperature, $\partial^i T$ is proportional to the heat current vector density and $\nabla^2 T$ is proportional to the heat input per unit volume and per unit time. Now it is well known for the steady case that T is determined in τ_3 if T is given on τ_2 and $\nabla^2 T$ is given in τ_3 and that T is determined to within a constant if $n^i\partial_i T$ is given on τ_2 and $\nabla^2 T$ is given in τ_3. These two cases are known as the *first* and *second boundary value problems* (*Randwertproblem*). There is a *third boundary value problem* where $hn^i\partial_i T + T$ is given on τ_2 and $\nabla^2 T$ in τ_3, h being a given positive constant.

The first boundary value problem can be solved in R_n by taking the function ψ in (8.4) in such a way that it is harmonic in τ_n and takes the value $\frac{1}{n-2}a^{-n+2}$ on τ_{n-1}. Then (8.4) takes the form

$$\int_{\tau_{n-1}} \left(\phi \partial^i \psi - \frac{1}{n-2} a^{-n+2} \partial^i \phi\right) d\mathfrak{f}_i + \int_{\tau_n} \psi \nabla^2 \phi \, d\mathfrak{f} = 0, \tag{8.20}$$

and from this equation and (8.11) it follows that

$$\underset{n}{\omega}\phi = \int_{\tau_n} \left(\psi - \frac{1}{n-2} a^{-n+2}\right) \nabla^2 \phi \, d\mathfrak{f} + \int_{\tau_{n-1}} \phi \partial^i \left(\psi - \frac{1}{n-2} a^{-n+2}\right) d\mathfrak{f}_i \tag{8.21}$$

expressing ϕ in $'x^h$ in terms of $\nabla^2\phi$ and ψ over τ_n and ϕ and $n^i\partial_i\psi$ over τ_{n-1}. $\psi - \frac{1}{n-2}a^{-n+2}$ is called a *Green's function*. So the first problem is reduced to the problem of the construction of a Green's function. Green's functions can also be constructed for the other boundary value problems of this kind.

If (8.11) is applied to all cartesian components of any quantity $P^{h_1 \ldots h_p}_{\cdot \cdot \; i_1 \ldots i_q}$ (suppressing the indices of P), we get

$$\underset{n}{\omega}P = -\frac{1}{n-2}\int_{\tau_n} a^{-n+2} \nabla^2 P \, d\mathfrak{f} + \frac{1}{n-2}\int_{\tau_{n-1}} a^{-n+2} \partial^i P \, d\mathfrak{f}_i + \int_{\tau_{n-1}} a^{-n} P a^i \, d\mathfrak{f}. \tag{8.22}$$

If we take a p-vector $w_{i_1 \ldots i_p}$, $p \geqslant 1$ for P, we get according to (IV, Ex. 1)

$$\underset{n}{\omega}w = -\frac{1}{n-2}\int_{\tau_n} a^{-n+2}(\operatorname{Div}\operatorname{Rot} + \operatorname{Rot}\operatorname{Div}) w \, d\mathfrak{f} +$$

$$+ \frac{1}{n-2}\int_{\tau_{n-1}} a^{-n+2} \partial^i w \, d\mathfrak{f}_i + \int_{\tau_{n-}} a^{-n} w a^i \, d\mathfrak{f}_i, \tag{8.23}$$

where the indices $i_1 \ldots i_p$ are suppressed.

If the components $w_{i_1 \ldots i_p}$ are potential functions (8.23) takes the form (cf. (8.13))

$$w = \operatorname{Pot}\operatorname{Div}\operatorname{Rot} w + \operatorname{Pot}\operatorname{Rot}\operatorname{Div} w. \tag{8.24}$$

If $w_{i_1 \ldots i_p}$ and its first derivatives are finite everywhere and if $w_{i_1 \ldots i_p}$ is zero at all points outside some finite sphere it can be proved that, besides (8.24), there is an identity

$$w = \operatorname{Div}\operatorname{Pot}\operatorname{Rot} w + \operatorname{Rot}\operatorname{Pot}\operatorname{Div} w. \tag{8.25}$$

The conditions are sufficient but not necessary. We use (8.25) to derive an integral expression for a field $w_{i_1 \ldots i_p}$ which only satisfies the condition that its first derivatives in τ_n and on τ_{n-1} are finite. By cutting off the part of the field outside τ_{n-1} we get a field which is discontinuous on τ_{n-1}. But the continuity can be restored by introducing a small layer of thickness h over τ_{n-1}, in which the field changes uniformly from the inside value to zero. Let $x^h = 0$ be chosen in a point of τ_{n-1}, $w_{i_1...i}$ be the field value in that point and $\underset{1}{i^h}$ be chosen in the direction perpendicular to τ_{n-1} from inside to outside. Then the field value on the x^1-axis between $x^1 = 0$ and $x^1 = h$ is

$$w_{i_1...i_p} = \underset{0}{w}_{i_1...i_p} - \frac{x^1}{h} \underset{0}{w}_{i_1...i_p} \tag{8.26}$$

and the field value vanishes for $x^1 > h$. Now if h is small enough the changes of $w_{i_1...i_p}$ in directions perpendicular to $\underset{1}{i^h}$ can be neglected, and consequently the values of Div w and Rot w on the x^1 axis are

$$\begin{aligned} &(a) \qquad \partial^{i_1} w_{i_1...i_p} = -\frac{1}{h} \underset{1}{i}^{i_1} \underset{0}{w}_{i_1...i_p}, \\ &(b) \qquad (p+1)\partial_{[i_0} w_{i_1...i_p]} = -\frac{1}{h}(p+1) \underset{1}{i}_{[i_0} \underset{0}{w}_{i_1...i_p]}. \end{aligned} \tag{8.27}$$

Now for the $d\mathfrak{f}$ of the layer we have to take $d\mathfrak{f} = hn^i d\mathfrak{f}_i$. Hence, by applying (8.25) to the field with the layer and proceeding to the limit $h \to 0$, we get the identity

$$\begin{aligned} -(n-2)\underset{n}{\omega} w_{i_1...i_p} &= \text{Div} \int_{\tau_n} a^{-n+2} \text{Rot}\, w\, d\mathfrak{f} + \text{Rot} \int_{\tau_n} a^{-n+2} \text{Div}\, w\, d\mathfrak{f} - \\ &-\partial^{i_0} \int_{\tau_{n-1}} a^{-n+2}(p+1) w_{[i_1...i_p} d\mathfrak{f}_{i_0]} - p\partial_{[i_1} \int_{\tau_{n-1}} a^{-n+2} w_{|j|i_2...i_p} d\mathfrak{f}^j, \end{aligned} \tag{8.28}$$

expressing w in $'x^h$ in terms of Div w and Rot w in τ_n and $w_{[i_1...i_p} n_{i_0]}$ and $n^j w_{ji_2...i_p}$ on τ_{n-1}.

The identity (8.11) that has been derived here by means of Green's identities could also have been obtained in the same way as (8.28). Then we should have had to start from the equation (8.12) that had to be proved first, but only for fields with finite values of field and first and second derivatives and field values zero in all points at the outside of some finite sphere. After cutting off the field outside τ_{n-1} finiteness of the first derivative could be restored by introducing one layer, and after that the second derivatives could be made finite by two other layers at the inside and at the outside of the first one (a so-called double layer). After this the process (8.12) could be applied. The second term to the right of (8.11) then comes from the first layer, and in fact it has the same form as the correcting terms in (8.28). The third term to the right is due to the double layer.

EXERCISES

V.1. Prove that in V_n

$$\nabla^2 w_{\lambda_1 \dots \lambda_p} \stackrel{\text{def}}{=} \nabla^\mu \nabla_\mu w_{\lambda_1 \dots \lambda_p} = \nabla^\mu (p+1) \nabla_{[\mu} w_{\lambda_1 \dots \lambda_p]} + p \partial_{[\lambda_1} \nabla^\mu w_{|\mu| \lambda_2 \dots \lambda_p]} +$$

$$+ K_{[\lambda_1}^{\cdot\;\;\kappa} w_{|\kappa| \lambda_2 \dots \lambda_p]} + \tfrac{1}{2}(p-1) K_{[\lambda_1 \lambda_2}^{\cdot\;\;\cdot\;\;\mu\kappa} w_{|\mu\kappa| \lambda_3 \dots \lambda_p]} \qquad (1\,\alpha)$$

for $p > 1$ and that the last term on the right-hand side disappears for $p = 1$.

V.2. For every vector field w_λ in a V_n with vanishing $G_{\lambda\kappa}$ we have

$$\nabla^2 w = \operatorname{Div} \operatorname{Rot} w + \operatorname{Grad} \operatorname{Div} w. \qquad (2\,\alpha)$$

V.3. $G_{\lambda\kappa}$ vanishes for $n = 2$.

V.4. A Weyl displacement is defined by

$$\nabla_\mu g_{\lambda\kappa} = -Q_\mu g_{\lambda\kappa}, \qquad S_{\mu\lambda}^{\cdot\cdot\;\kappa} = 0. \qquad (4\,\alpha)$$

Prove that

$$\Gamma_{\mu\kappa}^{\kappa} = \tfrac{1}{2}\partial_\mu \mathfrak{g} + \tfrac{1}{2} n Q_\mu. \qquad (4\,\beta)$$

V.5. Prove that for a Weyl displacement,

$$'\nabla_\mu {}'g_{\lambda\kappa} = -'Q_\mu {}'g_{\lambda\kappa}$$

if

$$'g_{\lambda\kappa} \stackrel{\text{def}}{=} \sigma g_{\lambda\kappa}, \qquad 'Q_\lambda \stackrel{\text{def}}{=} Q_\lambda - \partial_\lambda \log \sigma. \qquad (5\,\beta)$$

V.6. If $\Gamma_{\mu\lambda}^{\kappa}$ and $'\Gamma_{\mu\lambda}^{\kappa}$ are the parameters of two linear displacements, the differences $'\Gamma_{\mu\lambda}^{\kappa} - \Gamma_{\mu\lambda}^{\kappa}$ are components of an affinor.

V.7. If a V_n suffers a conformal transformation, i.e. if $g_{\lambda\kappa}$ is replaced by $\sigma g_{\lambda\kappa}$, $K_{\nu\mu\lambda}^{\cdot\cdot\cdot\;\kappa}$ transforms as follows:

$$'K_{\nu\mu\lambda}^{\cdot\cdot\cdot\;\kappa} = K_{\nu\mu\lambda}^{\cdot\cdot\cdot\;\kappa} + g_{[\nu[\lambda} s_{\mu]\rho]} g^{\rho\kappa},$$

where

$$s_{\mu\lambda} \stackrel{\text{def}}{=} 2\nabla_\mu s_\lambda - s_\mu s_\lambda + \tfrac{1}{2} s_\rho s^\rho g_{\mu\lambda}, \qquad (7\,\beta)$$

$$s_\mu \stackrel{\text{def}}{=} \partial_\mu \log \sigma.$$

V.8. Prove that, for an anholonomic coordinate system (h)

$$\partial_{[j} \partial_{i]} p = -\Omega_{ji}^{h} \partial_h p, \qquad (8\,\alpha)$$

$$\underset{[2\;\;1]}{d(d\xi)^h} = \Omega_{ji}^{h} \underset{2}{(d\xi)^j} \underset{1}{(d\xi)^i}. \qquad (8\,\beta)$$

V.9. Prove that for a general linear connexion

$$\underset{v}{\pounds}\, \Gamma_{\mu\lambda}^{\kappa} = \nabla_\mu v_\lambda^{\cdot\kappa} + v^\sigma R_{\sigma\mu\lambda}^{\cdot\cdot\cdot\;\kappa}, \qquad (9\,\alpha)$$

where

$$v_\lambda^{\cdot\kappa} = \nabla_\lambda v^\kappa - 2 S_{\lambda\mu}^{\cdot\cdot\;\kappa} v^\mu. \qquad (9\,\beta)$$

SUMMARY
OF CHAPTERS I–V

Principles of the kernel-index method

To every coordinate system there belongs a set of *running* indices and a corresponding set of *fixed* indices. These running and fixed indices are used for this coordinate system only. Every geometric object has its own *kernel letter*. If the coordinates are transformed the running and fixed indices change but the kernel letters remain unchanged. If an object is transformed the kernel letter changes but the running and fixed indices remain the same. We agree to use italic fixed indices with Greek running indices, e.g. $\kappa, \lambda = \mathit{1}, \ldots, n$ and vertical fixed indices with Roman running ones, e.g. $h, i = 1, \ldots, \mathrm{n}$.

In this book we prefer

$$\left.\begin{array}{r}\kappa, \lambda, \mu, \nu, \rho, \sigma, \tau \text{ and sometimes } \omega \\ = \mathit{1}, \ldots, n \\ \kappa', \lambda', \mu', \nu', \rho', \sigma', \tau' \text{ and sometimes } \omega' \\ = \mathit{1}', \ldots, n'\end{array}\right\} \text{for curvilinear coordinates in } X_n \text{ and } E_n \text{ and rectilinear coordinates in } R_n.$$

$$\left.\begin{array}{l}h, i, j, k, l \quad = 1, \ldots, \mathrm{n} \\ h', i', j', k', l' = 1', \ldots, \mathrm{n}'\end{array}\right\} \begin{array}{l}\text{for anholonomic coordinates in } X_n \text{ and} \\ \text{cartesian coordinates in } R_n.\end{array}$$

Other sets of running indices are, for instance, $a, b, c, d, e; x, y, z, u, v; \alpha, \beta, \gamma, \delta, \epsilon; \xi, \eta, \zeta, \theta, \iota; \mathfrak{a}, \mathfrak{b}, \mathfrak{c}, \mathfrak{d}, \mathfrak{e}$; and so on. A special set of fixed indices has to be chosen to go with each set of running indices.

For kernel letters we prefer:

Latin letters for scalars and affinors,
Gothic letters for densities and Δ-densities,
Capital Greek letters for more general geometric objects,
ξ for the coordinates in X_n,
x for rectilinear coordinates in E_n,

but we do not keep strictly to these rules. In physics there are often other considerations which govern the choice of a kernel.

A coordinate system is denoted by one of its running indices in round brackets, for instance (κ).

Components of a quantity with indices belonging to different coordinate systems are called *intermediate* (II, § 3). Quantities with indices belonging to different spaces are called *connecting quantities* (II, § 3).

Manifolds

An X_n is an n-dimensional manifold with coordinates ξ^κ and the convertible transformations

$$\xi^{\kappa'} = \xi^{\kappa'}(\xi^\kappa) \tag{IV.1.1}$$

with functions analytic in the region considered or having, in this region, a certain number of continuous derivatives. The coordinates are called *curvilinear*. To each coordinate system in X_n there belong n sets of ∞^{n-1} coordinate-X_1's (curves), $\binom{n}{2}$ sets of ∞^{n-2} coordinate-X_2's (surfaces), ..., and n sets of ∞^1 coordinate-X_{n-1}'s (hypersurfaces).

An E_n is an X_n in which certain coordinate systems are preferred which transform into each other by the invertible transformations of the linear group G_a

$$x^{\kappa'} = A^{\kappa'}_{\kappa} x^{\kappa} + a^{\kappa'}; \quad A^{\kappa'}_{\kappa} \text{ and } a^{\kappa'} \text{ constants.} \qquad \text{(I.1.1 } a, b)$$

These coordinates are called *rectilinear*. Curvilinear coordinates, however, can be used also. To each system of rectilinear coordinates in E_n there belong n sets of ∞^{n-1} coordinate-E_1's (straight lines), $\binom{n}{2}$ sets of ∞^{n-2} coordinate-E_2's (planes),..., and n sets of ∞^1 coordinate-E_{n-1}'s (hyperplanes).

A *centred* E_n is an E_n with a fixed origin. Its group is the group G_{ho} of all invertible linear homogeneous transformations

$$x^{\kappa'} = A^{\kappa'}_{\kappa} x^{\kappa}; \quad A^{\kappa'}_{\kappa} \text{ constants.}$$

A V_n is an X_n with a *real* fundamental tensor $g_{\lambda\kappa}$, $\text{Det}(g_{\lambda\kappa}) \neq 0$. In a V_n we only consider *real* coordinate transformations.

In a V_n orthogonal curvilinear coordinate systems exist always for $n \leqslant 3$ and in special cases also for $n > 3$ (V, § 4).

An R_n is an E_n with a real fundamental tensor with constant components with respect to rectilinear coordinates. In an R_n *orthogonal* rectilinear coordinate systems exist. They are called *cartesian* if the measuring vectors are unit vectors. Curvilinear and general rectilinear coordinates can, however, be used also.

An oriented E_n is an E_n with a fixed screw-sense. Every coordinate system fixes a screw-sense by the order of the coordinates.

Because of the linear transformation of the $d\xi^{\kappa}$

$$d\xi^{\kappa'} = A^{\kappa'}_{\kappa} d\xi^{\kappa}, \qquad A^{\kappa'}_{\kappa} \stackrel{\text{def}}{=} \partial_{\kappa} \xi^{\kappa'}, \qquad \partial_{\kappa} \stackrel{\text{def}}{=} \frac{\partial}{\partial \xi^{\kappa}} \qquad \text{(IV.2.2)}$$

to every point of an X_n there belongs a centred E_n, the *local* E_n of that point. In a V_n the local E_n is a centred R_n.

Flat sub-spaces in an E_n (I, § 3)

$n-p$ independent linear equations in rectilinear coordinates define a *proper* E_p if they have finite solutions and an *improper* E_p (or E_p at infinity) if this is not so. An improper E_p is also called a *(p+1)-direction*.

An E_p and an E_q in E_n either intersect in an E_s or they have no E_0 in common ($s = -1$). E_s is called the *section* of E_p and E_q. The join of E_p and E_q is the E_t with smallest dimension containing E_p and E_q. The relations between p, q, s, and t are

$$-1 \leqslant s \leqslant p, \qquad p+q-n \leqslant s \leqslant q,$$
$$t = p+q-s.$$

If a proper E_p and an $(n-p)$-direction, that has no direction in common with E_p, are given in E_n, the following processes are possible:

(1) Section of a figure with E_p;

(2) Reduction of a figure with respect to the $(n-p)$-direction;

(3) Projection of a figure on the E_p in the $(n-p)$-direction.

An E_p in E_n can have an *inner* orientation, i.e. a p-dimensional screw-sense *in* it and an *outer* orientation, i.e. an $(n-p)$-dimensional screw-sense *around* it.

Numbers necessary to fix

an E_p in E_n	$(p+1)(n-p)$
an E_p through a given E_q in E_n	$(p-q)(n-p)$
an E_p through O in a centred E_n	$p(n-p)$.

p linearly independent contravariant (covariant) vectors are said to *span* a contravariant (covariant) *domain* consisting of all vectors linearly dependent on them. They fix a p-direction ($(n-p)$-direction) called the *support of the domain*, which is said to be *spanned* by the p vectors. p is called the *dimension* of the domain.

Geometric objects

Geometric objects are distinguished by the law of transformation of their components with coordinate transformations. If the new components are linear homogeneous in the old components, homogeneous algebraical in the $A^{\kappa'}_{\lambda}$ and do not depend on the derivatives of the $A^{\kappa'}_{\lambda}$, the object is called a *geometric quantity* (II, § 1; IV, § 2). A quantity in a point of an X_n may always be considered as a quantity in the local E_n of this point (IV, § 2).

Transformation of an *affinor* of *valence* $p+q$, *contravariant valence* p, and *covariant valence* q:

$$P^{\kappa'_1 \dots \kappa'_p}{}_{\lambda'_1 \dots \lambda'_q} = A^{\kappa'_1 \dots \kappa'_p \lambda_1 \dots \lambda_q}_{\kappa_1 \dots \kappa_p \lambda'_1 \dots \lambda'_q} P^{\kappa_1 \dots \kappa_p}{}_{\lambda_1 \dots \lambda_q}. \qquad \text{(II.3.1)}$$

An affinor is a *scalar* if its valence is *0*, a *vector* if its valence is *1*, *contravariant* (*covariant*) if it has only upper (lower) indices, and *mixed* in all other cases.

An *affinor* Δ-*density* (*density*) of weight $\mathfrak{k}$ acquires an extra factor $\Delta^{-\mathfrak{k}}$ ($|\Delta|^{-\mathfrak{k}}$) (II, § 8).

A *W-affinor* acquires an extra factor $\Delta/|\Delta|$ (II, § 8).

Apart from a few exceptions upper and lower indices are never written in the same vertical line.

Algebraic processes for quantities (II, § 4):

Addition: $P^{\kappa}{}_{.\lambda\mu} + Q_{\mu}{}^{\cdot\kappa}{}_{.\lambda} = R_{\mu\lambda}{}^{\cdot\cdot\kappa}$.

Multiplication: $P^{\kappa\lambda}{}_{\cdot\cdot\mu}\, Q^{\nu}{}_{.\rho\sigma} = R^{\kappa\lambda\nu}{}_{\cdot\cdot\cdot\mu\rho\sigma}$.

Transvection: $P^{\kappa\lambda}{}_{\cdot\cdot\mu}\, Q^{\mu}{}_{.\rho\kappa} = R^{\lambda}{}_{.\rho}$.

Building of an isomer: $P^{\kappa\lambda}{}_{\cdot\cdot\mu} = Q^{\lambda}{}_{.\mu}{}^{\cdot\kappa}$.

Mixing: $P_{(\kappa\lambda\mu)} \overset{\text{def}}{=} \frac{1}{3!}(P_{\kappa\lambda\mu}+P_{\lambda\mu\kappa}+P_{\mu\kappa\lambda}+P_{\kappa\mu\lambda}+P_{\lambda\kappa\mu}+P_{\mu\lambda\kappa})$.

A *tensor* is symmetric in all indices:

$$P_{\kappa\lambda\mu} = P_{(\kappa\lambda\mu)}.$$

Alternation: $P_{[\kappa\lambda\mu]} \overset{\text{def}}{=} \frac{1}{3!}(P_{\kappa\lambda\mu}+P_{\lambda\mu\kappa}+P_{\mu\kappa\lambda}-P_{\kappa\mu\lambda}-P_{\lambda\kappa\mu}-P_{\mu\lambda\kappa})$.

A *multivector* is alternating in all indices:

$$P_{\kappa\lambda\mu} = P_{[\kappa\lambda\mu]}.$$

Mixed product of p vectors: $v^{\kappa_1\ldots\kappa_p} = \underset{1}{v}{}^{(\kappa_1}\ldots\underset{p}{v}{}^{\kappa_p)}$.

Alternating product of p vectors:

$$v^{\kappa_1\ldots\kappa_p} = \underset{1}{v}{}^{[\kappa_1}\ldots\underset{p}{v}{}^{\kappa_p]}.$$

A multivector is called *simple* if it is the alternating product of vectors. Necessary and sufficient conditions are

$$v^{[\kappa_1\ldots\kappa_p}v^{\lambda_1]\ldots\lambda_p} = 0 \qquad \text{(II.6.6)}$$

and also

$$v^{[\kappa_1\ldots\kappa_p}v^{\lambda_1\lambda_2]\ldots\lambda_p} = 0. \qquad \text{(II.6.7)}$$

Strangling of an index by transvection with a measuring vector:

$$\overset{(\kappa)}{P}{}^{\lambda}{}_{.\mu} = P^{(\kappa)\lambda}{}_{\cdot\cdot\mu} = \overset{\kappa}{e}_{\rho}\, P^{\rho\lambda}{}_{\cdot\cdot\mu}.$$

Running indices which do not transform with coordinate transformations are called *dead*. They are written right over or right under the central letter or marked in their own place by round brackets. Dead indices belong to the kernel. The other indices, called *living*, are always written to the right of the central letter.

If all indices of an affinor are strangled except, for example, the index κ, and if there are exactly r linearly independent vectors among those that come from this process, r is called the κ-rank of the quantity. These vectors span the κ-domain. The *support of the κ-domain* is an E_r (E_{n-r}) if κ is an upper (lower) index.

Quantities of valence *2* (II, § 9, 10, 11, 12)

To every quantity of valence *2* there always belongs a matrix. We agree that the first index numbers the columns and the second the rows and that if the two indices are vertically one above the other, the upper index is considered as to be the first. A matrix has rank r if it contains non-zero sub-determinants with r rows but none with $r+1$ rows. The rank of an affinor of valence *2* is equal to the rank of its matrix. The rank of $P^{\kappa}_{.\lambda}$ is r if and only if

$$s!\,P^{[\kappa_1}_{.\,[\lambda_1}\ldots P^{\kappa_s]}_{.\,\lambda_s]}\begin{cases}\neq 0 \text{ for } s \leqslant r\\ = 0 \text{ for } s > r.\end{cases} \tag{II.9.3}$$

The components of the quantity on the left-hand side are the s-rowed sub-determinants of the matrix of $P^{\kappa}_{.\lambda}$ and we have

$$\mathrm{Det}(P^{\kappa}_{.\lambda}) = n!\,P^{[1}_{.[1}\ldots P^{n]}_{.\,n]} = n!\,P^{[1}_{.1}\ldots P^{n]}_{.\,n} = n!\,P^{1}_{.[1}\ldots P^{n}_{.\,n]}. \tag{II.9.1}$$

A bivector, e.g. $v^{\kappa\lambda}$, has the rank r (always even) if and only if

$$v^{[\kappa_1\kappa_2}\ldots v^{\kappa_{2\rho-1}\kappa_{2\rho}]}\begin{cases}\neq 0 \text{ for } 2\rho = r\\ = 0 \text{ for } 2\rho > r\end{cases} \tag{II.9.6}$$

and for $r = n$ we have

$$\mathrm{Det}(v^{\kappa\lambda}) = \left\{\frac{n!}{2^{\frac{1}{2}n}(\frac{1}{2}n)!}\,v^{[12}\ldots v^{n-1,n]}\right\}^2. \tag{II.9.5}$$

If the rank is n there is an inverse. The inverse of $P^{\kappa}_{.\lambda}$, for example, is written $\overset{-1}{P}{}^{\kappa}_{.\lambda}$. For every value of κ and λ, the element $\overset{-1}{P}{}^{\kappa}_{.\lambda}$ is equal to the minor of the element $P^{\lambda}_{.\kappa}$ divided by $\mathrm{Det}(P^{\kappa}_{.\lambda})$ (II, § 3). The same holds *mutatis mutandis* for all quantities of valence *2* and their matrices.

If r is the rank of a *real tensor* (symmetric affinor) of valence *2* there is always a real coordinate system with respect to which the matrix takes the normal form consisting of s elements -1 and $r-s$ elements $+1$ in the main diagonal and all other elements zero. The *signature* $---\ldots+++\ldots$ is called even (odd) if the *index* s is even (odd). Signature and index are invariant for *real* coordinate transformations (II, § 10).

If r is the rank of a *real bivector* there is always a *real* coordinate system with respect to which the matrix takes the normal form with

$\frac{1}{2}r$ matrices $\begin{smallmatrix}0&1\\-1&0\end{smallmatrix}$ along the main diagonal and all other elements zero (II, § 10).

If
$$T^{\kappa}_{.\lambda}v^{\lambda} = \sigma v^{\kappa} \tag{II.14.1}$$
v^{κ} is called an *eigenvector* of $T^{\kappa}_{.\lambda}$ and σ an *eigenvalue*. The eigenvalues are the roots of the equation
$$\mathrm{Det}(T^{\kappa}_{.\lambda}-\sigma A^{\kappa}_{\lambda}) = 0, \tag{II.14.2}$$
or
$$\mathrm{Det}(T_{\mu\lambda}-\sigma g_{\mu\lambda}) = 0 \tag{II.14.4}$$
if there is a fundamental tensor. If $g_{\lambda\kappa}$ is *definite* and $T_{\lambda\kappa}$ *real and symmetric* the eigenvalues are always real. In this case there is always a real cartesian coordinate system (h) such that (*theorem of principal axes*)
$$T_{ji} = 0 \qquad (j \neq i; \quad i,j = 1,...,n). \tag{II.14.7}$$

If $f_{\mu\lambda}$ is a *real* bivector of rank r and $g_{\lambda\kappa}$ *definite* it is always possible to decompose $f_{\mu\lambda}$ into $\frac{1}{2}r$ mutually perpendicular real blades (*theorem of principal blades*) (II, § 14.)

Special quantities

A^{κ}_{λ}, $\tilde{\mathfrak{E}}^{\kappa_1...\kappa_n}$ and $\tilde{\mathfrak{e}}_{\lambda_1...\lambda_n}$ exist *a priori* in X_n (II, § 8)
$$A^{\kappa}_{\lambda} \overset{*}{=} \delta^{\kappa}_{\lambda},$$
$$\tilde{\mathfrak{E}}^{1...n} = +1, \qquad \tilde{\mathfrak{e}}_{1...n} = +1. \tag{II.8.3}$$
$\tilde{\mathfrak{E}}^{\kappa_1...\kappa_n}$ and $\tilde{\mathfrak{e}}_{\lambda_1...\lambda_n}$ establish one-to-one correspondences between alternating quantities, expressed in the simplest way by the equations
$$\begin{aligned}\tilde{\mathfrak{v}}_{\mu_1...\mu_p} &= (-1)^{\alpha_p}v^{\mu_{p+1}...\mu_n},\\ \tilde{\mathfrak{w}}^{\mu_{p+1}...\mu_n} &= (-1)^{\beta_p}w_{\mu_1...\mu_p},\\ \mathfrak{v}_{\mu_1...\mu_p} &= (-1)^{\gamma_p}\tilde{v}^{\mu_{p+1}...\mu_n},\\ \mathfrak{w}^{\mu_{p+1}...\mu_n} &= (-1)^{\delta_p}\tilde{w}_{\mu_1...\mu_p}.\end{aligned} \tag{II.8, 6, 14}$$
$\mu_1...\mu_n$ = even permutation of $1...n$, where α_p, β_p, γ_p, δ_p can be taken zero if we do not wish to introduce sub-groups of G_a and to identify quantities by means of the invariant quantities.

The most frequently used sub-groups of G_a and the invariant quantities belonging to them are (I, § 2; III, §§ 1–4)

G_{ho} (homogeneous): the origin $x^{\kappa} = 0$;

G_{eq} (equivoluminar, $\Delta = \pm 1$): the origin; a unit volume (density) $\mathfrak{q}$: $\mathfrak{q}[\kappa] = \pm 1$;

G_{sa} (special affine, $\Delta = +1$): the origin; $\mathfrak{q}$ and a screw-sense (W-scalar) $\tilde{\omega}$; $\tilde{\omega}[\kappa] = \pm 1$; $\underset{(\kappa)}{E}{}^{\kappa_1...\kappa_n}$, $\overset{(\kappa)}{e}_{\lambda_1...\lambda_n}$;

G_{or} (orthogonal, $\Delta = \pm 1$):	the origin; a fundamental tensor $g_{\lambda\kappa}$; $\mathfrak{g}$; $\mathfrak{i}$;
G_{ro} (rotational, $\Delta = +1$):	the origin; $g_{\lambda\kappa}$; a screw-sense $\tilde{\omega}$; $\mathfrak{g}$; $\mathfrak{i}$; $I^{\kappa_1 \dots \kappa_n}$; $i_{\lambda_1 \dots \lambda_n}$.

By means of these invariants, quantities, different by G_a, can be identified (III, §§ 1–4). When $n = 2\nu+1$ there remain from the $4n+2$ different quantities:

scalar	contr. vector	…	contr. n-vector
	cov. vector	…	cov. n-vector
W-scalar	contr. W-vector	…	contr. W-n-vector
	cov. W-vector	…	cov. W-n-vector

exactly $2n$ for G_{eq}:

Scalar	contr. vector	…	contr. ν-vector	each with 2 different geometric representations
	cov. vector	…	cov. ν-vector	
W-scalar	contr. W-vector	…	contr. W-ν-vector	
	cov. W-vector	…	cov. W-ν-vector	

exactly $n+1$ for G_{or}:

Scalar	vector	…	ν-vector	each with 4 different geometric representations,
W-scalar	W-vector	…	W-ν-vector	

and exactly $\nu+1$ for G_{ro}:

Scalar	vector	…	ν-vector	each with 8 different geometric representations

For $n = 2\nu$ we can only introduce identifications down to G_{or} if imaginary components are to be avoided. So we have for $n = 4$:

Group G_a:

Scalar	contr. vector	…	contr. 4-vector
	cov. vector	…	cov. 4-vector
W-scalar	contr. W-vector	…	contr. W-4-vector
	cov. W-vector	…	cov. W-4-vector

Group G_{eq}:

Scalar	contr. vector	…	contr. 4-vector	each with 2 different geometric interpretations
W-scalar	cov. vector	…	cov. 4-vector	

Group G_{or}:

Scalar, W-scalar	vector	…	4-vector	each with 4 different geometric interpretations.

For $n = 3$ we have only scalars and vectors for the group G_{ro}. This is the point of view of ordinary vector analysis. For the theory of elasticity (Ch. VII) it is best not to go farther than G_{or}. Then we have scalars, W-scalars (pseudo-scalars), vectors, and W-vectors (pseudo-vectors).

The *fundamental tensor* $g_{\lambda\kappa}$ (II, § 12):

Components with respect to cartesian coordinates

$$g_{11} = -1, \quad ..., \quad g_{ss} = -1, \quad g_{s+1s+1} = +1, \quad ..., \quad g_{nn} = +1;$$
$$g^{11} = -1, \quad ..., \quad g^{ss} = -1, \quad g^{s+1s+1} = +1, \quad ..., \quad g^{nn} = +1. \tag{II.12.7}$$

Length of vector v^κ:

$$|\sqrt{(g_{\lambda\kappa} v^\lambda v^\kappa)}| = |\sqrt{(-v^1v^1 - ... - v^sv^s + v^{s+1}v^{s+1} + ... + v^nv^n)}|. \tag{II.12.3, 15}$$

Null vector = vector of length zero.

Null cone in R_n:

$$g_{\lambda\kappa} x^\lambda x^\kappa = 0, \quad -x^1x^1 - ... - x^sx^s + x^{s+1}x^{s+1} + ... + x^nx^n = 0. \tag{II.12.4, 16}$$

$\pm$ region: $\qquad g_{\lambda\kappa} x^\lambda x^\kappa \gtrless 0.$

Unit hypersphere:

$$g_{\lambda\kappa} x^\lambda x^\kappa = \pm 1, \qquad -x^1x^1 - ... - x^sx^s + x^{s+1}x^{s+1} + ... + x^nx^n = \pm 1.$$

Angle between two vectors in the same region:

$$\cos\phi = \frac{\mp g_{\lambda\kappa} u^\lambda v^\kappa}{|\sqrt{\{(g_{\nu\rho} u^\nu u^\rho)(g_{\sigma\tau} v^\sigma v^\tau)\}}|} \quad \text{in the } \mp \text{ region.} \tag{II.12.8}$$

Co- and contravariant cartesian components:

$$v^h = \begin{cases} -v_h & (h = 1,..., s) \\ +v_h & (h = s+1,..., n). \end{cases} \tag{II.12.14}$$

Concomitants of $g_{\lambda\kappa}$ are

$$\mathfrak{g}[\kappa] = |\text{Det}(g_{\lambda\kappa})|, \qquad \mathfrak{i} = |\sqrt{\mathfrak{g}}|, \tag{II.12.20, 22}$$

and if a screw-sense is given by a cartesian coordinate system (h)

$$\underset{(h)}{I}^{\kappa_1...\kappa_n} = n!\, \underset{1}{i}^{[\kappa_1}...\underset{n}{i}^{\kappa_n]}, \qquad \overset{(h)}{i}_{\lambda_1...\lambda_n} = n!\, \overset{1}{i}_{[\lambda_1}...\overset{n}{i}_{\lambda_n]}. \tag{II.12.17}$$

Matrix calculus is an autonomous discipline which is not necessarily connected with covariant, contravariant, or mixed quantities of valence 2. But if we apply matrix calculus to these quantities in an E_n we have to distinguish the matrices of these three kinds of quantities. If only mixed affinors and vectors are used we have the translation table (cf. II, § 13)

$$PQ = R \qquad P^{\kappa}_{\cdot\mu}\, Q^{\mu}_{\cdot\lambda} = R^{\kappa}_{\cdot\lambda}$$

$$'v = Pv \qquad 'v^{\kappa} = P^{\kappa}_{\cdot\lambda} v^{\lambda}$$

$$'w = wP \qquad 'w_{\lambda} = w_{\kappa} P^{\kappa}_{\cdot\lambda}$$

$$\overset{\shortmid}{P} \qquad \overset{\shortmid}{P}{}_{\lambda}^{\cdot\kappa} = P^{\kappa}_{\cdot\lambda}$$

$$\overset{\shortmid}{Q}\overset{\shortmid}{P} = \overset{\shortmid}{R} \qquad \overset{\shortmid}{Q}{}_{\lambda}^{\cdot\mu}\overset{\shortmid}{P}{}_{\mu}^{\cdot\kappa} = \overset{\shortmid}{R}{}_{\lambda}^{\cdot\kappa}.$$

This calculus is very efficient in an R_n with a positive definite fundamental tensor because here there is no difference between co- and contravariant cartesian components. But if co- and contravariant affinors of valence *2* occur in an E_n, we have the additional translation table (cf. II, § 13)

$$f, \overset{\shortmid}{f} \qquad f^{\kappa\lambda}, \overset{\shortmid}{f}{}^{\kappa\lambda} = f^{\lambda\kappa}$$

$$h, \overset{\shortmid}{h} \qquad h_{\lambda\kappa}, \overset{\shortmid}{h}_{\lambda\kappa} = h_{\kappa\lambda}$$

$$hv = v\overset{\shortmid}{h} \qquad h_{\lambda\kappa} v^{\kappa} = v^{\kappa}\overset{\shortmid}{h}_{\kappa\lambda}$$

$$wf = \overset{\shortmid}{f}w \qquad w_{\lambda} f^{\lambda\kappa} = \overset{\shortmid}{f}{}^{\kappa\lambda} w_{\lambda}.$$

If this calculus is used the co- or contravariant character of all quantities must be memorized.

Invariant differential operators in X_n

1. *The natural derivatives* (IV, § 3):

$$Dp = \mathrm{Grad}\, p: \quad \partial_{\lambda} p$$

$$D\tilde{p} = \mathrm{Grad}\, \tilde{p}: \quad \partial_{\lambda} \tilde{p}$$

$$(q+1)Dw = \mathrm{Rot}\, w: \quad (q+1)\partial_{[\mu}\, w_{\lambda_1 \ldots \lambda_q]}$$

$$(q+1)D\tilde{w} = \mathrm{Rot}\, \tilde{w}: \quad (q+1)\partial_{[\mu}\, \tilde{w}_{\lambda_1 \ldots \lambda_q]}$$

$$\left.\begin{array}{ll} \mathrm{Div}\, \mathfrak{w}: & \partial_{\mu}\, \mathfrak{w}^{\mu\kappa_2 \ldots \kappa_p} \\ \mathrm{Div}\, \tilde{\mathfrak{w}}: & \partial_{\mu}\, \tilde{\mathfrak{w}}^{\mu\kappa_2 \ldots \kappa_p} \end{array}\right\} \text{weight } +1.$$

$$\mathrm{Rot}\,\mathrm{Grad}\, p = 0; \qquad \mathrm{Rot}\,\mathrm{Grad}\, \tilde{p} = 0,$$

$$\mathrm{Rot}\,\mathrm{Rot}\, w = 0; \qquad \mathrm{Rot}\,\mathrm{Rot}\, \tilde{w} = 0,$$

$$\mathrm{Div}\,\mathrm{Div}\, \mathfrak{w} = 0; \qquad \mathrm{Div}\,\mathrm{Div}\, \tilde{\mathfrak{w}} = 0.$$

The operator div in div $\bar{v}$ in ordinary vector analysis corresponds to the operator Div (Rot) if $\bar{v}$ is considered as a contravariant vector density (covariant bivector). The operator rot in rot $\bar{v}$ in ordinary vector analysis corresponds for odd (even) signature to the operator $\mp$Rot ($-$Div) if $\bar{v}$ is considered as a covariant vector (contravariant bivector density).

If the rotation of a covariant q-vector (W-q-vector) vanishes in some region of X_n, there is always a region where the q-vector (W-q-vector) for $q = 1$ can be expressed as the gradient of a scalar (W-scalar) and for $q > 1$ as the rotation of a covariant $(q-1)$-vector (W-$(q-1)$-vector). This theorem can also be expressed in terms of contravariant multivector densities (Δ-densities) of weight $+1$ (IV, § 4).

2. *Stokes's theorem* (IV, § 4):

$$\int_{\tau_{q+1}} \partial_{[\mu} v_{\lambda_1 \ldots \lambda_q]} \, df^{\mu\lambda_1 \ldots \lambda_q} = \int_{\tau_q} v_{\lambda_1 \ldots \lambda_q} \, df^{\lambda_1 \ldots \lambda_q}. \qquad \text{(IV.4.1)}$$

There are 9 other forms of this theorem. In an X_3 the most important formulae of this kind are

$$\int_{\tau_3} \partial_{[\mu} v_{\lambda\kappa]} \, df^{\mu\lambda\kappa} = \int_{\tau_2} v_{\lambda\kappa} \, df^{\lambda\kappa}, \qquad \text{(IV.4.21)}$$

$$\int_{\tau_3} \partial_\mu \mathfrak{v}^\mu \, d\mathfrak{f} = \int_{\tau_2} \mathfrak{v}^\mu \, d\mathfrak{f}_\mu, \qquad \text{(IV.4.27)}$$

$$\int_{\tau_2} \partial_{[\mu} v_{\lambda]} \, df^{\mu\lambda} = \int_{\tau_1} v_\lambda \, df^\lambda, \qquad \text{(IV.4.29)}$$

$$2\int_{\tau_2} \partial_\mu \mathfrak{v}^{\lambda\mu} \, d\mathfrak{f}_\lambda = \int_{\tau_1} \mathfrak{v}^{\lambda\mu} \, d\mathfrak{f}_{\lambda\mu}, \qquad \text{(IV.4.35)}$$

each of which occurs in four other forms also.

3. *The Lie derivative with respect to a field* v^κ (IV, § 5):

$$\underset{v}{\pounds} \mathfrak{P}^{\kappa\lambda}_{\cdot\cdot\nu} = v^\mu \partial_\mu \mathfrak{P}^{\kappa\lambda}_{\cdot\cdot\nu} + \mathfrak{P}^{\kappa\lambda}_{\cdot\cdot\rho} \partial_\nu v^\rho - \mathfrak{P}^{\sigma\lambda}_{\cdot\cdot\nu} \partial_\sigma v^\kappa -$$
$$- \mathfrak{P}^{\kappa\sigma}_{\cdot\cdot\nu} \partial_\sigma v^\lambda + \mathfrak{k}\mathfrak{P}^{\kappa\lambda}_{\cdot\cdot\nu} \partial_\mu v^\mu. \qquad \text{(IV.5.22)}$$

The Lie derivatives of v^κ, A^κ_λ, $\mathfrak{E}^{\kappa_1 \ldots \kappa_n}$, and $\mathfrak{e}_{\lambda_1 \ldots \lambda_n}$ are zero. If a field is dragged along over $v^\kappa dt$ the increase of the field at the fixed point ξ^κ is equal to the negative Lie differential at this point.

A field is called *absolutely invariant* with respect to the field v^κ if its Lie derivative with respect to v^κ vanishes. Its field value does not change if it is dragged along by any one of the point transformations of the group generated by the infinitesimal point transformation $v^\kappa dt$. Its components in ξ^κ with respect to (κ) are equal to the components of the field dragged along over $v^\kappa dt$ in $\xi^\kappa + v^\kappa dt$ with respect to (κ'), if (κ') arises from (κ) by dragging along over $v^\kappa dt$.

A field that is absolutely invariant with respect to σv^κ for every scalar field σ is called *absolutely invariant with respect to the streamlines of* v^κ. It is invariant for all point transformations which leave the streamlines of v^κ invariant. The additional condition for this stronger invariance for a covariant p-vector $w_{\lambda_1 \dots \lambda_p}$ or a p-vector density of weight $+1$ is

$$v^\mu w_{\mu\lambda_2 \dots \lambda_p} = 0, \tag{IV.5.23}$$

$$\mathfrak{w}^{[\kappa_1 \dots \kappa_p} v^{\kappa]} = 0. \tag{IV.5.24}$$

4. *The Lagrange derivative* (IV, § 6)

If Φ_Λ ($\Lambda = 1, \dots, N$) are N functions of ξ^κ and if $\mathfrak{L}$ is a function of the Φ_Λ and their derivatives $\Phi_\mu = \partial_\mu \Phi$, $\Phi_{\nu\mu} = \partial_\nu \partial_\mu \Phi, \dots$ (indices Λ suppressed) with respect to the ξ^κ up to a certain order, the expression

$$[\mathfrak{L}] \overset{\text{def}}{=} \frac{\partial \mathfrak{L}}{\partial \Phi} - \partial_\mu \frac{\partial \mathfrak{L}}{\partial \Phi_\mu} + \partial_\nu \partial_\mu \frac{\partial^2 \mathfrak{L}}{\partial \Phi_{\nu\mu}} - \dots \tag{IV.6.5}$$

is called the Lagrange derivative of $\mathfrak{L}$. If Φ is subject to a variation $\overset{v}{d}\Phi$ such that the variation of Φ and of all its derivatives occurring in $\mathfrak{L}$ vanish at the boundary of τ_n, for the derivative $[\mathfrak{L}]$ the following holds:

$$\overset{v}{d} \int_{\tau_n} \mathfrak{L}\, d\xi^1 \dots d\xi^n = \int_{\tau_n} [\mathfrak{L}]\, \overset{v}{d}\Phi d\xi^1 \dots d\xi^n. \tag{IV.6.6}$$

If Φ_Λ is an affinor with the valences p, q and if $\tilde{\mathfrak{L}}$ is a scalar Δ-density of weight $+1$, $[\tilde{\mathfrak{L}}]$ is an affinor Δ-density of weight $+1$ with the valences q, p.

In this case there is a relation between the Φ_Λ, $[\tilde{\mathfrak{L}}]$, and their first derivatives. If Φ_Λ is the affinor $P_{\lambda\kappa}$ this identity has the form

$$[\tilde{\mathfrak{L}}]^{\lambda\kappa}(\partial_\mu P_{\lambda\kappa} - \partial_\lambda P_{\mu\kappa} - \partial_\kappa P_{\lambda\mu}) - P_{\mu\kappa} \partial_\lambda [\tilde{\mathfrak{L}}]^{\lambda\kappa} - P_{\lambda\mu} \partial_\kappa [\tilde{\mathfrak{L}}]^{\lambda\kappa} = 0. \tag{IV.6.11}$$

If the field Φ_Λ is subjected to a variation $\overset{v}{d}\Phi_\Lambda$ such that the variation of the Φ_Λ and of all their derivatives which occur in $\mathfrak{L}$ vanish at the boundary τ_{n-1} of a region τ_n of X_n, and if $[\mathfrak{L}]$ is zero in that region, we have

$$\overset{v}{d} \int_{\tau_n} \mathfrak{L}\, d\xi^1 \dots d\xi^n = 0.$$

Anholonomic coordinates (IV, § 7)

If an arbitrary set of measuring vectors $\underset{i}{e}^\kappa$, $\overset{h}{e}_\lambda$ ($h, i = 1, \dots, n$) is introduced in every local E_n in a region of X_n and if $\partial_{[\mu} \overset{h}{e}_{\lambda]}$ does not vanish for every value of h, there is no ordinary coordinate system with these measuring vectors in this region. In this case the system of all these measuring vectors is called the *anholonomic* coordinate system (h) in the region considered. The necessary and sufficient condition is that the components of the *object of anholonomity* with respect to (h)

$$\Omega_{ji}^{\cdot\cdot h} = A_{ji}^{\mu\lambda}\, \partial_{[\mu} A_{\lambda]}^{h}, \qquad A_{\lambda}^{h} \overset{\text{def}}{=} \overset{i}{e}_{\lambda}\, \underset{i}{e}^{h}, \qquad A_{i}^{\kappa} \overset{\text{def}}{=} \overset{h}{e}_{i}\, \underset{h}{e}^{\kappa} \qquad \text{(IV.7.8)}$$

do not all vanish. Instead of $d\xi^h$ we must now write

$$(d\xi)^h \overset{\text{def}}{=} A_{\kappa}^{h}\, d\xi^{\kappa} \qquad \text{(IV.7.4)}$$

because there are no variables ξ^h.

Equations with respect to anholonomic coordinates nearly always contain correction terms with $\Omega_{ij}^{\cdot\cdot h}$, for instance

$$A_{ji}^{\mu\lambda} \partial_{[\mu}\, w_{\lambda]} = \partial_{[j}\, w_{i]} + \Omega_{ji}^{\cdot\cdot h}\, w_h, \qquad \partial_j \overset{\text{def}}{=} A_j^{\mu}\, \partial_{\mu}, \qquad \text{(IV.7.7)}$$

$$\Delta^{-1}\, \partial_{\mu}\, \mathfrak{v}^{\mu} = \partial_j\, \mathfrak{v}^j + 2\Omega_{ji}^{\cdot\cdot i}\mathfrak{v}^j, \qquad \Delta = \operatorname{Det}(A_{\lambda}^{h}); \quad \text{weight } +1. \qquad \text{(IV.7.9)}$$

Manifolds with a given linear displacement (Ch. V)

A *linear displacement* or *pseudo-parallelism* in X_n is given by a geometric object $\Gamma_{\mu\lambda}^{\kappa}$ with the transformation

$$\begin{aligned} \Gamma_{\mu'\lambda'}^{\kappa'} &= A_{\kappa\mu'\lambda'}^{\kappa'\mu\lambda}\, \Gamma_{\mu\lambda}^{\kappa} - A_{\mu'\lambda'}^{\mu\lambda}\, \partial_{\mu}\, A_{\lambda}^{\kappa'} \\ &= A_{\kappa\mu'\lambda'}^{\kappa'\mu\lambda}\, \Gamma_{\mu\lambda}^{\kappa} + A_{\lambda}^{\kappa'}\, \partial_{\mu'}\, A_{\lambda'}^{\lambda}. \end{aligned} \qquad \text{(V.1.9)}$$

$\Gamma_{\mu} \overset{\text{def}}{=} \Gamma_{\mu\lambda}^{\lambda}$ is a geometric object with the transformation

$$\Gamma_{\mu'} = A_{\mu'}^{\mu}\, \Gamma_{\mu} - A_{\lambda'}^{\lambda}\, \partial_{\mu'}\, A_{\lambda}^{\lambda'} = A_{\mu'}^{\mu}\, \Gamma_{\mu} - \partial_{\mu'} \log \Delta. \qquad \text{(V.1.13)}$$

The *covariant derivative* and the *covariant differential* of a quantity, for instance $\mathfrak{P}^{\kappa}_{\cdot\lambda}$ of weight $\mathfrak{k}$, are

$$\nabla_{\mu}\, \mathfrak{P}^{\kappa}_{\cdot\lambda} = \partial_{\mu}\, \mathfrak{P}^{\kappa}_{\cdot\lambda} + \Gamma_{\mu\nu}^{\kappa}\, \mathfrak{P}^{\nu}_{\cdot\lambda} - \Gamma_{\mu\lambda}^{\nu}\, \mathfrak{P}^{\kappa}_{\cdot\nu} - \mathfrak{k}\mathfrak{P}^{\kappa}_{\cdot\lambda}\, \Gamma_{\mu},$$

$$\delta\mathfrak{P}^{\kappa}_{\cdot\lambda} = d\mathfrak{P}^{\kappa}_{\cdot\lambda} + \Gamma_{\mu\nu}^{\kappa}\, \mathfrak{P}^{\nu}_{\cdot\lambda}\, d\xi^{\mu} - \Gamma_{\mu\lambda}^{\nu}\, \mathfrak{P}^{\kappa}_{\cdot\nu}\, d\xi^{\mu} - \mathfrak{k}\mathfrak{P}^{\kappa}_{\cdot\lambda}\, \Gamma_{\mu}\, d\xi^{\mu}. \qquad \text{(V.1.18)}$$

There is a term with a positive (negative) sign for every upper (lower) index and an extra term, containing Γ_{μ} and a factor $\mathfrak{k}$.

If a quantity is displaced in such a way that its covariant differential vanishes the displacement is called *pseudo-parallel.*

An X_n with a linear displacement is called L_n. If the affinor

$$S_{\mu\lambda}^{\cdot\cdot\kappa} \overset{\text{def}}{=} \Gamma_{[\mu\lambda]}^{\kappa} \qquad \text{(V.1.22)}$$

is zero the displacement is said to be *symmetric* and the X_n is called A_n. In an A_n infinitesimal parallelograms are possible in all *2*-directions.

The natural derivatives and the Lie derivative can be written in an

A_n with ∇_μ instead of ∂_μ, for instance

$$\nabla_\mu p = \partial_\mu p,$$

$$\nabla_{[\mu} w_{\lambda]} = \partial_{[\mu} w_{\lambda]}, \qquad \text{(V.1.26 a)}$$

$$\nabla_\mu \mathfrak{v}^\mu = \partial_\mu \mathfrak{v}^\mu \quad \text{(weight } +1), \qquad \text{(V.1.26 b)}$$

$$\underset{v}{£} \mathfrak{P}^{\kappa}_{.\lambda} = v^\mu \nabla_\mu \mathfrak{P}^{\kappa}_{.\lambda} + \mathfrak{P}^{\kappa}_{.\rho} \nabla_\lambda v^\rho - \mathfrak{P}^{\sigma}_{.\lambda} \nabla_\sigma v^\kappa + \mathfrak{k}\mathfrak{P}^{\kappa}_{.\lambda} \nabla_\mu v^\mu. \qquad \text{(V.1.27)}$$

A curve in L_n is a *geodesic* if its tangent vector remains tangent under pseudo-parallel displacements along the curve. On every geodesic there exists a *natural parameter* t, determined to within affine transformations with constant coefficients, such that the equation of the geodesic takes the form

$$\frac{\delta}{dt}\frac{d\xi^\kappa}{dt} = \frac{d^2\xi^\kappa}{dt^2} + \Gamma^\kappa_{\mu\lambda} \frac{d\xi^\mu}{dt}\frac{d\xi^\lambda}{dt} = 0. \qquad \text{(V.2.8)}$$

In an A_n the coordinates can be chosen in such a way that the $\Gamma^\kappa_{\mu\lambda}$ vanish at some given point ξ^κ. Certain coordinates of this kind are called *normal coordinates* with respect to ξ^κ. Normal coordinates can also be defined with respect to a given curve and in special cases also with respect to a given X_m in A_n (V, § 3).

A V_n is an A_n because the equation $\nabla_\mu g_{\lambda\kappa} = 0$ fixes a symmetric displacement

$$\Gamma^\kappa_{\mu\lambda} = \{^{\,\kappa}_{\mu\lambda}\} \stackrel{\text{def}}{=} \tfrac{1}{2} g^{\kappa\rho}(\partial_\mu g_{\lambda\rho} + \partial_\lambda g_{\mu\rho} - \partial_\rho g_{\mu\lambda}). \qquad \text{(V.4.7)}$$

$\{^{\,\kappa}_{\mu\lambda}\}$ is the Christoffel symbol. For this displacement the following equation holds:

$$\Gamma^\lambda_{\mu\lambda} = \tfrac{1}{2} g^{\lambda\rho}\, \partial_\mu g_{\lambda\rho} = \tfrac{1}{2}\partial_\mu \log \mathfrak{g}, \qquad \text{(V.4.8, 10)}$$

$$\nabla_\mu \mathfrak{g} = 0. \qquad \text{(V.4.11)}$$

The geodesics in a V_n are not only the straightest curves but also the longest or shortest ones. The length s is a natural parameter and accordingly the equation of the geodesic is

$$\delta \frac{d\xi^\kappa}{ds} = 0. \qquad \text{(V.4.20)}$$

The *curvature affinor* of an A_n is

$$R_{\nu\mu\lambda}^{\cdot\cdot\cdot\kappa} \stackrel{\text{def}}{=} 2\partial_{[\nu} \Gamma^\kappa_{\mu]\lambda} + 2\Gamma^\kappa_{[\nu|\rho|} \Gamma^\rho_{\mu]\lambda}. \qquad \text{(V.5.12)}$$

The following identities hold:

$$\nabla_{[\nu} \nabla_{\mu]} v^\kappa = \tfrac{1}{2} R_{\nu\mu\lambda}^{\cdot\cdot\cdot\kappa} v^\lambda, \qquad \text{(V.5.13)}$$

$$\nabla_{[\nu} \nabla_{\mu]} w_\lambda = -\tfrac{1}{2} R_{\nu\mu\lambda}^{\cdot\cdot\cdot\kappa} w_\kappa, \qquad \text{(V.5.14)}$$

$$\nabla_{[\nu} \nabla_{\mu]} \mathfrak{P}^{\kappa}_{.\lambda} = \tfrac{1}{2} R_{\nu\mu\rho}^{\cdot\cdot\cdot\kappa} \mathfrak{P}^{\rho}_{.\lambda} - \tfrac{1}{2} R_{\nu\mu\lambda}^{\cdot\cdot\cdot\rho} \mathfrak{P}^{\kappa}_{.\rho} - \tfrac{1}{2}\mathfrak{k} R_{\nu\mu\rho}^{\cdot\cdot\cdot\rho} \mathfrak{P}^{\kappa}_{.\lambda}. \qquad \text{(V.5.17)}$$

For every upper (lower) index there is a term with a positive (negative) sign and there is an extra term, containing $R_{\mu\lambda\rho}^{\cdot\cdot\cdot\rho}$ and a factor $\mathfrak{k}$.

In a V_n we write $K_{\nu\mu\lambda}^{\cdot\cdot\cdot\kappa}$ instead of $R_{\nu\mu\lambda}^{\cdot\cdot\cdot\kappa}$. $K_{\nu\mu\lambda}^{\cdot\cdot\cdot\kappa}$ has $\frac{1}{12}n^2(n^2-1)$ independent components. Identities for the curvature affinor are

I. $R_{(\nu\mu)\lambda}^{\cdot\cdot\cdot\kappa} = 0.$ (V.5.18)

II. $R_{[\nu\mu\lambda]}^{\cdot\cdot\cdot\kappa} = 0$ can be replaced in V_n by $K_{[\nu\mu\lambda\kappa]} = 0.$ (V.5.19, 23)

III. $K_{\nu\mu(\lambda\kappa)} = 0.$ (V.5.20)

IV. $K_{\nu\mu\lambda\kappa} = K_{\lambda\kappa\nu\mu}.$ (V.5.22)

Bianchi's identity: $\nabla_{[\omega} R_{\nu\mu]\lambda}^{\cdot\cdot\cdot\kappa} = 0.$ (V.5.25)

If

$$R_{\mu\lambda} \stackrel{\text{def}}{=} R_{\kappa\mu\lambda}^{\cdot\cdot\cdot\kappa}, \quad K_{\mu\lambda} \stackrel{\text{def}}{=} K_{\kappa\mu\lambda}^{\cdot\cdot\cdot\kappa}, \quad G_{\mu\lambda} \stackrel{\text{def}}{=} K_{\mu\lambda}-\tfrac{1}{2}Kg_{\mu\lambda}, \quad K \stackrel{\text{def}}{=} g^{\mu\lambda}K_{\mu\lambda}$$

we have

$$\nabla_{\omega} R_{\nu\mu\lambda}^{\cdot\cdot\cdot\omega}+2\nabla_{[\mu} R_{\nu]\lambda} = 0, \tag{V.5.34}$$

$$\nabla_{\mu} G_{\lambda}^{\cdot\mu} = 0. \tag{V.5.37}$$

The tensor density of weight $+1$

$$\mathfrak{G}^{\kappa\lambda} \stackrel{\text{def}}{=} \mathfrak{g}^{\frac{1}{2}}\, G^{\kappa\lambda}$$

is the Lagrange derivative of $-\mathfrak{g}^{\frac{1}{2}}K$ as a function of $g_{\lambda\kappa}$ and its derivatives up to the second order

$$-[\mathfrak{g}^{\frac{1}{2}}K]^{\kappa\lambda} = \mathfrak{G}^{\kappa\lambda}. \tag{V.5.39}$$

For anholonomic coordinates (h) in L_n we have

$$\Gamma_{ji}^{h} = A_{\kappa ji}^{h\mu\lambda}\,\Gamma_{\mu\lambda}^{\kappa}-A_{ji}^{\mu\lambda}\,\partial_{\mu} A_{\lambda}^{h}, \tag{V.7.1}$$

$$\Gamma_{[ji]}^{h} = S_{ji}^{\cdot\cdot h}-\Omega_{ji}^{\cdot\cdot h}, \tag{V.7.2}$$

and in A_n

$$\Gamma_{[ji]}^{h} = -\Omega_{ji}^{\cdot\cdot h}, \tag{V.7.3}$$

$$\nabla_{[j}\, w_{i]} = \partial_{[j}\, w_{i]}+\Omega_{ji}^{\cdot\cdot h}w_h, \tag{V.7.4 a}$$

$$\nabla_j\, \mathfrak{v}^j = \partial_j\, \mathfrak{v}^j+2\Omega_{ji}^{\cdot\cdot i}\mathfrak{v}^j, \tag{V.7.4 b}$$

$$R_{kji}^{\cdot\cdot\cdot h} = 2\partial_{[k}\, \Gamma_{j]i}^{h}+2\Gamma_{[k|l|}^{h}\, \Gamma_{j]i}^{l}+2\Omega_{kj}^{\cdot\cdot l}\, \Gamma_{li}^{h}. \tag{V.7.8}$$

In V_n we have

$$\Gamma_{ji}^{h} = \left\{ {h \atop ji} \right\}-\Omega_{ji}^{\cdot\cdot h}+g_{jl}\,g^{hk}\,\Omega_{ik}^{\cdot\cdot l}+g_{il}\,g^{hk}\,\Omega_{jk}^{\cdot\cdot l}, \tag{V.7.5}$$

$$\Gamma_{ji}^{i} = \left\{ {i \atop ji} \right\} = \tfrac{1}{2}\partial_j \log \mathfrak{g}, \tag{V.7.7}$$

and in the special case when the anholonomic system consists of mutually perpendicular unit vectors at every point

$$\Gamma_{ji}^{h} = -\Omega_{ji}^{\cdot\cdot h}+g_{jl}\,g^{hk}\,\Omega_{ik}^{\cdot\cdot l}+g_{il}\,g^{hk}\,\Omega_{jk}^{\cdot\cdot l}, \tag{V.7.6}$$

$$\Gamma_{ji}^{i} = 0.$$

Integral formulae in V_n and R_n (V, § 8)

Green's first identity for V_n

$$\int_{\tau_{n-}} \mathfrak{g}^{\frac{1}{2}}\phi\nabla^{\mu}\psi\, d\mathfrak{f}_{\mu} = \int_{\tau_n} \mathfrak{g}^{\frac{1}{2}}(\nabla_{\mu}\phi)(\nabla^{\mu}\psi)\, d\mathfrak{f}+\int_{\tau_n} \mathfrak{g}^{\frac{1}{2}}\phi\nabla_{\mu}\,\nabla^{\mu}\psi\, d\mathfrak{f}. \tag{V.8.3}$$

Green's second identity for V_n

$$\int_{\tau_{n-1}} \mathfrak{g}^{\frac{1}{2}}(\phi\nabla^{\mu}\psi-\psi\nabla^{\mu}\phi)\,d\mathfrak{f}_{\mu} = \int_{\tau_n} \mathfrak{g}^{\frac{1}{2}}(\phi\nabla_{\mu}\nabla^{\mu}\psi-\psi\nabla_{\mu}\nabla^{\mu}\phi)\,d\mathfrak{f}. \qquad (\text{V.8.4})$$

In an R_n with a positive definite fundamental tensor we have

$$\underset{n}{\omega}\phi = -\frac{1}{n-2}\int_{\tau_n} a^{-n+2}\nabla^2\phi\,d\mathfrak{f}+\frac{1}{n-2}\int_{\tau_{n-1}} a^{-n+2}\,\partial^i\phi\,d\mathfrak{f}_i+\int_{\tau_{n-1}} a^{-n}\phi a^i\,d\mathfrak{f}_i. \qquad (\text{V.8.11})$$

On the right-hand side ϕ is considered as a function of the x^h, on the left-hand side ϕ appears as a function of the $'x^h$ and a is the length of the vector $a^h = x^h - x^{h'}$. $\underset{n}{\omega}$ is the $(n-1)$-dimensional area of a sphere with radius R divided by R^{n-1} (V.8.10).

ϕ is a *potential function* if (conditions only sufficient):

(1) ϕ and its first and second derivatives are finite everywhere;

(2) the average value of ϕ on a sphere with radius ρ tends to zero for $\rho \to \infty$;

(3)
$$\lim_{\rho\to\infty}\int_{\tau_n} \rho^{-n+2}\nabla^2\phi\,d\mathfrak{f} = 0. \qquad (\text{V.8.15})$$

The conditions (2) and (3) are always satisfied if (1) holds and if the function is zero at all points outside some finite sphere. For a potential function we have

$$\phi = \operatorname{Pot}\nabla^2\phi, \qquad (\text{V.8.13})$$

where Pot is the integral operator

$$\operatorname{Pot} \overset{\text{def}}{=} -\frac{1}{(n-2)\underset{n}{\omega}}\int_{\infty} \ldots a^{-n+2}\,d\mathfrak{f}. \qquad (\text{V.8.14})$$

For a *harmonic* function ϕ, i.e. a function with finite first and second derivatives satifying certain conditions of continuity and $\nabla^2\phi = 0$, we have

$$\underset{n}{\omega}\rho^{n-1}\phi = \int_{\tau_{n-1}} \phi n^i\,d\mathfrak{f}_i, \qquad (\text{V.8.18})$$

expressing the fact that the value of ϕ at $'x^h$ is the mean of its values on any sphere with centre at $'x^h$ and radius ρ.

The first *boundary value problem* is solved by the equation

$$\underset{n}{\omega}\phi = \int_{\tau_n}\left(\psi-\frac{1}{n-2}a^{-n+2}\right)\nabla^2\phi\,d\mathfrak{f}+\int_{\tau_{n-1}}\phi\partial^i\left(\psi-\frac{1}{n-2}a^{-n+2}\right)d\mathfrak{f}_i \qquad (\text{V.8.21})$$

expressing ϕ in $'x^h$ in terms of $\nabla^2\phi$ and ψ over τ_n and $n^i\partial_i\psi$ over τ_{n-1}.

Green's function $\psi - \dfrac{1}{n-2}a^{-n+2}$ is a function that is harmonic in τ_n and takes the value zero on τ_{n-1}.

If the components of the p-vector $w_{i_1 \ldots i_p}$ are potential functions, we have

$$w = \operatorname{Pot}\operatorname{Div}\operatorname{Rot} w + \operatorname{Pot}\operatorname{Rot}\operatorname{Div} w. \tag{V.8.24}$$

If $w_{i_1 \ldots i_p}$ and its first derivatives are finite everywhere and if $w_{i_1 \ldots i_p}$ is zero at all points outside some finite sphere we have

$$w = \operatorname{Div}\operatorname{Pot}\operatorname{Rot} w + \operatorname{Rot}\operatorname{Pot}\operatorname{Div} w. \tag{V.8.25}$$

VI

PHYSICAL OBJECTS AND THEIR DIMENSIONS†

1. Physical objects

Quantities such as scalars, vectors, densities, etc., occurring in physics are not by any means identical with the quantities introduced in Chapter II. For instance, though a velocity may be represented by an arrow, it is not true that it is simply a contravariant vector. In order to draw the vector belonging to a velocity it is necessary to introduce a unit of time and if this unit is changed the figure of the velocity changes. From this we see that quantities in physics have a property that geometric quantities do not have. Their components change not only with transformations of coordinates but also with transformations of certain units.

As to these units we remark that there are always certain *fundamental units* and that all other units, the *derived units*, can be derived from them. The choice of the fundamental units is free; sometimes three fundamental units are chosen and sometimes four, five, or six. For illustrative purposes in this chapter we use four fundamental units of mass, length, time, and electric charge (M, L, T, and Q). The results can, however, be interpreted immediately if any other choice of fundamental units is preferred.

The fundamental units are always supposed to be subject to transformations of the form

$$M' = m^{-1}M, \quad L' = l^{-1}L, \quad T' = t^{-1}T, \quad Q' = q^{-1}Q, \qquad (1.1)$$

where m, l, t, and q are arbitrary *constants*. All sets of units which can be derived in this way from one set which is given originally will be called *allowable systems of fundamental units*.

We now define *physical objects* in a V_n ($n = 3$, $V_3 = R_3$ in ordinary space and $n = 4$ in the space of general relativity) as follows (cf. II, § 1 and IV, § 2):

If at a definite point ξ^κ of V_n there is a correspondence between ordered sets of N numbers Φ_Λ ($\Lambda = 1,..., N$) on the one hand and the allowable coordinate systems (κ) in a neighbourhood of ξ^κ together with the allowable systems of fundamental units M, L, T, Q on the other hand, such that

† Cf. Dorgelo and Schouten 1946. 1, also for references to literature.

(1) *to every* (κ) *and every allowable system of fundamental units there corresponds one and only one set* Φ_Λ;

(2) *the set* $\Phi_{\Lambda'}$ *corresponding to* (κ'), M', L', T', Q' *can be expressed in terms of* Φ_Λ, *the values of* $A_\kappa^{\kappa'}$, $\partial_\lambda A_\kappa^{\kappa'}$, $\partial_\mu \partial_\lambda A_\kappa^{\kappa'}$,... *at* ξ^κ *and* m, l, t, *and* q *only*;

(3) *in these expressions*, m, l, t, *and* q *occur only in the form of a factor* $m^\alpha l^\beta t^\gamma q^\delta$ *which is the same for each of the numbers* $\Phi_{\Lambda'}$;†

then the Φ_Λ *are said to be the components of a physical object with respect to* (κ), M, L, T, Q *at the point* ξ^κ.

If the $\Phi_{\Lambda'}$ are linear and homogeneous in the Φ_Λ, are homogeneous algebraical in the $A_\kappa^{\kappa'}$, and do not involve the derivatives of the $A_\kappa^{\kappa'}$, then the Φ_Λ are the components of a *physical quantity* at ξ^κ.

If the physical object is defined for each point of a definite region of V_n we have a *field* of physical objects in that region.

A set of components may constitute a physical object with respect to certain coordinate transformations and not with respect to other coordinate transformations. For instance, the three components of the electric field strength form a physical quantity with respect to orthogonal transformations in x, y, z. But these three components have to be taken together with the three components of magnetic induction to form a physical quantity with respect to the Lorentz transformations in x, y, z, t.

If p, q, and r are components of a physical object, then $\log p$, q^2, and $\cos r$, for example, are also components, although they transform in quite another way. This freedom in the choice of the components enables us to choose them in such a way that their transformation is as simple as possible.

If M, L, T, and Q are fixed, only the transformations of coordinates remain, and this means that for every definite choice of M, L, T, and Q one and only one geometric object belongs to every physical object (cf. IV, § 2). This object acquires (i.e. all its components acquire) a factor $m^\alpha l^\beta t^\gamma q^\delta$ if another set of allowable fundamental units is introduced. We call this geometric object the *geometric image* of the physical object and the symbol $m^\alpha l^\beta t^\gamma q^\delta$ the *dimension*‡ *of the geometric image* and also the ABSOLUTE DIMENSION *of the physical object*. A dimension is usually written in the form $[m^\alpha l^\beta t^\gamma q^\delta]$.

The absolute dimension is defined with respect to the set of all trans-

† The case when the factors differ for different components (e.g. a velocity in polar coordinates) can always be reduced to the case mentioned here.

‡ This notion has nothing to do with the dimension of a space used in Ch. I.

formations of the form (IV.1.1) or with respect to the group G_a if we work with rectilinear coordinates in an E_n. This dimension is *not* the dimension usually mentioned in physics. Take, for example, a velocity in R_3. Its geometric image is a segment of a straight line described by a point moving uniformly in unit time and is provided with an arrow in the direction of motion. If T changes, this contravariant vector v^κ acquires a factor t^{-1}. Hence the absolute dimension of a velocity is $[t^{-1}]$. But the components v^h with respect to a cartesian coordinate system (h) in R_3, based on the unit of length L, acquire the factor lt^{-1} if the fundamental units are changed. This leads us to the definition

If the components of a physical object with respect to a local cartesian coordinate system based on the unit of length have to be multiplied by the factor $m^\alpha l^\beta t^\gamma q^\delta$ if the fundamental units are changed, we call $[m^\alpha l^\beta t^\gamma q^\delta]$ the RELATIVE DIMENSION *or briefly* DIMENSION *of the physical object.*†

This is the dimension generally used in physics. Obviously the difference between absolute and relative dimension is due to the fact that a general coordinate system is not connected in any way with the unit of length and that a local cartesian coordinate system changes if another unit of length is introduced. Consequently the relation between the relative dimension and the absolute dimension of a physical object depends only on the transformation of the components under coordinate transformations. In fact, the transformation of general components of a quantity into cartesian components and vice versa is effected by transvection with $A^h_\lambda \overset{*}{=} \overset{h}{i}_\lambda$ and $A^\kappa_i \overset{*}{=} \underset{i}{i}^\kappa$ $(h, i = 1,\dots, n)$. Now with a change of the unit of length the unit vectors $\underset{i}{i}^\kappa$ acquire a factor l^{-1} and the unit vectors $\overset{h}{i}_\lambda$ a factor l. Hence the relative dimension acquires a factor l^{-1} for every factor A^κ_i and a factor l for every factor A^h_λ occurring in the formulae of transformation of general components into cartesian components. Accordingly, in order to derive the relative dimensions of a quantity from the absolute ones we have to add a factor l for every contravariant index, a factor l^{-1} for every covariant index, and a factor $l^{-n\mathfrak{k}}$ for a weight $\mathfrak{k}$.

This leads to the table on p. 129 for the quantities in R_3 introduced in Ch. II, § 8.

In a V_n, the linear element $d\xi^\kappa$ has the absolute dimension $[1]$. Hence

† If, for example, the velocity vector in R_3 is fixed by two angles and its length, two components have a dimension $[1]$ and the third a dimension $[LT^{-1}]$. But for the physical objects which usually occur it is always possible to find components which all have the same dimension. In what follows we make use of components of this kind only.

$g_{\lambda\kappa}$ and $g^{\kappa\lambda}$ have the absolute dimension $[l^2]$ and $[l^{-2}]$ respectively. This is in accordance with the fact that the relative dimensions of g_{ih} and g^{hi} (which take only the values $+1$, -1, or 0) are $[1]$. Differentiation with respect to the ξ^κ does not change the absolute dimension. Hence the absolute dimension of $\Gamma^\kappa_{\mu\lambda}$ is $[1]$ and the absolute dimension of the covariant derivative of a quantity, for instance $\nabla_\mu P^{\kappa}_{\cdot\lambda}$, is the same as the absolute dimension of $P^{\kappa}_{\cdot\lambda}$.

Absolute and Relative Dimensions in R_3.

	Absolute	Relative		Absolute	Relative
s, $\tilde{s}$		$[D]$	$\mathfrak{s}_{\mu\lambda\kappa}$, $\tilde{\mathfrak{s}}_{\mu\lambda\kappa}$		$[D]$
			$\mathfrak{S}^{\kappa\lambda\mu}$, $\tilde{\mathfrak{S}}^{\kappa\lambda\mu}$		$[D]$
v^κ, $\tilde{v}^\kappa$		$[Dl]$	$\mathfrak{v}_{\lambda\kappa}$, $\tilde{\mathfrak{v}}_{\lambda\kappa}$		$[Dl]$
w_λ, $\tilde{w}_\lambda$		$[Dl^{-1}]$	$\mathfrak{w}^{\kappa\lambda}$, $\tilde{\mathfrak{w}}^{\kappa\lambda}$		$[Dl^{-1}]$
$f^{\kappa\lambda}$, $\tilde{f}^{\kappa\lambda}$	$[D]$	$[Dl^2]$	$\mathfrak{f}_\lambda$, $\tilde{\mathfrak{f}}_\lambda$	$[D]$	$[Dl^2]$
$h_{\lambda\kappa}$, $\tilde{h}_{\lambda\kappa}$		$[Dl^{-2}]$	$\mathfrak{h}^\kappa$, $\tilde{\mathfrak{h}}^\kappa$		$[Dl^{-2}]$
$p^{\kappa\lambda\mu}$, $\tilde{p}^{\kappa\lambda\mu}$		$[Dl^3]$	$\mathfrak{p}$, $\tilde{\mathfrak{p}}$		$[Dl^3]$
$q_{\mu\lambda\kappa}$, $\tilde{q}_{\mu\lambda\kappa}$		$[Dl^{-3}]$	$\mathfrak{q}$, $\tilde{\mathfrak{q}}$		$[Dl^{-3}]$

(1.2)

From (V.5.12), (V.5.28), (V.5.29), and (V.5.36) we see that the absolute dimension of $K_{\nu\mu\lambda}^{\cdot\cdot\cdot\kappa}$, $K_{\mu\lambda}$, and $G_{\mu\lambda}$ is $[1]$. But the absolute dimension of the scalar curvature $\dfrac{1}{n(n-1)}K$ is $[l^{-2}]$ because of (V.5.29), and the relative dimension has the same value.

If $\mathfrak{L}$ is a scalar density of weight $+1$ and absolute dimension $[d_1]$ and Φ is an affinor (indices suppressed) of absolute dimension $[d_2]$ we see from (IV.6.6) that the absolute dimension of the Lagrange derivative is $[d_1/d_2]$. For example, we have the following dimensions for $\mathfrak{g}^{\frac{1}{2}}K$ and its Lagrange derivative $\mathfrak{G}^{\kappa\lambda}$ (cf. (V.5.38))

	Absolute	Relative
$\mathfrak{g}^{\frac{1}{2}}K$	$[l^{n-2}]$	$[l^{-2}]$
$\mathfrak{G}_{\lambda\kappa} = \mathfrak{g}^{\frac{1}{2}}G_{\lambda\kappa}$	$[l^n]$	$[l^{-2}]$
$\mathfrak{G}^{\kappa\lambda} = [\mathfrak{g}^{\frac{1}{2}}K]^{\kappa\lambda}$	$[l^{n-4}] = [l^{n-2}/l^2]$	$[l^{-2}]$
$g_{\lambda\kappa}$	l^2	$[1]$

(1.3)

In fact, according to (II.12.20), $\mathfrak{g}$ has the absolute dimension $[l^{2n}]$ and the relative dimension $[1]$ because the component of $\mathfrak{g}$ with respect to a cartesian coordinate system is always $+1$.

We see from IV, § 5, that the absolute dimensions of a quantity and its Lie derivative are equal. The relative dimensions are also equal because a quantity and its Lie derivative have the same law of transformation.

We give below some examples of dimensions of mechanical quantities in R_3:

	Absolute	*Relative*
Velocity	v^κ: $[t^{-1}]$ v_λ: $[l^2t^{-1}]$	$\left.\begin{matrix} v^h \\ v_i \end{matrix}\right\}$: $[lt^{-1}]$
Force	K^κ: $[mt^{-2}]$ K_λ: $[ml^2t^{-2}]$	$\left.\begin{matrix} K^h \\ K_i \end{matrix}\right\}$: $[mlt^{-2}]$
Momentum	$V^\kappa = mv^\kappa$: $[mt^{-1}]$	V^h: $[mlt^{-1}]$
Moment of force	$D^{\kappa\lambda} = 2r^{[\kappa}K^{\lambda]}$: $[mt^{-2}]$	D^{hi}, D^h: $[ml^2t^{-2}]$
Energy	P: $[ml^2t^{-2}]$	P: $[ml^2t^{-2}]$
Mass density	$\mu[\kappa]$: $[m]$	$\mu[h]$: $[ml^{-3}]$
Material current density	$\mathfrak{v}^\kappa = \mu v^\kappa$: $[mt^{-1}]$	$\mathfrak{v}^h$: $[ml^{-2}t^{-1}]$

(1.4)

2. The absolute dimension and the construction of the geometric image

Absolute dimensions are simpler in some ways than relative ones because factors occur in the relative dimensions which only arise because a cartesian coordinate system itself depends on the choice of the unit of length. But there is still another reason which makes them useful. An absolute dimension is the dimension of the geometric image and accordingly it always contains just those fundamental units that are necessary for the construction of this image. On the other hand, the relative dimension contains just those fundamental units that are necessary for the determination of the cartesian components. We give a number of examples.

(1) Let v^κ be the velocity of a fluid and μ its density. μ is a scalar density of weight $+1$. Its general component $\mu[\kappa]$ is the mass per measuring parallelepiped (parallelepiped of the measuring vectors $\underset{\lambda}{e^\kappa}$) and its cartesian component $\mu[h]$ is the mass per unit volume. Accordingly its absolute dimension is $[m]$ and its relative dimension is $[ml^{-3}]$. The contravariant vector density $\mathfrak{v}^\kappa = \mu v^\kappa$ is the current density of the fluid. Its geometric image is a cylinder with an inner orientation. If at any point, $\mathfrak{q}$ is the mass of the fluid pumped in from outside per measuring parallelepiped and per unit time, the equation of continuity for the most general case

$$\partial_\mu \mathfrak{v}^\mu + \partial_t \mu[\kappa] = \mathfrak{q}[\kappa] \quad \text{(absolute dimension } [mt^{-1}]) \tag{2.1}$$

expresses the fact that the divergence of the current density equals the sum of the decrease of mass density per unit time and the mass pumped in per measuring parallelepiped and per unit time. In cartesian coordinates the equation has the form

$$\nabla_j \mathfrak{v}^j + \partial_t \mu[h] = \mathfrak{q}[h] \quad \text{(relative dimension } [ml^{-3}t^{-1}]). \tag{2.2}$$

The geometric image of $\mathfrak{v}^\kappa$ is a cylinder carrying exactly unit mass in unit time. The geometric image of μ is the volume that would contain unit mass. The geometric image of $\mathfrak{q}$ is the volume into which unit mass would be pumped in unit time. Hence the geometric images of $\mathfrak{v}^\kappa$ and $\mathfrak{q}$ can be constructed by using the fundamental units M and T. Only the unit M is needed for the geometric image of μ.

(2) Maxwell's equations, written in ordinary vector analysis, are†

$$
\begin{aligned}
&(a) & \nabla\times\mathbf{F}+\dot{\mathbf{B}} &= 0, \\
&(b) & \nabla.\mathbf{B} &= 0, \\
&(c) & \dot{\mathbf{D}}-\nabla\times\mathbf{H} &= -\rho\mathbf{u}, \\
&(d) & \nabla.\mathbf{D} &= \rho, \\
&(e) & \mathbf{D} &= \epsilon\mathbf{F}, \\
&(f) & \mathbf{B} &= \overset{*}{\mu}\mathbf{H},
\end{aligned}
\qquad (2.3)
$$

where

$\mathbf{F}$ = electric field strength,
$\mathbf{D}$ = dielectric displacement,
$\mathbf{H}$ = magnetic field strength,
$\mathbf{B}$ = magnetic induction,
ρ = electric charge density,
ϵ = dielectric constant $(\epsilon = \epsilon_0$ in vacuum),
$\overset{*}{\mu}$ = permeability $(\overset{*}{\mu} = \overset{*}{\mu}_0$ in vacuum).

If these equations are written in cartesian components we have

$$
\begin{aligned}
&(a) & \partial_2 F^3-\partial_3 F^2+c\partial_4 B^1 &= 0, \qquad \partial_4 \overset{\text{def}}{=} \frac{1}{c}\frac{\partial}{\partial t}, \\
&(b) & \partial_1 B^1+\partial_2 B^2+\partial_3 B^3 &= 0, \qquad c \overset{\text{def}}{=} \text{velocity of light}, \\
&(c) & c\partial_4 D^1-\partial_2 H^3+\partial_3 H^2 &= -\rho u^1, \\
&(d) & \partial_1 D^1+\partial_2 D^2+\partial_3 D^3 &= \rho, \\
&(e) & D^1 &= \epsilon F^1, \\
&(f) & B^1 &= \overset{*}{\mu} H^1, \qquad \text{cycl. } 1, 2, 3.
\end{aligned}
\qquad (2.4)
$$

In the equations (2.3) and (2.4) **F**, **D**, **H**, and **B** are 'vectors' in the sense of Ch. III, § 4, after the introduction of a fundamental tensor and a screw-sense (cf. (III.4.21) and Fig. 12).

† This is the form of the equations in the electromagnetic system, the electrostatic system, and the system of Giorgi if these systems are rationalized (cf. Dorgelo and Schouten 1946. 1). We use here the signature $---$ in space, corresponding to the signature $---+$ in space-time.

Now we try to bring the equations (2.4) into a form which is invariant for affine transformations. The only affine invariant derivatives available are the gradient of a scalar or W-scalar, the rotation of a covariant q-vector or W-q-vector, and the divergence of a contravariant p-vector density or W-p-vector density of weight $+1$. Hence we see from (2.4 a) that **F** can only be a covariant vector or W-vector and **B** only a covariant bivector or W-bivector. But since the orientation in an electric field is along the streamlines and not around them, **F** must be a covariant vector and **B** a covariant bivector. This is in agreement with (2.4 b). We see from (2.4 d) that **D** must be an ordinary contravariant vector density or Δ-vector density of weight $+1$ and ρ an ordinary scalar density or scalar Δ-density of weight $+1$. But since an electric charge has no screw-sense,† **D** must be an ordinary vector density and ρ an ordinary scalar density. Hence it follows from (2.4 c) that **H** is an ordinary bivector density of weight $+1$.

Now that this is settled we get the following form of the equations (2.4) (cf. Ch. III, Fig. 8):

$$\begin{aligned}
&(a) \quad -\partial_2 F_3+\partial_3 F_2+c\partial_4 B_{23} = 0,\\
&(b) \quad \partial_1 B_{23}+\partial_2 B_{31}+\partial_3 B_{12} = 0,\\
&(c) \quad c\partial_4 \mathfrak{D}^1-\partial_2 \mathfrak{H}^{12}+\partial_3 \mathfrak{H}^{31} = -\rho u^1,\\
&(d) \quad \partial_1 \mathfrak{D}^1+\partial_2 \mathfrak{D}^2+\partial_3 \mathfrak{D}^3 = \rho,\\
&(e) \quad \mathfrak{D}^1 = -\epsilon F_1,\\
&(f) \quad B_{23} = \overset{*}{\mu}\mathfrak{H}^{23}, \qquad \text{cycl. } 1,2,3;
\end{aligned} \tag{2.5}$$

or

$$\begin{aligned}
&(a) \quad -\frac{2}{c}\partial_{[c} F_{b]}+\partial_4 B_{cb} = 0,\\
&(b) \quad \partial_{[c} B_{ba]} = 0,\\
&(c) \quad \partial_4 \mathfrak{D}^a+\frac{1}{c}\partial_b \mathfrak{H}^{ba} = -\frac{\rho}{c}u^a,\\
&(d) \quad \partial_c \mathfrak{D}^c = \rho,\\
&(e) \quad \mathfrak{D}^a = -\epsilon \mathfrak{g}^{\frac{1}{2}} g^{ab} F_b,\\
&(f) \quad B_{cb} = \overset{*}{\mu}\mathfrak{g}^{-\frac{1}{2}} g_{ca} g_{bd} \mathfrak{H}^{ad} \quad (a,b,c,d = 1,2,3),
\end{aligned} \tag{2.6}$$

which contain only affine invariant operations, and obviously this is the only way to do it. Since the equations (2.6) have the affine invariant

† A spinning electron has no screw-sense either!

form they can be written in general coordinates:

$$
\begin{aligned}
&(a) && -\frac{2}{c}\partial_{[\gamma}F_{\beta]}+\partial_4 B_{\gamma\beta} = 0, \\
&(b) && \partial_{[\gamma}B_{\beta\alpha]} = 0, \\
&(c) && \partial_4 \mathfrak{D}^{\alpha}+\frac{1}{c}\partial_{\beta}\mathfrak{H}^{\beta\alpha} = -\frac{\rho}{c}u^{\alpha}, \\
&(d) && \partial_{\gamma}\mathfrak{D}^{\gamma} = \rho, \\
&(e) && \mathfrak{D}^{\alpha} = -\epsilon\mathfrak{g}^{\frac{1}{2}}g^{\alpha\beta}F_{\beta}, \\
&(f) && B_{\gamma\beta} = \overset{*}{\mu}\mathfrak{g}^{-\frac{1}{2}}g_{\gamma\alpha}g_{\beta\delta}\mathfrak{H}^{\alpha\delta} \quad (\alpha,\beta,\gamma,\delta = 1,2,3).
\end{aligned}
\tag{2.7}
$$

If the medium is not isotropic we get equations of the form

$$
\begin{aligned}
&(a) && \mathfrak{D}^{\alpha} = \epsilon^{\alpha\beta}F_{\beta}, \\
&(b) && B_{\gamma\beta} = \overset{*}{\mu}_{\gamma\beta\alpha\delta}\mathfrak{H}^{\alpha\delta}
\end{aligned}
\tag{2.8}
$$

instead of (2.7 e, f), where $\epsilon^{\alpha\beta}$ and $\overset{*}{\mu}_{\gamma\beta\alpha\delta}$ are affinor densities of weight $+1$ and -1 respectively, and are characteristic for the electromagnetic properties of the medium.

The geometric images of **F**, **D**, **H**, and **B** are

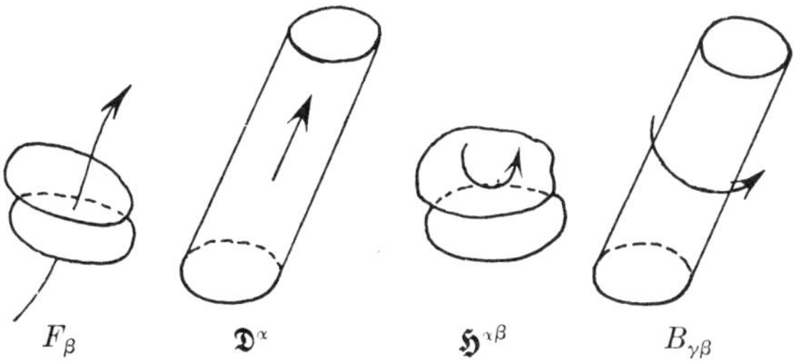

FIG. 23.

Also, from these figures, it is obvious that the relations between **F** and **D** and those between **H** and **B** in an isotropic medium can only be established by means of the fundamental tensor, because they require the notion 'perpendicular'.

In order to derive the dimensions of **F**, **D**, **H**, and **B** we make use of the fact that $e\mathbf{F}$ is the force acting on a charge e in the field. Now the relative dimension of a force is $[mlt^{-2}]$ and consequently the relative dimension of **F** is $[mlt^{-2}q^{-1}]$. Accordingly the covariant vector F_{β} has the absolute dimension $[ml^2t^{-2}q^{-1}]$ or $[ht^{-1}q^{-1}]$ if $[h]$ denotes the dimension (absolute and relative) of the quantum. From (2.7 a) it follows that

$B_{\gamma\beta}$ has the absolute dimension $[hq^{-1}]$ and the relative dimension $[hl^{-2}q^{-1}] = [mt^{-1}q^{-1}]$. The absolute dimension of ρ is $[q]$. Hence, according to (2.7 c), $\mathfrak{D}^{\alpha}$ has the absolute dimension $[q]$ and $\mathfrak{H}^{\gamma\alpha}$ the absolute dimension $[t^{-1}q]$.

We give a table of dimensions

	Absolute		Relative	
electric charge	Q	$[q]$	Q	$[q]$
electric density	ρ	$[q]$	ρ	$[ql^{-3}]$
electric field strength	F_{β}	$[ht^{-1}q^{-1}]$	F_{b}	$[hl^{-1}t^{-1}q^{-1}] = [mlt^{-2}q^{-1}]$
dielectric displacement	$\mathfrak{D}^{\alpha}$	$[q]$	$\mathfrak{D}^{a}$	$[l^{-2}q]$
magnetic field strength	$\mathfrak{H}^{\alpha\beta}$	$[t^{-1}q]$	$\mathfrak{H}^{ab}$	$[l^{-1}t^{-1}q]$
magnetic induction	$B_{\gamma\beta}$	$[hq^{-1}]$	B_{cb}	$[hl^{-2}q^{-1}] = [mt^{-1}q^{-1}]$
magnetic flux	Φ	$[hq^{-1}]$	Φ	$[hq^{-1}] = [ml^{2}t^{-1}q^{-1}]$
dielectric constant	ϵ	$[h^{-1}l^{-1}tq^{2}]$	ϵ	$[h^{-1}l^{-1}tq^{2}] = [m^{-1}l^{-3}t^{2}q^{2}]$
permeability	$\overset{*}{\mu}$	$[hl^{-1}tq^{-2}]$	$\overset{*}{\mu}$	$[hl^{-1}tq^{-2}] = [mlq^{-2}]$
dielectric affinor density	$\epsilon^{\alpha\beta}$	$[h^{-1}tq^{2}]$	ϵ^{ab}	$[h^{-1}l^{-1}tq^{2}] = [m^{-1}l^{-3}t^{2}q^{2}]$
affinor density of permeability	$\overset{*}{\mu}_{\delta\gamma\beta\alpha}$	$[htq^{-2}]$	$\overset{*}{\mu}_{dcba}$	$[hl^{-1}tq^{-2}] = [mlq^{-2}]$
quantum		$[h]$		$[h] = [ml^{2}t^{-1}]$

(2.9)

In order to prove that for the construction of the geometric image of F_{β} with the absolute dimension $[ht^{-1}q^{-1}]$ we only need the units of quantum, time, and electric charge, we consider an electron (charge e) at rest at O in a field F_{p} in a gas with a radiation frequency ν (dimension $[t^{-1}]$). The radiation begins at all those places where the energy of the electron is $h\nu$. If a slowly variable magnetic field is applied, this field does not change the energy of the electron because the force due to it is always perpendicular to the velocity. Only the path of the electron changes. Hence the locus of the points where radiation sets in is exactly the second plane of the covariant vector $\frac{e}{h\nu} F_{\beta}$ if the first plane is taken through O. This means that *the radiation itself draws the covariant vector* $\frac{e}{h\nu} F_{\beta}$ *in space. F_{β} can be derived from this vector if the units of quantum, time, and charge are known.*

Since the absolute dimension of $B_{\gamma\beta}$ is $[hq^{-1}]$ we only need here the units of quantum and electric charge. A moving electron in a magnetic field is subjected to the force

$$\mathbf{K} = e\mathbf{v}\times\mathbf{B} \tag{2.10}$$

perpendicular to the velocity $\mathbf{v}$. According to the theorem of Huygens this force is $(mv^{2})/R$ if R is the radius of curvature of the path. This implies that the electron describes a helical curve with an axis in the direction of **B**.

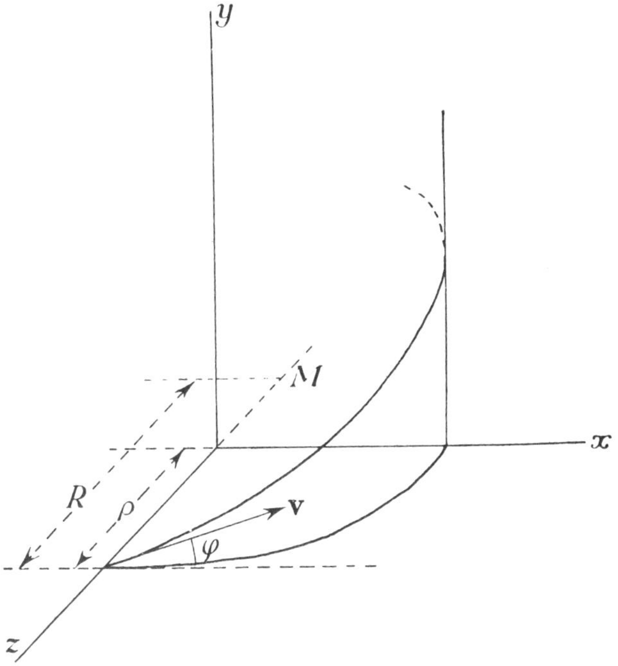

FIG. 24.

If the y-axis is this axis and if ρ is the radius of the cylinder of the curve, ϕ the angle between **v** and the xz-plane, and M the centre of curvature belonging to the point of the curve on the z-axis, it is well known that

$$R = \frac{\rho}{\cos^2\phi}. \tag{2.11}$$

It follows from (2.10) that

$$evB\cos\phi = \frac{mv^2}{R} = \frac{mv^2\cos^2\phi}{\rho}. \tag{2.12}$$

Hence the moment of momentum with respect to the y-axis is

$$m\rho v\cos\phi = eB\rho^2 = \frac{1}{\pi}eBO, \qquad O \overset{\text{def}}{=} \pi\rho^2. \tag{2.13}$$

Now according to the quantum theory this moment of momentum must be ψ times $h/2\pi$ if ψ is the quantum number of the motion. Hence

$$2eBO = \psi h, \tag{2.14}$$

and this implies that the cylinder represents the covariant bivector $\frac{2e}{\psi h} B_{\gamma\beta}$. Hence *the moving electron itself draws in space the bivector* $B_{\gamma\beta}$ *provided with a factor that can be computed if the quantum number of the motion and the unit of quantum and charge are known.*

If we write $[r]$ for the dimension $[hq^{-2}]$ of a resistance, the absolute dimension of $B_{\gamma\beta}$ can be written in the form $[rq]$ and this implies that the bivector $B_{\gamma\beta}$ can be constructed if the units of resistance and electric charge are known. Suppose we have a flat coil connected with a ballistic galvanometer. The sense indicating the positive sense of the galvanometer makes this coil into a contravariant bivector $f^{\alpha\beta}$. The scalar

$$\tfrac{1}{2}f^{\alpha\beta}B_{\alpha\beta} = \frac{\text{area of coil}}{\text{area of section of } B_{\alpha\beta} \text{ with plane of coil}} \tag{2.15}$$

is the magnetic flux through the coil. This flux is positive if the orientation of $f^{\alpha\beta}$ is the same as the orientation of $B_{\alpha\beta}$ and negative otherwise. If R is the resistance of the coil, N its number of turns, and I the current in the coil we have

$$\frac{R}{N}\int I\,dt = \tfrac{1}{2}f^{\alpha\beta}B_{\alpha\beta} \tag{2.16}$$

if the coil is withdrawn suddenly from the field. Accordingly

$$\text{area of section of } B_{\alpha\beta} \text{ with plane of coil} = \frac{N}{R}\,\frac{\text{area of coil}}{\int I\,dt}.$$

Hence this method enables us to construct the section of the cylinder of $B_{\alpha\beta}$ with any plane. In fact *it requires only the units of resistance and electric charge.*

H and **D** are merely quantities used for calculation and they cannot be measured directly because a moving charge is only subjected to forces depending on **F** and **B**. Hence the determination of **H** and **D** has always to be done by measuring some directly measurable quantities such as mass, length, time, charge, **F** and **B**, and calculating **H** and **D** from the results of these measurements. *Despite this the construction of the geometric images of* **H** *and* **D** *only requires the units occurring in their absolute dimensions.*

For the construction of the vector density $\mathfrak{D}^\alpha$ (absolute dimension $[q]$) at a point of the electric field of a condenser of any form we need only measure the charge Ψ of the condenser by means of a ballistic galvanometer. The lines of force can now be determined by calculation and this only requires knowledge of the *form* of the condenser and not of its absolute size. Hence no unit of length is needed. If a small tube in this diagram corresponds to the charge $\Delta\Psi$, a little piece of this tube represents the vector density $\frac{1}{\Delta\Psi}\,\mathfrak{D}^{\alpha}$, with better and better accuracy as $\Delta\Psi$ is reduced. Of course for the determination of the *cartesian* components of **D** the unit of length is also required. This agrees with the relative dimension $[l^{-2}q]$ of **D**.

Similarly the construction of the geometric image of $\mathfrak{H}^{\alpha\beta}$ of a given current only requires the unit of electric current (dimension $[t^{-1}q]$) and this is in accordance with the absolute dimension $[t^{-1}q]$ of **H**.

EXERCISES

VI.1. Usually [*1*] is chosen for the dimension of a temperature. But since it is possible to define temperature as 'heat per particle' it is also possible to choose $[ml^2t^{-2}]$ as its dimension. Check the following dimensional formulae for the quantities in Chapter VII, §§ 1–3, writing $[\theta]$ for the dimension of a temperature.

	Absolute	*Relative*
u^{κ}	$[1]$	$[l]$
u_{λ}	$[l^2]$	$[l]$
$S_{\mu\lambda}$	$[l^2]$	$[1]$
$\mathfrak{f}_{\lambda}\, d\sigma$	$[1]$	$[l^2]$
$p^{\kappa}\, d\sigma$	$[mt^{-2}]$	$[mlt^{-2}]$
$\mathfrak{T}^{\kappa\lambda}$	$[mt^{-2}]$	$[ml^{-1}t^{-2}]$
$\mathfrak{T}^{\mu\lambda}\, dS_{\mu\lambda}$	$[ml^2t^{-2}]$	$[ml^{-1}t^{-2}]$
$d\tau$	$[1]$	$[l^3]$
$\mathfrak{c}^{\kappa\lambda\mu\nu}$	$[ml^{-2}t^{-2}]$	$[ml^{-1}t^{-2}]$
$\mathfrak{s}_{\kappa\lambda\mu\nu}$	$[m^{-1}l^2t^2]$	$[m^{-1}lt^2]$
T, θ	$[\theta]$	$[\theta]$
$\mathfrak{E}$, η	$[ml^2t^{-2}\theta^{-1}]$	$[ml^{-1}t^{-2}\theta^{-1}]$
$\mathfrak{Q}$	$[ml^2t^{-2}]$	$[ml^{-1}t^{-2}]$
ρ	$[m]$	$[ml^{-3}]$
C	$[l^2t^{-2}\theta^{-1}]$	$[l^2t^{-2}\theta^{-1}]$
$\gamma^{\kappa\lambda}$	$[l^{-2}]$	$[1]$
$\alpha_{\mu\lambda}$	$[l^2\theta^{-1}]$	$[\theta^{-1}]$

(Dead indices omitted.)

VI.2. Check the following dimensional formulae for the quantities in Chapter VII, § 4:

	Absolute	*Relative*
$h_{\mu}^{\cdot\kappa\lambda}$	$[mt^{-2}q^{-1}]$	$[mlt^{-2}q^{-1}]$
$\beta_{\lambda\mu}$	$[ml^2t^{-2}q^{-2}]$	$[ml^3t^{-2}q^{-2}]$
$\mathfrak{q}_{\lambda}$	$[q^{-1}]$	$[l^2q^{-1}]$
$d^{\kappa}_{\cdot\lambda\mu}$	$[m^{-1}t^2q]$	$[m^{-1}l^{-1}t^2q]$
$\epsilon^{\kappa\lambda}$	$[m^{-1}l^{-2}t^2q^2]$	$[m^{-1}l^{-3}t^2q^2]$
$\mathfrak{p}^{\kappa}$	$[q\theta^{-1}]$	$[l^{-2}q\theta^{-1}]$
$\mathfrak{e}^{\kappa\lambda\mu}$	$[l^{-2}q]$	$[l^{-2}q]$
$\mathfrak{g}_{\mu\lambda\kappa}$	$[l^2q^{-1}]$	$[l^2q^{-1}]$

(Dead indices omitted.)

VI.3. Check the following dimensional formulae for the quantities in Chapter IX.

Quantities in three dimensions:

	Absolute	*Relative*
s^{α}	$[t^{-1}]$	$[lt^{-1}]$
ρ	$[q]$	$[l^{-3}q]$
μ	$[m]$	$[ml^{-3}]$
$\mathfrak{s}^{\alpha}$	$[l^{-1}q]$	$[l^{-3}q]$
$\mathfrak{f}_{\beta}\,d\sigma$	$[1]$	$[l^2]$
$E\,d\sigma$	$[ml^2t^{-3}]$	$[ml^2t^{-3}]$
$p^{\alpha}\,d\sigma$	$[mt^{-2}]$	$[mlt^{-2}]$
$\mathfrak{E}^{\alpha}$	$[ml^2t^{-3}]$	$[mt^{-3}]$
$\mathfrak{T}^{\alpha\beta}$	$[mt^{-2}]$	$[ml^{-1}t^{-2}]$
$\mathfrak{M}^{\alpha}$	$[mt^{-1}]$	$[ml^{-2}t^{-1}]$
$\mathfrak{E}, \underset{0}{\mathfrak{E}}, \mathfrak{p}$	$[ml^2t^{-2}]$	$[ml^2t^{-2}]$
$\mathfrak{Q}^{\alpha}$	$[ml^2t^{-3}]$	$[mt^{-3}]$

$(\alpha, \beta = 1, 2, 3)$

Quantities in four dimensions:

	Absolute	*Relative*
$F_{\lambda\kappa}$	$[ml^2t^{-1}q^{-1}] = [hq^{-1}]$	$[mt^{-1}q^{-1}]$
$\mathfrak{F}^{\kappa\lambda}$	$[q]$	$[l^{-2}q]$
$\mathfrak{s}^{\kappa}$	$[q]$	$[l^{-3}q]$
$\underset{0}{\rho}$	$[lq]$	$[l^{-3}q]$
$\overset{4}{K}_{\lambda}$	$[ml^2t^{-2}]$	$[mlt^{-2}]$
$\underset{I}{v}^{\kappa}, \underset{I}{s}^{\kappa}$	$[l^{-1}]$	$[1]$
$\mathfrak{k}_{\lambda}$	$[ml^3t^{-2}]$	$[ml^{-2}t^{-2}]$
$\mathfrak{k}^{\kappa}$	$[mlt^{-2}]$	$[ml^{-2}t^{-2}]$
$\underset{0}{\mu}$	$[ml]$	$[ml^{-3}]$
$\underset{m}{\mathfrak{P}}^{\kappa\lambda}, \underset{e}{\mathfrak{P}}^{\kappa\lambda}$	$[mlt^{-2}]$	$[ml^{-1}t^{-2}]$
κ	$[m^{-1}l]$	$[m^{-1}l]$
	$[m^{-1}l^3t^{-2}]$	$[m^{-1}l^3t^{-2}]$
$\mathfrak{B}_{\lambda}\,d\omega$	$[l^{-1}t]$	$[l^2t]$
$M^{\kappa}\,d\omega$	$[mt^{-1}]$	$[mlt^{-1}]$

VII

APPLICATIONS TO THE THEORY OF ELASTICITY†

1. Deformation and strain

LET the particles of a piece of matter which occupy a region of ordinary space having the cartesian coordinates x^h ($h = 1, 2, 3$) and the curvilinear or general rectilinear coordinates ξ^κ ($\kappa = 1, 2, 3$), undergo a small displacement u^κ. Let v^κ be an infinitesimal vector at a point ξ, which is fixed in the matter. Then if (κ') is the coordinate system arising from (κ) by dragging along over the infinitesimal vector u^κ, the coordinates $\xi^{\kappa'}$ of the displaced point ξ^κ and the components $v^{\kappa'}$ of the displaced vector v^κ in that displaced point have the same values as ξ^κ and v^κ respectively. Now we know that (cf. (IV.5.9))

$$A^\kappa_{\kappa'} \overset{*}{=} \delta^\lambda_{\kappa'}(A^\kappa_\lambda + \partial_\lambda u^\kappa), \tag{1.1}$$

and consequently the displaced vector v^κ at $\xi^\kappa + u^\kappa$ has the components

$$v^\kappa + v^\lambda \partial_\lambda u^\kappa \tag{1.2}$$

with respect to (κ).

If the vector v^κ at ξ^κ is displaced parallel‡ to $\xi^\kappa + u^\kappa$ the components of the displaced vector are (cf. V, § 1)

$$v^\kappa - \Gamma^\kappa_{\mu\lambda} v^\lambda u^\mu. \tag{1.3}$$

Hence the variation of the vector v^κ, seen from the point of view of an observer whose system of reference undergoes parallel displacement, is

$$\overset{v}{d} v^\kappa = v^\lambda \partial_\lambda u^\kappa + \Gamma^\kappa_{\mu\lambda} u^\mu v^\lambda = v^\mu \nabla_\mu u^\kappa. \tag{1.4}$$

This variation can be split up into two parts

$$\overset{v}{d} v^\kappa = v^\mu g^{\kappa\lambda} \nabla_{[\mu} u_{\lambda]} + v^\mu g^{\kappa\lambda} \nabla_{(\mu} u_{\lambda)}. \tag{1.5}$$

The first part represents a rotation with $\nabla_{[\mu} u_{\lambda]}$ as bivector of rotation. If the second part is zero, an infinitesimal piece of matter at ξ^κ only undergoes a translation over u^κ and a rotation without deformation. The second part represents a pure deformation without rotation. Accordingly we call

$$S_{\mu\lambda} \overset{\text{def}}{=} \mp \nabla_{(\mu} u_{\lambda)} \quad \text{or} \quad S_{ji} = \mp \partial_{(j} u_{i)} \quad (i, j = 1, 2, 3)§ \tag{1.6}$$

† General references: Voigt 1898. 1; Tedone 1907. 1; Brillouin 1938. 1; Love 1944. 1; Cady 1946. 2. Brillouin considered also finite deformations.

‡ As we are working in an R_3 'parallel displacement' has the ordinary sense here.

§ The components S_{11}, S_{12} are often written x_x, $\frac{1}{2}x_y = \frac{1}{2}y_x$.

for signature $\mp\mp\mp$, the *strain tensor* of the deformation.† Every tensor can be split up into a scalar part and a deviator, i.e. a scalar-free tensor:

$$S_{\mu\lambda} = \tfrac{1}{3}S_{\nu}^{\cdot\nu}g_{\mu\lambda}+(S_{\mu\lambda}-\tfrac{1}{3}S_{\nu}^{\cdot\nu}g_{\mu\lambda}). \tag{1.7}$$

The scalar of $S_{\mu\lambda}$ is the divergence $\nabla_{\mu} u^{\mu}$ or $\partial_j u^j$. If this scalar is zero there is no change in the volume of an infinitesimal piece of matter.

2. Forces and stress

Every two-dimensional element or facet in R_3 with an outer orientation is a covariant vector density of weight -1. If $d\sigma$ is the area (in m.[2]) we may denote it by $\mathfrak{f}_\lambda\, d\sigma$. Then $\mathfrak{f}_\lambda$ represents a facet with area 1. Let the two sides of $\mathfrak{f}_\lambda$ be called *front* and *back*, the arrow of the orientation being directed from back to front. Now in matter under deformation the matter at the front of $\mathfrak{f}_\lambda\, d\sigma$ will exert a force on the matter at the back. If this force be denoted by $p^\kappa\, d\sigma$, p^κ stands for a force per unit of area.

If $\overset{1}{\mathfrak{f}}_\lambda\, \overset{1}{d\sigma}$, $\overset{2}{\mathfrak{f}}_\lambda\, \overset{2}{d\sigma}$, $\overset{3}{\mathfrak{f}}_\lambda\, \overset{3}{d\sigma}$, and $\mathfrak{f}_\lambda\, d\sigma$ are the facets of an infinitesimal tetrahedron with their fronts to the outside we have the identity

$$\mathfrak{f}_\lambda\, d\sigma = -\overset{1}{\mathfrak{f}}_\lambda\, \overset{1}{d\sigma}-\overset{2}{\mathfrak{f}}_\lambda\, \overset{2}{d\sigma}-\overset{3}{\mathfrak{f}}_\lambda\, \overset{3}{d\sigma}. \tag{2.1}$$

If $\underset{1}{p^\kappa}\, \overset{1}{d\sigma}$, $\underset{2}{p^\kappa}\, \overset{2}{d\sigma}$, $\underset{3}{p^\kappa}\, \overset{3}{d\sigma}$, and $p^\kappa\, d\sigma$ are the forces exerted through these facets by the outside matter on the inside matter, the sum of these forces is equal to the total mass of the inside matter multiplied by the acceleration of its centre of gravity. But since this mass is of the third order in the differentials $d\xi^\kappa$ and the forces are only of second order, we have

$$p^\kappa\, d\sigma = -\underset{1}{p^\kappa}\, \overset{1}{d\sigma}-\underset{2}{p^\kappa}\, \overset{2}{d\sigma}-\underset{3}{p^\kappa}\, \overset{3}{d\sigma}. \tag{2.2}$$

Hence, since $\overset{1}{\mathfrak{f}}_\lambda$, $\overset{2}{\mathfrak{f}}_\lambda$, and $\overset{3}{\mathfrak{f}}_\lambda$ are linearly independent, there is one and only one affinor density $\mathfrak{T}^{\kappa\lambda}$ of weight $+1$, which satisfies the equations

$$\underset{1}{p^\kappa} = \mathfrak{T}^{\kappa\lambda}\overset{1}{\mathfrak{f}}_\lambda,\quad \underset{2}{p^\kappa} = \mathfrak{T}^{\kappa\lambda}\overset{2}{\mathfrak{f}}_\lambda,\quad \underset{3}{p^\kappa} = \mathfrak{T}^{\kappa\lambda}\overset{3}{\mathfrak{f}}_\lambda,\quad p^\kappa = \mathfrak{T}^{\kappa\lambda}\mathfrak{f}_\lambda.\ddagger \tag{2.3}$$

Now let σ be a closed surface in the region occupied by the matter, τ the space bounded by σ, x^h the radius vector of a point in cartesian coordinates, $\mathfrak{f}_i\, d\sigma$ an element of σ with its front to the outside, ρ the mass density, $p^h\, d\sigma$ the force exerted through $\mathfrak{f}_i\, d\sigma$ by the outside matter

† In this chapter, from § 2 onwards, we use the signature $+++$.

‡ The components $\mathfrak{T}^{11}$, $\mathfrak{T}^{12}$ are often written as $-X_x$, $-X_y$; $+X_x, +X_y$ or σ_x, τ_{xy}. There are many other notations.

on the inside matter and $\mathfrak{k}^h\,d\tau$ the resultant of some other external forces acting on the matter in the volume element $d\tau$. Then the total force acting on the matter within σ is

$$K^h \overset{*}{=} \int\limits_\sigma p^h\,d\sigma + \int\limits_\tau \mathfrak{k}^h\,d\tau = \int\limits_\sigma \mathfrak{T}^{hi}\mathfrak{f}_i\,d\sigma + \int\limits_\tau \mathfrak{k}^h\,d\tau.\dagger \tag{2.4}$$

Using the theorem of Stokes we get

$$K^h \overset{*}{=} \int\limits_\tau (\partial_i\,\mathfrak{T}^{hi} + \mathfrak{k}^h)\,d\tau. \tag{2.5}$$

This force must be equal to the resultant of all first time derivatives of the momenta of all particles in σ. Hence

$$K^h \overset{*}{=} \int\limits_\tau \rho\ddot{x}^h\,d\tau. \tag{2.6}$$

Since (2.5) and (2.6) are valid for every choice of τ we get

$$\rho\ddot{x}^h \overset{*}{=} \partial_i\,\mathfrak{T}^{hi} + \mathfrak{k}^h \overset{*}{=} \nabla_i\,\mathfrak{T}^{hi} + \mathfrak{k}^h. \tag{2.7}$$

But since this equation has the invariant form, it can be written in general coordinates

$$\rho\ddot{\xi}^\kappa = \nabla_\mu\,\mathfrak{T}^{\kappa\mu} + \mathfrak{k}^\kappa. \tag{2.8}$$

The total moment about the origin of all forces acting on the matter in σ is

$$\begin{aligned} M^{hi} &\overset{*}{=} 2\int\limits_\sigma x^{[h}p^{i]}\,d\sigma + 2\int\limits_\tau x^{[h}\mathfrak{k}^{i]}\,d\tau \\ &\overset{*}{=} 2\int\limits_\sigma x^{[h}\mathfrak{T}^{i]j}\mathfrak{f}_j\,d\sigma + 2\int\limits_\tau x^{[h}\mathfrak{k}^{i]}\,d\tau \\ &\overset{*}{=} 2\int\limits_\tau (\partial_j\,x^{[h}\mathfrak{T}^{i]j} + x^{[h}\mathfrak{k}^{i]})\,d\tau \\ &\overset{*}{=} 2\int\limits_\tau (\mathfrak{T}^{[ih]} + x^{[h}\partial_j\,\mathfrak{T}^{i]j} + x^{[h}\mathfrak{k}^{i]})\,d\tau. \end{aligned} \tag{2.9}$$

This moment must be equal to the resultant of all first time derivatives of the moments of momentum of all particles within σ. Hence

$$M^{hi} \overset{*}{=} 2\int\limits_\tau \rho x^{[h}\ddot{x}^{i]}\,d\tau. \tag{2.10}$$

From (2.9) and (2.10) we get, using (2.7),

$$\int\limits_\tau \mathfrak{T}^{[ih]}\,d\tau \overset{*}{=} 0, \tag{2.11}$$

† These equations are only valid in rectilinear coordinates. They are not valid in curvilinear coordinates because in general it is not then possible to sum vectors at different points. We therefore use the sign $\overset{*}{=}$ (cf. II, § 2).

which is valid for every choice of τ. Hence

$$\mathfrak{T}^{[\kappa\lambda]} = 0 \tag{2.12}$$

at all points. Consequently $\mathfrak{T}^{\kappa\lambda}$ is a *tensor density of weight +1*. $\mathfrak{T}^{\kappa\lambda}$ is called the *stress tensor density* and

$$T^{\kappa\lambda} \overset{\text{def}}{=} \mathfrak{g}^{-\frac{1}{2}}\mathfrak{T}^{\kappa\lambda} \tag{2.13}$$

the *stress tensor* of the deformation. If we use cartesian coordinates the difference between tensors and tensor densities vanishes.

The expression

$$\partial_i \mathfrak{T}^{hi} \overset{*}{=} \nabla_i \mathfrak{T}^{hi} = \nabla_i \mathfrak{T}^{ih} \tag{2.14}$$

is the force per unit volume due to the *internal* forces. In general coordinates, $\nabla_\mu \mathfrak{T}^{\mu\kappa}$ is the force per measuring parallelepiped due to the internal forces. A force per measuring parallelepiped is a vector density of weight $+1$.

3. The elastic coefficients†

If we consider a piece of matter at rest which undergoes an infinitesimal deformation du^κ and if there are no thermal processes, the energy due to all the forces acting on the matter in σ will be equal to the increase of the elastic potential energy. According to (2.3) this increase of energy is

$$\int_\sigma \mathfrak{p}^\mu\, du_\mu\, d\sigma + \int_\tau \mathfrak{k}^\mu\, du_\mu\, d\tau = \int_\sigma \mathfrak{T}^{\mu\lambda}\mathfrak{f}_\mu\, du_\lambda\, d\sigma + \int_\tau \mathfrak{k}^\mu\, du_\mu\, d\tau.\ddagger \tag{3.1}$$

Hence, using the theorem of Stokes and the relations (1.6) and (2.8),

$$\int_\tau \{(\nabla_\mu \mathfrak{T}^{\mu\lambda})\, du_\lambda\, d\tau + \mathfrak{T}^{\mu\lambda}\, d\nabla_\mu u_\lambda + \mathfrak{k}^\mu\, du_\mu\}\, d\tau = \int_\tau \mathfrak{T}^{\mu\lambda}\, dS_{\mu\lambda}\, d\tau \tag{3.2}$$

for this increase of energy, and this proves that $\mathfrak{T}^{\mu\lambda}\, dS_{\mu\lambda}$ is the increase of the potential energy per measuring parallelepiped and $T^{ji}\, dS_{ji}$ is the increase of the potential energy per unit volume.

If there are thermal processes the increase of the total energy per measuring parallelepiped is§

$$d\mathfrak{W} = \mathfrak{T}^{\mu\lambda}\, dS_{\mu\lambda} + T\, d\mathfrak{E} = \mathfrak{T}^{\mu\lambda}\, dS_{\mu\lambda} + d\mathfrak{Q}, \tag{3.3}$$

where T is the temperature, $\mathfrak{E}$ the entropy per measuring parallelepiped, and $d\mathfrak{Q} = T\, d\mathfrak{E}$ the increase of heat energy per measuring parallelepiped.

We suppose now that

(1) the matter is a homogeneous crystal;

(2) originally $\mathfrak{T}^{\kappa\lambda} = 0$, $S_{\mu\lambda} = 0$, $T = \underset{0}{T}$, $\mathfrak{E} = \underset{0}{\mathfrak{E}}$;

† Cf. Voigt 1898. 1; Tedone 1907. 1; Love 1944. 1; Mason 1947. 4.

‡ Since the integrand is a scalar this equation can be written in general coordinates.

§ This is the first law of thermodynamics when no convection of interior energy is considered.

(3) we consider a state with $\mathfrak{T}^{\kappa\lambda}$, $S_{\mu\lambda}$, $T = \underset{0}{T}+\theta$, $\mathfrak{E} = \underset{0}{\mathfrak{E}}+\eta$ when $\mathfrak{T}^{\kappa\lambda}$, $S_{\mu\lambda}$, θ and η (a density) are small;

(4) $\mathfrak{T}^{\kappa\lambda}$ and θ are uniquely determined by $S_{\mu\lambda}$ and η and vice versa;

(5) the process is reversible and *only rectilinear coordinates* (either affine or cartesian) *are used*, hence

$$d\mathfrak{W} = \mathfrak{T}^{\mu\lambda}\, dS_{\mu\lambda} + \underset{0}{T}\, d\eta$$

is a total differential.

Since we use rectilinear coordinates only, the equations in this and in the following section are valid only for this kind of coordinates. The only exceptions are the equations that have the invariant form and are marked with an asterisk.

If $\mathfrak{T}^{\kappa\lambda}$ and θ are expressed in terms of $S_{\mu\lambda}$ and η we have

$$\begin{aligned} &(a) \qquad d\mathfrak{T}^{\kappa\lambda} = \frac{\partial \mathfrak{T}^{\kappa\lambda}}{\partial S_{\mu\nu}}\, dS_{\mu\nu} + \frac{\partial \mathfrak{T}^{\kappa\lambda}}{\partial \eta}\, d\eta, \\ &(b) \qquad d\theta = \frac{\partial \theta}{\partial S_{\mu\nu}}\, dS_{\mu\nu} + \frac{\partial \theta}{\partial \eta}\, d\eta. \end{aligned} \tag{3.4}$$

The term $\dfrac{\partial \theta}{\partial \eta}\, d\eta$ in (3.4 *b*) represents the change of the temperature at constant strain. Hence, if $\underset{S}{C}$ (a density) is the heat capacity per unit mass at constant strain we have

$$\frac{\partial \theta}{\partial \eta}\, d\eta = d\mathfrak{Q}/\rho \underset{S}{C}, \tag{3.5}$$

and this expression is a total differential if $\underset{S}{C}$ is constant.

To a first approximation we may now suppose that $\partial \mathfrak{T}^{\kappa\lambda}/\partial S_{\mu\lambda}$, $\partial \mathfrak{T}^{\kappa\lambda}/\partial \eta$, and $\partial \theta/\partial S_{\mu\nu}$ in (3.4) are constant (Hooke's law) and that $\underset{S}{C}$ is constant (we remember that only rectilinear coordinates are used). Then $d\mathfrak{W}$ is a total differential if and only if

$$\begin{aligned} &(a) \qquad \frac{\partial \mathfrak{T}^{\kappa\lambda}}{\partial S_{\mu\nu}} = \frac{\partial \mathfrak{T}^{\mu\nu}}{\partial S_{\kappa\lambda}}, \\ &(b) \qquad \frac{\partial \theta}{\partial S_{\kappa\lambda}} = \frac{\partial \mathfrak{T}^{\kappa\lambda}}{\partial \eta}. \end{aligned} \tag{3.6}$$

If these conditions are satisfied and if we write

$$\begin{aligned} &(a) \qquad \underset{\eta}{\mathfrak{c}}^{\kappa\lambda\mu\nu} \overset{\text{def}}{=} \frac{\partial \mathfrak{T}^{\kappa\lambda}}{\partial S_{\mu\nu}} = \frac{\partial \mathfrak{T}^{\mu\nu}}{\partial S_{\kappa\lambda}}, \\ &(b) \qquad \underset{S}{\gamma}^{\kappa\lambda} \overset{\text{def}}{=} -\frac{1}{\underset{0}{T}} \frac{\partial \mathfrak{T}^{\kappa\lambda}}{\partial \eta} = -\frac{1}{\underset{0}{T}} \frac{\partial \theta}{\partial S_{\kappa\lambda}}, \end{aligned} \tag{3.7}$$

the equations (3.4) take the form

$$\boxed{\begin{aligned} &(a) \quad \mathfrak{T}^{\kappa\lambda} = \underset{\eta}{\mathfrak{c}}^{\kappa\lambda\mu\nu} S_{\mu\nu} - \underset{S}{\gamma}^{\kappa\lambda} \Delta\mathfrak{Q} \\ &(b) \quad \theta = -\underset{0}{T}\underset{S}{\gamma}^{\mu\nu} S_{\mu\nu} + \frac{\Delta\mathfrak{Q}}{\rho \underset{S}{C}} \end{aligned}} \tag{3.8*}$$

which is valid for small deformations $S_{\mu\lambda}$ and small amounts of heat flow. With respect to rectilinear coordinates the $\underset{\eta}{\mathfrak{c}}^{\kappa\lambda\mu\nu}$ are the *adiabatic elastic constants* and $-\underset{S}{\gamma}^{\kappa\lambda}$ is the stress per unit of heat necessary to prevent deformation.

If $S_{\mu\lambda}$ and η are expressed in terms of $\mathfrak{T}^{\kappa\lambda}$ and θ we get

$$\begin{aligned} &(a) \qquad dS_{\mu\nu} = \frac{\partial S_{\mu\nu}}{\partial \mathfrak{T}^{\kappa\lambda}} d\mathfrak{T}^{\kappa\lambda} + \frac{\partial S_{\mu\nu}}{\partial\theta} d\theta, \\ &(b) \qquad d\eta = \frac{\partial\eta}{\partial \mathfrak{T}^{\kappa\lambda}} d\mathfrak{T}^{\kappa\lambda} + \frac{\partial\eta}{\partial\theta} d\theta. \end{aligned} \tag{3.9}$$

The expression $\dfrac{\partial\eta}{d\theta} d\theta$ in (3.9 b) is the change of entropy at constant stress. Hence, if $\overset{\mathfrak{T}}{C}$ is the heat capacity per unit mass at constant stress we have

$$\underset{0}{T} \frac{\partial\eta}{\partial\theta} = \rho \overset{\mathfrak{T}}{C}, \tag{3.10}$$

and the expression

$$\frac{\partial\eta}{\partial\theta} d\theta = \frac{d\theta}{\underset{0}{T}} \rho \overset{\mathfrak{T}}{C} \tag{3.11}$$

is a total differential if $\overset{\mathfrak{T}}{C}$ is a constant.

Now we suppose for a first approximation that $\partial S_{\mu\nu}/\partial\mathfrak{T}^{\kappa\lambda}$, $\partial S_{\mu\nu}/\partial\theta$, $\partial\eta/\partial\mathfrak{T}^{\kappa\lambda}$, and $\overset{\mathfrak{T}}{C}$ are constants. Then $d\mathfrak{W}$ is a total differential if and only if

$$S_{\mu\lambda} d\mathfrak{T}^{\mu\lambda} + \eta\, d\theta \tag{3.12}$$

is a total differential, and this is the case if and only if

$$\begin{aligned} \frac{\partial S_{\mu\nu}}{\partial\mathfrak{T}^{\kappa\lambda}} &= \frac{\partial S_{\kappa\lambda}}{\partial\mathfrak{T}^{\mu\nu}}, \\ \frac{\partial\eta}{\partial\mathfrak{T}^{\kappa\lambda}} &= \frac{\partial S_{\kappa\lambda}}{\partial\theta}. \end{aligned} \tag{3.13}$$

Hence, if we write

$$\begin{aligned} &(a) \qquad \overset{\theta}{\mathfrak{s}}_{\kappa\lambda\mu\nu} \overset{\text{def}}{=} \frac{\partial S_{\mu\nu}}{\partial\mathfrak{T}^{\kappa\lambda}} = \frac{\partial S_{\kappa\lambda}}{\partial\mathfrak{T}^{\mu\nu}}, \\ &(b) \qquad \overset{\mathfrak{T}}{\alpha}_{\mu\lambda} \overset{\text{def}}{=} \frac{\partial S_{\mu\lambda}}{\partial\theta} = \frac{\partial\eta}{\partial\mathfrak{T}^{\mu\lambda}}, \end{aligned} \tag{3.14}$$

the equations (3.9) take the form

$$\boxed{\begin{aligned} &(a) \quad S_{\mu\nu} = \overset{\theta}{\mathfrak{s}}_{\kappa\lambda\mu\nu}\,\mathfrak{T}^{\kappa\lambda} + \overset{\mathfrak{T}}{\alpha}_{\mu\nu}\,\theta, \\ &(b) \quad \Delta\mathfrak{Q} = \underset{0}{T}\overset{\mathfrak{T}}{\alpha}_{\kappa\lambda}\,\mathfrak{T}^{\kappa\lambda} + \rho\overset{\mathfrak{T}}{C}\theta, \end{aligned}} \tag{3.15*}$$

which is valid for small deformations and small changes of temperature. With respect to rectilinear coordinates the $\overset{\theta}{\mathfrak{s}}_{\kappa\lambda\mu\nu}$ are *isothermal elastic constants* and the $\overset{\mathfrak{T}}{\alpha}_{\mu\lambda}$ are the temperature coefficients of strain without stress.

We see from (3.7 a) and (3.14 a) that $\underset{\eta}{\mathfrak{c}}^{\kappa\lambda\mu\nu}$ and $\overset{\theta}{\mathfrak{s}}_{\kappa\lambda\mu\nu}$ are not only symmetrical in the first two and the last two indices but are also invariant for the interchange of the first two with the last two indices. Hence these quantities have, apart from their density character, the properties of a *tensor-tensor*, that is a quantity that plays the same role for tensors of valence *2* as a tensor does for vectors and so has $\frac{6.7}{1.2} = 21$ independent components. (Cf. the bivector-tensor in V, § 5.)

To get the relations between $\underset{\eta}{\mathfrak{c}}^{\kappa\lambda\mu\nu}$, $\underset{S}{\gamma}^{\kappa\lambda}$, $\underset{S}{C}$ and $\overset{\theta}{\mathfrak{s}}_{\kappa\lambda\mu\nu}$, $\overset{\mathfrak{T}}{\alpha}_{\mu\lambda}$, $\overset{\mathfrak{T}}{C}$ we introduce the inverse of $\underset{\eta}{\mathfrak{c}}^{\kappa\lambda\mu\nu}$ (as a tensor-tensor) by means of the equation

$$\underset{\eta}{\mathfrak{s}}_{\kappa\lambda\mu\nu}\,\underset{\eta}{\mathfrak{c}}^{\mu\nu\rho\sigma} = A^{\rho}_{(\kappa}A^{\sigma}_{\lambda)}. \tag{3.16*}$$

Then (3.8 a) can be written in the form

$$S_{\mu\nu} = \underset{\eta}{\mathfrak{s}}_{\kappa\lambda\mu\nu}\,\mathfrak{T}^{\kappa\lambda} + \underset{\eta}{\mathfrak{s}}_{\kappa\lambda\mu\nu}\,\underset{S}{\gamma}^{\kappa\lambda}\Delta\mathfrak{Q}. \tag{3.17*}$$

θ can be eliminated from (3.15*):

$$S_{\mu\nu} = \overset{\theta}{\mathfrak{s}}_{\kappa\lambda\mu\nu}\,\mathfrak{T}^{\kappa\lambda} - (\underset{0}{T}/\rho\overset{\mathfrak{T}}{C})\overset{\mathfrak{T}}{\alpha}_{\mu\nu}\,\overset{\mathfrak{T}}{\alpha}_{\kappa\lambda}\,\mathfrak{T}^{\kappa\lambda} + (1/\rho\overset{\mathfrak{T}}{C})\overset{\mathfrak{T}}{\alpha}_{\mu\nu}\,\Delta\mathfrak{Q}, \tag{3.18*}$$

and from this and (3.17*) we see that

$$\begin{aligned} &(a) \quad \underset{\eta}{\mathfrak{s}}_{\kappa\lambda\mu\nu} = \overset{\theta}{\mathfrak{s}}_{\kappa\lambda\mu\nu} - (\underset{0}{T}/\rho\overset{\mathfrak{T}}{C})\overset{\mathfrak{T}}{\alpha}_{\mu\nu}\,\overset{\mathfrak{T}}{\alpha}_{\kappa\lambda}, \\ &(b) \quad \overset{\mathfrak{T}}{\alpha}_{\mu\nu} = \rho\overset{\mathfrak{T}}{C}\underset{\eta}{\mathfrak{s}}_{\kappa\lambda\mu\nu}\,\underset{S}{\gamma}^{\kappa\lambda}. \end{aligned} \tag{3.19*}$$

In the same way, by writing (3.15* a) in the form

$$\mathfrak{T}^{\kappa\lambda} = \overset{\theta}{\mathfrak{c}}^{\kappa\lambda\mu\nu}S_{\mu\nu} - \overset{\theta}{\mathfrak{c}}^{\kappa\lambda\mu\nu}\overset{\mathfrak{T}}{\alpha}_{\mu\nu}\,\theta, \tag{3.20*}$$

by introducing the inverse $\overset{\theta}{\mathfrak{c}}{}^{\kappa\lambda\mu\nu}$ of $\overset{\theta}{\mathfrak{s}}{}_{\kappa\lambda\mu\nu}$ (as a tensor-tensor), and eliminating $\Delta\mathfrak{Q}$ from (3.8*) we get

$$\mathfrak{T}^{\kappa\lambda} = \underset{\eta}{\mathfrak{c}}{}^{\kappa\lambda\mu\nu}S_{\mu\nu} - \rho \underset{S}{C}\underset{0}{T}\underset{S}{\gamma}{}^{\kappa\lambda}\underset{S}{\gamma}{}^{\mu\nu}S_{\mu\nu} - \rho\underset{S}{C}\underset{S}{\gamma}{}^{\kappa\lambda}\theta, \tag{3.21*}$$

and from this and (3.20*)

$$\begin{aligned} &(a) \qquad \overset{\theta}{\mathfrak{c}}{}^{\kappa\lambda\mu\nu} = \underset{\eta}{\mathfrak{c}}{}^{\kappa\lambda\mu\nu} - \rho\underset{S}{C}\underset{0}{T}\underset{S}{\gamma}{}^{\kappa\lambda}\underset{S}{\gamma}{}^{\mu\nu}, \\ &(b) \qquad \rho\underset{S}{C}\underset{S}{\gamma}{}^{\kappa\lambda} = \overset{\theta}{\mathfrak{c}}{}^{\kappa\lambda\mu\nu}\overset{\mathfrak{T}}{\alpha}_{\mu\nu}. \end{aligned} \tag{3.22*}$$

It follows from (3.19* b) that

$$\rho\underset{S}{\overset{\mathfrak{T}}{C}}\gamma^{\kappa\lambda} = \underset{\eta}{\mathfrak{c}}{}^{\kappa\lambda\mu\nu}\overset{\mathfrak{T}}{\alpha}_{\mu\nu}, \tag{3.23*}$$

hence $$\rho(\overset{\mathfrak{T}}{C} - \underset{S}{C})\underset{S}{\gamma}{}^{\kappa\lambda} = \rho\underset{S}{C}\underset{0}{T}\underset{S}{\gamma}{}^{\kappa\lambda}\underset{S}{\gamma}{}^{\mu\nu}\overset{\mathfrak{T}}{\alpha}_{\mu\nu}, \tag{3.24*}$$

or, using (3.22* b),

$$\overset{\mathfrak{T}}{C} - \underset{S}{C} = \underset{S}{C}\underset{0}{T}\underset{S}{\gamma}{}^{\mu\nu}\overset{\mathfrak{T}}{\alpha}_{\mu\nu} = (1/\rho)\underset{0}{T}\overset{\theta}{\mathfrak{c}}{}^{\kappa\lambda\mu\nu}\overset{\mathfrak{T}}{\alpha}_{\kappa\lambda}\overset{\mathfrak{T}}{\alpha}_{\mu\nu}. \tag{3.25*}$$

The differences between $\overset{\theta}{\mathfrak{s}}{}^{\kappa\lambda\mu\nu}$ and $\underset{\eta}{\mathfrak{s}}{}^{\kappa\lambda\mu\nu}$ are very small. For instance, the values of $\overset{\mathfrak{T}}{\alpha}_{\mu\lambda}$ for α-quartz, with respect to the usual system of orthogonal axes, measured in °K. (degrees Kelvin), and the values of ρ and $\overset{\mathfrak{T}}{C}$ are

$$\begin{gathered} \overset{\mathfrak{T}}{\alpha}_{11} = 14{\cdot}3\times 10^{-6}, \qquad \overset{\mathfrak{T}}{\alpha}_{22} = 14{\cdot}3\times 10^{-6}, \qquad \overset{\mathfrak{T}}{\alpha}_{33} = 7{\cdot}8\times 10^{-6}\ (1/°\text{K.}), \\ \overset{\mathfrak{T}}{\alpha}_{23} = \overset{\mathfrak{T}}{\alpha}_{31} = \overset{\mathfrak{T}}{\alpha}_{12} = 0, \\ \rho = 2{,}650\ \text{kg./m.}^3, \qquad \overset{\mathfrak{T}}{C} = 7{\cdot}35\times 10^5\ \text{joule/kg.} \end{gathered} \tag{3.26}$$

Hence according to (3.19* a) only the values of $\mathfrak{s}_{1111} = \mathfrak{s}_{2222}$, $\mathfrak{s}_{1122}$, $\mathfrak{s}_{1133} = \mathfrak{s}_{2233}$, and $\mathfrak{s}_{3333}$ show any difference. For the values of $\underset{\eta}{\mathfrak{s}}{}_{hijk}$ given by Mason† in m.²/newton:

$$\begin{aligned} \underset{\eta}{\mathfrak{s}}{}_{1111} &= 127{\cdot}9\times 10^{-13}, \qquad & \underset{\eta}{\mathfrak{s}}{}_{1122} &= -15{\cdot}35\times 10^{-13}, \\ \underset{\eta}{\mathfrak{s}}{}_{1133} &= -11{\cdot}0\times 10^{-13}, \qquad & \underset{\eta}{\mathfrak{s}}{}_{3333} &= 95{\cdot}6\times 10^{-13}, \end{aligned} \tag{3.27}$$

this gives

$$\begin{aligned} \overset{\theta}{\mathfrak{s}}{}_{1111} &= 128{\cdot}2\times 10^{-13}, \qquad & \overset{\theta}{\mathfrak{s}}{}_{1122} &= -15{\cdot}04\times 10^{-13}, \\ \overset{\theta}{\mathfrak{s}}{}_{1133} &= -10{\cdot}83\times 10^{-13}, \qquad & \overset{\theta}{\mathfrak{s}}{}_{3333} &= 95{\cdot}7\times 10^{-13}. \end{aligned} \tag{3.28}$$

† 1947. 4.

for $\overset{\theta}{\mathfrak{s}}_{hijk}$. The differences are probably smaller than the accuracy of measurement.†

4. Dielectric and piezo-electric constants‡

Now we suppose that there is also an electric field F_λ and a dielectric displacement $\mathfrak{D}^\kappa$, both originally zero, which take small values during the process. We also suppose that $\mathfrak{T}^{\kappa\lambda}$, θ, and F_λ are uniquely determined by $S_{\mu\lambda}$, η, and $\mathfrak{D}^\kappa$ and vice versa, and that the process is reversible. The difference between $\overset{\theta}{\mathfrak{c}}^{\kappa\lambda\mu\nu}$ and $\underset{\eta}{\mathfrak{c}}^{\kappa\lambda\mu\nu}$ will be neglected.

The increase of the total energy per measuring parallelepiped is now

$$d\mathfrak{W} = \mathfrak{T}^{\mu\lambda}\, dS_{\mu\lambda} + F_\lambda\, d\mathfrak{D}^\lambda + T\, d\eta. \tag{4.1}$$

If $\mathfrak{T}^{\kappa\lambda}$, θ, and F_λ are expressed in terms of $S_{\mu\lambda}$, η, and $\mathfrak{D}^\kappa$ we have

$$\begin{aligned} d\mathfrak{T}^{\kappa\lambda} &= \frac{\partial \mathfrak{T}^{\kappa\lambda}}{\partial S_{\mu\nu}}\, dS_{\mu\nu} + \frac{\partial \mathfrak{T}^{\kappa\lambda}}{\partial \mathfrak{D}^\mu}\, d\mathfrak{D}^\mu + \frac{\partial \mathfrak{T}^{\kappa\lambda}}{\partial \eta}\, d\eta, \\ dF_\lambda &= \frac{\partial F_\lambda}{\partial S_{\mu\nu}}\, dS_{\mu\nu} + \frac{\partial F_\lambda}{\partial \mathfrak{D}^\mu}\, d\mathfrak{D}^\mu + \frac{\partial F_\lambda}{\partial \eta}\, d\eta, \\ d\theta &= \frac{\partial \theta}{dS_{\mu\nu}}\, \partial S_{\mu\nu} + \frac{\partial \theta}{d\mathfrak{D}^\mu}\, \partial \mathfrak{D}^\mu + \frac{\partial \theta}{\partial \eta}\, d\eta. \end{aligned} \tag{4.2}$$

For a first approximation we may suppose that all derivatives, except $\partial\theta/\partial\eta$, that occur in (4.2) are constants, and that the heat capacity $\underset{S,\mathfrak{D}}{C}$ per unit mass at constant strain and constant dielectric displacement is a constant. (We remember that only rectilinear coordinates are used.) Then

$$\frac{\partial \theta}{\partial \eta}\, d\eta = d\mathfrak{Q}/\rho \underset{S,\mathfrak{D}}{C} \tag{4.3}$$

is a total differential. $d\mathfrak{W}$ is a total differential if and only if

$$\begin{aligned} &(a)\quad \frac{\partial \mathfrak{T}^{\kappa\lambda}}{\partial S_{\mu\nu}} = \frac{\partial \mathfrak{T}^{\mu\nu}}{\partial S_{\kappa\lambda}}, &&(b)\quad \frac{\partial \mathfrak{T}^{\kappa\lambda}}{\partial \mathfrak{D}^\mu} = \frac{\partial F_\mu}{\partial S_{\kappa\lambda}}, \\ &(c)\quad \frac{\partial \mathfrak{T}^{\kappa\lambda}}{\partial \eta} = \frac{\partial \theta}{\partial S_{\kappa\lambda}}, &&(d)\quad \frac{\partial F_\lambda}{\partial \mathfrak{D}^\mu} = \frac{\partial F_\mu}{\partial \mathfrak{D}^\lambda}, \\ &&(e)\quad \frac{\partial F_\lambda}{\partial \eta} = \frac{\partial \theta}{\partial \mathfrak{D}^\lambda}. \end{aligned} \tag{4.4}$$

† Mason 1947. 4.

‡ Cf. Voigt 1898. 1; Cady 1946. 2; Mason 1947. 4.

If these conditions are satisfied and if we write

$$
\begin{aligned}
&(a) \qquad \underset{\mathfrak{D}}{\mathfrak{c}}^{\kappa\lambda\mu\nu} \overset{\text{def}}{=} \frac{\partial \mathfrak{T}^{\kappa\lambda}}{\partial S_{\mu\nu}} = \frac{\partial \mathfrak{T}^{\mu\nu}}{\partial S_{\kappa\lambda}}, \\
&(b) \qquad \underset{\eta}{h}{}_{\mu}^{\cdot\kappa\lambda} \overset{\text{def}}{=} -\frac{\partial \mathfrak{T}^{\kappa\lambda}}{\partial \mathfrak{D}^{\mu}} = -\frac{\partial F_{\mu}}{\partial S_{\kappa\lambda}}, \\
&(c) \qquad \underset{\mathfrak{D}}{\gamma}^{\kappa\lambda} \overset{\text{def}}{=} -\frac{1}{\underset{0}{T}} \frac{\partial \mathfrak{T}^{\kappa\lambda}}{\partial \eta} = -\frac{1}{\underset{0}{T}} \frac{\partial \theta}{\partial S_{\kappa\lambda}}, \\
&(d) \qquad \underset{S,\eta}{\beta}_{\lambda\mu} \overset{\text{def}}{=} \frac{\partial F_{\lambda}}{\partial \mathfrak{D}^{\mu}} = \frac{\partial F_{\mu}}{\partial \mathfrak{D}^{\lambda}}, \\
&(e) \qquad \underset{S}{\mathfrak{q}}_{\lambda} \overset{\text{def}}{=} -\frac{1}{\underset{0}{T}} \frac{\partial F_{\lambda}}{\partial \eta} = -\frac{1}{\underset{0}{T}} \frac{\partial \theta}{\partial \mathfrak{D}^{\lambda}},
\end{aligned}
\tag{4.5}
$$

the equations (4.2) take the form

$$
\boxed{
\begin{aligned}
&(a) \quad \mathfrak{T}^{\kappa\lambda} = \underset{\mathfrak{D}}{\mathfrak{c}}^{\kappa\lambda\mu\nu} S_{\mu\nu} - \underset{\eta}{h}{}_{\mu}^{\cdot\kappa\lambda} \mathfrak{D}^{\mu} - \underset{\mathfrak{D}}{\gamma}^{\kappa\lambda} \Delta \mathfrak{Q}, \\
&(b) \quad F_{\lambda} = -\underset{\eta}{h}{}_{\lambda}^{\cdot\mu\nu} S_{\mu\nu} + \underset{S,\eta}{\beta}_{\lambda\mu} \mathfrak{D}^{\mu} - \underset{S}{\mathfrak{q}}_{\lambda} \Delta \mathfrak{Q}, \\
&(c) \quad \theta = -\underset{0}{T} \underset{\mathfrak{D}}{\gamma}^{\mu\nu} S_{\mu\nu} - \underset{0}{T} \underset{S}{\mathfrak{q}}_{\mu} \mathfrak{D}^{\mu} + \frac{\Delta \mathfrak{Q}}{\rho \underset{S,\mathfrak{D}}{C}}
\end{aligned}
}
\tag{4.6*}
$$

which is valid for small deformations, small electric fields, and small amounts of heat flow. With respect to rectilinear coordinates $\underset{\mathfrak{D}}{\mathfrak{c}}^{\kappa\lambda\mu\nu}$ are the elastic constants at constant dielectric displacement, the $\underset{\eta}{h}{}_{\mu}^{\cdot\kappa\lambda}$ are adiabatic piezo-electric constants, the $\underset{\mathfrak{D}}{\gamma}^{\kappa\lambda}$ are thermo-elastic constants at constant dielectric displacement, the $\underset{S}{\mathfrak{q}}_{\lambda}$ are pyro-electric constants at constant strain, and the $\underset{S,\eta}{\beta}_{\lambda\mu}$ are adiabatic dielectric constants at constant strain.

If $S_{\mu\lambda}$, η, and $\mathfrak{D}^{\kappa}$ are expressed in terms of $\mathfrak{T}^{\kappa\lambda}$, θ, and F_{λ} we have

$$
\begin{aligned}
dS_{\mu\nu} &= \frac{\partial S_{\mu\nu}}{\partial \mathfrak{T}^{\kappa\lambda}} d\mathfrak{T}^{\kappa\lambda} + \frac{\partial S_{\mu\nu}}{\partial F_{\lambda}} dF_{\lambda} + \frac{\partial S_{\mu\nu}}{\partial \theta} d\theta, \\
d\mathfrak{D}^{\kappa} &= \frac{\partial \mathfrak{D}^{\kappa}}{\partial \mathfrak{T}^{\nu\lambda}} d\mathfrak{T}^{\nu\lambda} + \frac{\partial \mathfrak{D}^{\kappa}}{\partial F_{\lambda}} dF_{\lambda} + \frac{\partial \mathfrak{D}^{\kappa}}{\partial \theta} d\theta, \\
d\eta &= \frac{\partial \eta}{\partial^{\kappa} \mathfrak{T}^{\kappa\lambda}} d\mathfrak{T}^{\kappa\lambda} + \frac{\partial \eta}{\partial F_{\lambda}} dF_{\lambda} + \frac{\partial \eta}{\partial \theta} d\theta.
\end{aligned}
\tag{4.7}
$$

For a first approximation we suppose that the partial derivatives, except $\partial\eta/\partial\theta$, that occur in (4.7) are constants and that the heat capacity $\overset{\mathfrak{T},F}{C}$ at constant stress and constant electric field defined by

$$\underset{0}{T}\frac{\partial\eta}{\partial\theta} = \rho\overset{\mathfrak{T},F}{C} \tag{4.8}$$

is constant. Then the expression

$$\frac{\partial\eta}{\partial\theta}d\theta = \frac{d\theta}{\underset{0}{T}}\rho\overset{\mathfrak{T},F}{C} \tag{4.9}$$

is a total differential. $d\mathfrak{W}$ is a total differential if and only if

$$S_{\mu\lambda}\,d\mathfrak{T}^{\mu\lambda}+\mathfrak{D}^{\lambda}\,dF_{\lambda}+\eta\,d\theta \tag{4.10}$$

is a total differential. This is so if and only if

$$\begin{aligned}
&(a)\quad \frac{\partial S_{\mu\nu}}{\partial\mathfrak{T}^{\kappa\lambda}} = \frac{\partial S_{\kappa\lambda}}{\partial\mathfrak{T}^{\mu\nu}}, &&(b)\quad \frac{\partial S_{\mu\nu}}{\partial F_{\lambda}} = \frac{\partial\mathfrak{D}^{\lambda}}{\partial\mathfrak{T}^{\mu\nu}},\\
&(c)\quad \frac{\partial S_{\mu\nu}}{\partial\theta} = \frac{\partial\eta}{\partial\mathfrak{T}^{\mu\nu}}, &&(d)\quad \frac{\partial\mathfrak{D}^{\lambda}}{\partial F_{\mu}} = \frac{\partial\mathfrak{D}^{\mu}}{\partial F_{\lambda}},\\
&&&\hspace{-6em}(e)\quad \frac{\partial\mathfrak{D}^{\lambda}}{\partial\theta} = \frac{\partial\eta}{\partial F_{\lambda}}.
\end{aligned} \tag{4.11}$$

If these conditions are satisfied and if we write

$$\begin{aligned}
\overset{F}{\mathfrak{s}}_{\kappa\lambda\mu\nu} &\overset{\text{def}}{=} \frac{\partial S_{\mu\nu}}{\partial\mathfrak{T}^{\kappa\lambda}} = \frac{\partial S_{\kappa\lambda}}{\partial\mathfrak{T}^{\mu\nu}},\\
\overset{\theta}{d}{}^{\lambda}_{\cdot\mu\nu} &\overset{\text{def}}{=} \frac{\partial S_{\mu\nu}}{\partial F_{\lambda}} = \frac{\partial\mathfrak{D}^{\lambda}}{\partial\mathfrak{T}^{\mu\nu}},\\
\overset{F}{\alpha}_{\mu\nu} &\overset{\text{def}}{=} \frac{\partial S_{\mu\nu}}{\partial\theta} = \frac{\partial\eta}{\partial\mathfrak{T}^{\mu\nu}},\\
\overset{\mathfrak{T},\theta}{\epsilon}{}^{\kappa\lambda} &\overset{\text{def}}{=} \frac{\partial\mathfrak{D}^{\kappa}}{\partial F_{\lambda}} = \frac{\partial\mathfrak{D}^{\lambda}}{\partial F_{\kappa}},\\
\overset{\mathfrak{T}}{\mathfrak{p}}{}^{\kappa} &\overset{\text{def}}{=} \frac{\partial\mathfrak{D}^{\kappa}}{\partial\theta} = \frac{\partial\eta}{\partial F_{\kappa}},
\end{aligned} \tag{4.12}$$

the equations (4.7) take the form

$$\boxed{\begin{aligned}
&(a)\quad S_{\mu\nu} = \overset{F}{\mathfrak{s}}_{\kappa\lambda\mu\nu}\mathfrak{T}^{\kappa\lambda}+\overset{\theta}{d}{}^{\lambda}_{\cdot\mu\nu}F_{\lambda}+\overset{F}{\alpha}_{\mu\nu}\theta,\\
&(b)\quad \mathfrak{D}^{\kappa} = \overset{\theta}{d}{}^{\kappa}_{\cdot\nu\lambda}\mathfrak{T}^{\nu\lambda}+\overset{\mathfrak{T},\theta}{\epsilon}{}^{\kappa\lambda}F_{\lambda}+\overset{\mathfrak{T}}{\mathfrak{p}}{}^{\kappa}\theta,\\
&(c)\quad \Delta\mathfrak{Q} = \underset{0}{T}\overset{F}{\alpha}_{\kappa\lambda}\mathfrak{T}^{\kappa\lambda}+\underset{0}{T}\overset{\mathfrak{T}}{\mathfrak{p}}{}^{\lambda}F_{\lambda}+\rho\overset{\mathfrak{T},F}{C}\theta,
\end{aligned}} \tag{4.13*}$$

which is valid for small deformations, small electric fields, and small changes of temperature. With respect to rectilinear coordinates the $\overset{F}{\mathfrak{s}}$ are elastic constants at constant electric field, the $\overset{\theta}{d}{}^{\lambda}_{\cdot\mu\nu}$ are isothermal piezo-electric constants, the $\overset{F}{\alpha}_{\mu\nu}$ are thermo-elastic constants at constant electric field, the $\overset{\mathfrak{T},\theta}{\epsilon}{}^{\kappa\lambda}$ are isothermal dielectric constants at constant stress, and the $\overset{\mathfrak{T}}{p}{}^{\kappa}$ are pyro-electric constants at constant stress.

Now in problems concerning vibrations in crystals we may assume that there is no interchange of heat between adjacent elements. Hence the equations (4.6*) reduce to

$$\boxed{\begin{aligned} &(a) \quad \mathfrak{T}^{\kappa\lambda} = \underset{\mathfrak{D}}{\mathfrak{c}}{}^{\kappa\lambda\mu\nu}S_{\mu\nu} - h_{\mu}^{\cdot\kappa\lambda}\mathfrak{D}^{\mu}, \\ &(b) \quad F_{\lambda} = -h_{\lambda}^{\cdot\mu\nu}S_{\mu\nu} + \underset{S}{\beta}_{\lambda\mu}\mathfrak{D}^{\mu}, \\ &(c) \quad \theta = -\underset{0}{T}\underset{\mathfrak{D}}{\gamma}{}^{\mu\nu}S_{\mu\nu} - \underset{0}{T}\underset{S}{\mathfrak{q}}_{\mu}\mathfrak{D}^{\mu}, \end{aligned}} \tag{4.14*}$$

where the subscript η has been dropped because we are only considering adiabatic processes. In (4.13* c) we have $\Delta\mathfrak{Q} = 0$, hence θ can be solved from (4.13* c) and substituted in (4.13* a, b). This leads to equations of the form

$$\boxed{\begin{aligned} &(a) \quad S_{\mu\nu} = \overset{F}{\mathfrak{s}}_{\kappa\lambda\mu\nu}\mathfrak{T}^{\kappa\lambda} + d^{\lambda}_{\cdot\mu\nu}F_{\lambda}, \\ &(b) \quad \mathfrak{D}^{\kappa} = d^{\kappa}_{\cdot\nu\lambda}\mathfrak{T}^{\nu\lambda} + \overset{\mathfrak{T}}{\epsilon}{}^{\kappa\lambda}F_{\lambda}, \end{aligned}} \tag{4.15*}$$

where $\overset{F}{\mathfrak{s}}_{\kappa\lambda\mu\nu}$, $d^{\lambda}_{\cdot\mu\nu}$, and $\overset{\mathfrak{T}}{\epsilon}{}^{\kappa\lambda}$ differ slightly from $\overset{F}{\mathfrak{s}}_{\kappa\lambda\mu\nu}$, $\overset{\theta}{d}{}^{\lambda}_{\cdot\mu\nu}$, and $\overset{\mathfrak{T},\theta}{\epsilon}{}^{\kappa\lambda}$ in (4.13*). (4.14* a, b) and (4.15* a, b) are the equations, free from θ and η, which are to be used for vibrations in crystals.

In order to get the relations between the coefficients in (4.14*) and (4.15*) we introduce the inverses $\underset{S}{\epsilon}{}^{\kappa\lambda}$ of $\underset{S}{\beta}_{\lambda\kappa}$ and $\overset{\mathfrak{T}}{\beta}_{\lambda\kappa}$ of $\overset{\mathfrak{T}}{\epsilon}{}^{\kappa\lambda}$. Then (4.14* b) and (4.15* b) can be written in the form

$$\begin{aligned} &(a) \qquad \mathfrak{D}^{\kappa} = \underset{S}{\epsilon}{}^{\kappa\lambda}F_{\lambda} + \underset{S}{\epsilon}{}^{\kappa\lambda}h_{\lambda}^{\cdot\mu\nu}S_{\mu\nu}, \\ &(b) \qquad F_{\lambda} = \overset{\mathfrak{T}}{\beta}_{\lambda\kappa}\mathfrak{D}^{\kappa} - \overset{\mathfrak{T}}{\beta}_{\lambda\kappa}d^{\kappa}_{\cdot\nu\mu}\mathfrak{T}^{\nu\mu}. \end{aligned} \tag{4.16*}$$

Substituting this in (4.14* a) and (4.15* a) we get

$$\begin{aligned} \mathfrak{T}^{\kappa\lambda} &= (\underset{\mathfrak{D}}{\mathfrak{c}}{}^{\kappa\lambda\mu\nu} - h_{\rho}^{\cdot\kappa\lambda}\underset{S}{\epsilon}{}^{\rho\sigma}h_{\sigma}^{\cdot\mu\nu})S_{\mu\nu} - h_{\rho}^{\cdot\kappa\lambda}\underset{S}{\epsilon}{}^{\rho\mu}F_{\mu}, \\ S_{\mu\nu} &= (\overset{F}{\mathfrak{s}}_{\kappa\lambda\mu\nu} - d^{\rho}_{\cdot\mu\nu}\overset{\mathfrak{T}}{\beta}_{\rho\sigma}d^{\sigma}_{\cdot\kappa\lambda})\mathfrak{T}^{\kappa\lambda} + d^{\lambda}_{\cdot\mu\nu}\overset{\mathfrak{T}}{\beta}_{\lambda\kappa}\mathfrak{D}^{\kappa}. \end{aligned} \tag{4.17*}$$

Now by introducing the inverses $\underset{\mathfrak{D}}{\mathfrak{s}}_{\kappa\lambda\mu\nu}$ of $\underset{\mathfrak{D}}{\mathfrak{c}}^{\kappa\lambda\mu\nu}$ and $\overset{F}{\mathfrak{c}}^{\kappa\lambda\mu\nu}$ of $\overset{F}{\mathfrak{s}}_{\kappa\lambda\mu\nu}$ we get from (4.14* a) and (4.15* a)

$$\begin{aligned} S_{\mu\nu} &= \underset{\mathfrak{D}}{\mathfrak{s}}_{\kappa\lambda\mu\nu}\mathfrak{T}^{\kappa\lambda}+\underset{\mathfrak{D}}{\mathfrak{s}}_{\kappa\lambda\mu\nu}h_{\rho}^{\cdot\kappa\lambda}\mathfrak{D}^{\rho}, \\ \mathfrak{T}^{\kappa\lambda} &= \overset{F}{\mathfrak{c}}^{\kappa\lambda\mu\nu}S_{\mu\nu}-\overset{F}{\mathfrak{c}}^{\kappa\lambda\mu\nu}d^{\rho}_{\cdot\mu\nu}F_{\rho}, \end{aligned} \tag{4.18*}$$

and it follows from (4.17*) and (4.18*) that

$$\begin{aligned} &(a) \quad \overset{F}{\mathfrak{c}}^{\kappa\lambda\mu\nu} = \underset{\mathfrak{D}}{\mathfrak{c}}^{\kappa\lambda\mu\nu}-h_{\rho}^{\cdot\kappa\lambda}\underset{S}{\epsilon}^{\rho\sigma}h_{\sigma}^{\cdot\mu\nu} = \underset{\mathfrak{D}}{\mathfrak{c}}^{\kappa\lambda\mu\nu}-\mathfrak{e}^{\rho\kappa\lambda}h_{\rho}^{\cdot\mu\nu}, \\ &(b) \quad \underset{\mathfrak{D}}{\mathfrak{s}}_{\kappa\lambda\mu\nu} = \overset{F}{\mathfrak{s}}_{\kappa\lambda\mu\nu}-d^{\rho}_{\cdot\mu\nu}\overset{\mathfrak{T}}{\beta}_{\rho\sigma}d^{\sigma}_{\cdot\kappa\lambda} = \overset{F}{\mathfrak{s}}_{\kappa\lambda\mu\nu}-\mathfrak{g}_{\rho\mu\nu}d^{\rho}_{\cdot\kappa\lambda}, \\ &(c) \quad \mathfrak{e}^{\mu\kappa\lambda} \overset{\text{def}}{=} h_{\rho}^{\cdot\kappa\lambda}\underset{S}{\epsilon}^{\rho\mu} = \overset{F}{\mathfrak{c}}^{\kappa\lambda\rho\nu}d^{\mu}_{\cdot\rho\nu}, \\ &(d) \quad \mathfrak{g}_{\kappa\mu\nu} \overset{\text{def}}{=} d^{\lambda}_{\cdot\mu\nu}\overset{\mathfrak{T}}{\beta}_{\lambda\kappa} = \underset{\mathfrak{D}}{\mathfrak{s}}_{\rho\lambda\mu\nu}h_{\kappa}^{\cdot\rho\lambda}. \end{aligned} \tag{4.19*}$$

By substituting (4.14* a) and (4.15* a) in (4.16*) we get

$$\begin{aligned} \mathfrak{D}^{\kappa} &= (\underset{S}{\epsilon}^{\kappa\lambda}+\underset{S}{\epsilon}^{\kappa\rho}h_{\rho}^{\cdot\mu\nu}d^{\lambda}_{\cdot\mu\nu})F_{\lambda}+\underset{S}{\epsilon}^{\kappa\lambda}h_{\lambda}^{\cdot\mu\nu}\overset{F}{\mathfrak{s}}_{\rho\sigma\mu\nu}\mathfrak{T}^{\rho\sigma}, \\ F_{\lambda} &= (\overset{\mathfrak{T}}{\beta}_{\lambda\kappa}+\overset{\mathfrak{T}}{\beta}_{\lambda\rho}d^{\rho}_{\cdot\nu\mu}h_{\kappa}^{\cdot\nu\mu})\mathfrak{D}^{\kappa}-\overset{\mathfrak{T}}{\beta}_{\lambda\kappa}d^{\kappa}_{\cdot\nu\mu}\underset{\mathfrak{D}}{\mathfrak{c}}^{\nu\mu\rho\sigma}S_{\rho\sigma}, \end{aligned} \tag{4.20*}$$

and this gives because of (4.14* b) and (4.15* b)

$$\begin{aligned} \overset{\mathfrak{T}}{\epsilon}^{\kappa\lambda} &= \underset{S}{\epsilon}^{\kappa\lambda}+\underset{S}{\epsilon}^{\kappa\rho}h_{\rho}^{\cdot\mu\nu}d^{\lambda}_{\cdot\mu\nu} = \underset{S}{\epsilon}^{\kappa\lambda}+\mathfrak{e}^{\kappa\mu\nu}d^{\lambda}_{\cdot\mu\nu}, \\ \underset{S}{\beta}_{\lambda\kappa} &= \overset{\mathfrak{T}}{\beta}_{\lambda\kappa}+\overset{\mathfrak{T}}{\beta}_{\lambda\rho}d^{\rho}_{\cdot\nu\mu}h_{\kappa}^{\cdot\nu\mu} = \overset{\mathfrak{T}}{\beta}_{\lambda\kappa}+\mathfrak{g}_{\lambda\nu\mu}h_{\kappa}^{\cdot\nu\mu}. \end{aligned} \tag{4.21*}$$

With the abridged notations (4.19* c, d) the equations (4.16*) and (4.18*) take the form

$$\boxed{\begin{aligned} &(a) \quad \mathfrak{D}^{\kappa} = \underset{S}{\epsilon}^{\kappa\lambda}F_{\lambda}+\mathfrak{e}^{\kappa\mu\nu}S_{\mu\nu}, \\ &(b) \quad F_{\lambda} = \overset{\mathfrak{T}}{\beta}_{\lambda\kappa}\mathfrak{D}^{\kappa}-\mathfrak{g}_{\lambda\mu\nu}\mathfrak{T}^{\mu\nu}, \\ &(c) \quad S_{\mu\nu} = \underset{\mathfrak{D}}{\mathfrak{s}}_{\kappa\lambda\mu\nu}\mathfrak{T}^{\kappa\lambda}+\mathfrak{g}_{\kappa\mu\nu}\mathfrak{D}^{\kappa}, \\ &(d) \quad \mathfrak{T}^{\kappa\lambda} = \overset{F}{\mathfrak{c}}^{\kappa\lambda\mu\nu}S_{\mu\nu}-\mathfrak{e}^{\mu\kappa\lambda}F_{\mu}. \end{aligned}} \tag{4.22*}$$

These are the equations generally used.

We give a table of results:

$$(a)\quad \overset{F}{\mathfrak{c}}{}^{\kappa\lambda\mu\nu} = \underset{\mathfrak{D}}{\mathfrak{c}}{}^{\kappa\lambda\mu\nu} - \mathfrak{e}^{\rho\kappa\lambda} h_{\rho}^{\cdot\mu\nu}$$

$$(b)\quad \overset{F}{\mathfrak{c}}{}^{\kappa\lambda\rho\sigma} \overset{F}{\mathfrak{s}}_{\rho\sigma\mu\nu} = A^{\kappa}_{(\mu} A^{\lambda}_{\nu)}$$

$$(c)\quad \underset{\mathfrak{D}}{\mathfrak{s}}_{\kappa\lambda\mu\nu} = \overset{F}{\mathfrak{s}}_{\kappa\lambda\mu\nu} - \mathfrak{g}_{\rho\mu\nu} d^{\rho}_{\cdot\kappa\lambda}$$

$$(d)\quad \underset{\mathfrak{D}}{\mathfrak{c}}{}^{\kappa\lambda\rho\sigma} \underset{\mathfrak{D}}{\mathfrak{s}}_{\rho\sigma\mu\nu} = A^{\kappa}_{(\mu} A^{\lambda}_{\nu)}$$

$$(e)\quad \mathfrak{e}^{\mu\kappa\lambda} = h_{\rho}^{\cdot\kappa\lambda} \underset{S}{\epsilon}{}^{\rho\mu} =$$

$$(f)\quad = \overset{F}{\mathfrak{c}}{}^{\kappa\lambda\rho\nu} d^{\mu}_{\cdot\rho\nu}$$

$$(g)\quad \mathfrak{g}_{\kappa\mu\nu} = d^{\lambda}_{\cdot\mu\nu} \overset{\mathfrak{T}}{\beta}_{\kappa\lambda} =$$

$$(h)\quad = \underset{\mathfrak{D}}{\mathfrak{s}}_{\rho\lambda\mu\nu} h_{\kappa}^{\cdot\rho\lambda}$$

$$(i)\quad \overset{\mathfrak{T}}{\epsilon}{}^{\kappa\lambda} = \underset{S}{\epsilon}{}^{\kappa\lambda} + \mathfrak{e}^{\kappa\mu\nu} d^{\lambda}_{\cdot\mu\nu}$$

$$(j)\quad \overset{\mathfrak{T}}{\epsilon}{}^{\kappa\mu} \overset{\mathfrak{T}}{\beta}_{\mu\lambda} = A^{\kappa}_{\lambda}$$

$$(k)\quad \underset{S}{\beta}_{\lambda\kappa} = \overset{\mathfrak{T}}{\beta}_{\lambda\kappa} + \mathfrak{g}_{\lambda\nu\mu} h_{\kappa}^{\cdot\nu\mu}$$

$$(l)\quad \underset{S}{\epsilon}{}^{\kappa\mu} \underset{S}{\beta}_{\mu\lambda} = A^{\kappa}_{\lambda}$$

(4.23*)

5. Crystal classes†

The chief property of a crystal is that it is invariant for a finite number of orthogonal point transformations. These transformations form a finite group, the group of the crystal. As an illustration of a finite group we consider the group of all orthogonal transformations in

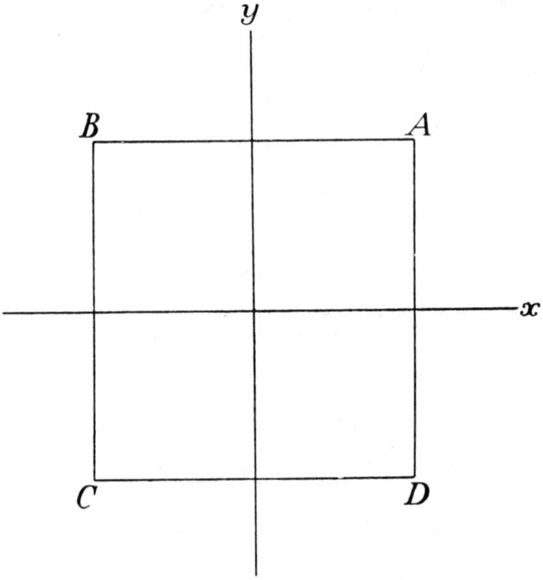

FIG. 25.

the xy-plane which leave the square $ABCD$ invariant. Writing $\underset{4}{z}$ for a rotation through $90°$ from x to y, $\underset{2}{z}$ for a rotation through $180°$, and I for the identical transformation, we have first the transformations $\underset{4}{z}$,

† Cf. Voigt 1898. 1; Tedone 1907. 1; Love 1944. 1; Cady 1946. 2.

$z_4^2 = z_2$, $z_4^3 = z_4^{-1}$, $z_4^4 = I$. All these transformations have $\Delta = +1$. The reflections E_x at the y-axis and the reflection E_y at the x-axis have $\Delta = -1$, and the same holds for the reflection F_A at the line BD and F_B at the line AC. We give a table for multiplication:

↙	I	z_4	z_4^2	z_4^3	E_x	E_y	F_A	F_B
I	I	z_4	z_4^2	z_4^3	E_x	E_y	F_A	F_B
z_4	z_4	z_4^2	z_4^3	I	F_B	F_A	E_x	E_y
z_4^2	z_4^2	z_4^3	I	z_4	E_y	E_x	F_B	F_A
z_4^3	z_4^3	I	z_4	z_4^2	F_A	F_B	E_y	E_x
E_x	E_x	F_A	E_y	F_B	I	z_4^2	z_4	z_4^3
E_y	E_y	F_B	E_x	F_A	z_4^2	I	z_4^3	z_4
F_A	F_A	E_y	F_B	E_x	z_4^3	z_4	I	z_4^2
F_B	F_B	E_x	F_A	E_y	z_4	z_4^3	z_4^2	I

(5.1)

This table is easily deduced from the schemes:

$$I\colon \begin{matrix} x & y \\ x & y \end{matrix}, \quad z_4\colon \begin{matrix} x & y \\ y & -x \end{matrix}, \quad z_4^2\colon \begin{matrix} x & y \\ -x & -y \end{matrix}, \quad z_4^3\colon \begin{matrix} x & y \\ -y & x \end{matrix},$$
$$E_x\colon \begin{matrix} x & y \\ -x & y \end{matrix}, \quad E_y\colon \begin{matrix} x & y \\ x & -y \end{matrix}, \quad F_A\colon \begin{matrix} x & y \\ -y & -x \end{matrix}, \quad F_B\colon \begin{matrix} x & y \\ y & x \end{matrix}, \tag{5.2}$$

We see from the table that the eight transformations form a group and that every transformation of the group can be obtained, for instance, from z_4 and E_x or from z_4 and F_A or from E_x and F_A. Accordingly we call z_4 and E_x a *generating set of transformations*. This is not the only generating set, z_4 and F_A or E_x and F_B also form generating sets.

In order to investigate crystals in space it has been found that it is sufficient to consider the following twenty-one transformations:

I: $x, y, z \to x, y, z$; identical transformation. $\Delta = +1$.

x_2: $x, y, z \to x, -y, -z$

x_3: $x, y, z \to x, -\tfrac{1}{2}y + \tfrac{1}{2}z\sqrt{3}, -\tfrac{1}{2}y\sqrt{3} - \tfrac{1}{2}z$

x_4: $x, y, z \to x, z, -y$

x_6: $x, y, z \to x, \tfrac{1}{2}y + \tfrac{1}{2}z\sqrt{3}, -\tfrac{1}{2}y\sqrt{3} + \tfrac{1}{2}z$

(for x_2, x_3, x_4, x_6:) cycl.; rotations about the x-, y-, and z-axis over *180°*, *120°*, *90°*, and *60°*. $\Delta = 1$.

$S\colon x, y, z \to y, z, x$, $S^2\colon x, y, z \to z, x, y$ — cyclical permutations of the axes. $\Delta = +1$.

$C\colon x, y, z \to -x, -y, -z$; reflection at the origin. $\Delta = -1$.

$E_x\colon x, y, z \to -x, y, z$ cycl.; reflections at the yz-, zx-, and xy-planes. $\Delta = -1$.

$S_x = E_x x_4 = C x_4^{-1}\colon x, y, z \to -x, z, -y$ cycl.; reflexotations (Drehspiegelungen), being rotations about the x-, y-, and z-axis over $90°$ followed by a reflection at the plane perpendicular to the axis. $\Delta = -1$. (5.3)

By using these transformations it has been proved that exactly thirty-two different groups exist and that there are thirty-two different crystal classes corresponding to them. We now set out a table of all classes with their groups given by a generating set of transformations.

System			*Generating set*	*Generating set with regard to central symmetric properties*	*Enantiomorphous*	*Pyro-electric*	*Pyro-magnetic*	*Piezo-electric*	*Piezo-magnetic*
Triclinic	1. (2)	holohedral	C	No transformations			×		×
	2. (1)	hemihedral	No transformations		×	×	×	×	×
Monoclinic	3. (5)	holohedral	$C, z_2 \equiv C, E_z \equiv z_2, E_z$	z_2			×		×
	4. (4)	hemihedral	E_z			×	×	×	×
	5. (3)	hemimorphic	z_2		×	×	×	×	×
Rhombic	6. (8)	holohedral	$C, z_2, x_2 \equiv C, z_2, E_x$	$z_2, x_2, (y_2)$					×
	7. (6)	hemihedral	$z_2, x_2, (y_2)$†		×			×	×
	8. (7)	hemimorphic	z_2, E_x			×		×	×
Trigonal I	9. (20)	holohedral	C, z_3, x_2	z_3, x_2					×
	10. (18)	enantiomorphous hemihedral	z_3, x_2		×			×	×
	11. (19)	hemimorphic hemihedral	z_3, E_x			×		×	×
Trigonal II	12. (17)	paramorphic hemihedral	C, z_3	z_3			×		×
	13. (16)	tetartohedral	z_3		×	×	×	×	×
Tetragonal I	14. (15)	holohedral	$C, z_4, x_2 \equiv C, z_4, E_x$	z_4, x_2					×
	15. (12)	enantiomorphous hemihedral	z_4, x_2		×			×	×
	16. (14)	hemimorphic hemihedral	z_4, E_x			×		×	×
	19. (11)	hemihedral with inversion axis	S_z, x_2					×	×

† The transformations in parentheses belong to the group, but it is not necessary to take them in the generating set because $z_4 = x_4 y_4 x_4^{-1}$; $S_z = S_y^{-2} S_x^{-2} S_y^{-1} S_x S_y$; $z_2 = x_2 y_2$.

System			*Generating set*	*Generating set with regard to central symmetric properties*	*Enantiomorphous*	*Pyro-electric*	*Pyro-magnetic*	*Piezo-electric*	*Piezo-magnetic*
Tetragonal II	*17.* (*13*)	paramorphic hemihedral	C, z_4	z_4			×		×
	18. (*10*)	tetartohedral	z_4		×	×	×	×	×
	20. (*9*)	tetartohedral with inversion axis	S_z				×	×	×
Hexagonal I	*21.* (*27*)	holohedral	C, z_6, x_2	z_6, x_2					×
	22. (*24*)	enantiomorphous hemihedral	z_6, x_2		×			×	×
	23. (*26*)	hemimorphic hemihedral	z_6, E_x			×		×	×
	26. (*22*)	hemihedral with threefold axis	z_3, x_2, E_z					×	×
Hexagonal II	*24.* (*25*)	paramorphic hemihedral	C, z_6	z_6			×		×
	25. (*23*)	tetartohedral	z_6		×	×	×	×	×
	27. (*21*)	tetartohedral with threefold axis	z_3, E_z				×	×	×
Cubic I	*28.* (*32*)	holohedral	$C, x_4, y_4, (z_4)$†	$x_4, y_4, (z_4)$					
	29. (*29*)	enantiomorphous hemihedral	$x_4, y_4, (z_4)$		×				
	30. (*31*)	hemimorphic hemihedral	$S_x, S_y, (S_z)$†					×	
Cubic II	*31.* (*30*)	paramorphic hemihedral	$C, x_2, y_2, (z_2), S$†	$x_2, y_2, (z_2), S$					×
	32. (*28*)	tetartohedral	$x_2, y_2, (z_2), S$		×			×	×

(5.4)

The classes are indicated by their number as given by Voigt, with the number assigned to them by Cady in parentheses. There are seven systems, four of which contain two sub-systems. A transformation T is equivalent to CT for any property with central symmetry. Consequently the generating set for these properties is much simpler. The systems are sub-divided by these simpler generating sets into sub-systems. A class is called *enantiomorphous* if it can contain crystals that can be transformed into each other by a reflection but not by a rotation. The necessary and sufficient condition for this case to arise is that the generating set does not contain a transformation with $\Delta = -1$. The enantiomorphous classes are *2*, *5*, *7*, *10*, *13*, *15*, *18*, *22*, *25*, *29*, and *32*.

† See note opposite.

The *principle of Neumann* connects the crystal classes with the possible physical properties. According to this principle the symmetry that characterizes the outward appearance of the crystal, i.e. the generating set of its group, also fixes the symmetry of all its physical properties. So a physical quantity without central symmetry, for instance a vector, can never occur in the eleven classes whose groups contain C: *1, 3, 6, 9, 12, 14, 17, 21, 24, 28, 31*. But a W-vector, since it has central symmetry, may occur in these classes if nothing else prevents its doing so.

In order to investigate whether certain quantities may occur in the different classes we have only to find out which quantities are invariant for the transformations (5.3). For instance a vector v_1, v_2, v_3 is transformed by $\underset{x}{E}$ into $-v_1, v_2, v_3$; hence only a vector $0, v_2, v_3$ is invariant for $\underset{x}{E}$. But a W-vector $\tilde{v}_1, \tilde{v}_2, \tilde{v}_3$ is transformed into $\tilde{v}_1, -\tilde{v}_2, -\tilde{v}_3$, hence an invariant W-vector has the form $\tilde{v}_1, 0, 0$.

An affinor of valence *2* can be split up into a tensor and a bivector. A tensor can be split up into a scalar and a deviator, and a bivector is equivalent to a W-vector for the group of rotations and reflections. Hence the decomposition is

$$\text{scalar}+W\text{-vector}+\text{deviator} \quad (9 = 1+3+5) \tag{5.5}$$

and all these parts are centro-symmetric. For a W-affinor of valence *2* we get

$$W\text{-scalar}+\text{vector}+W\text{-deviator} \tag{5.6}$$

and none of these parts are centro-symmetric.

An affinor P^{hij} of valence *3* contains a trivector $P^{[hij]}$, which is equivalent to a W-scalar, and a tensor $P^{(hij)}$ (*10* components). A vector $P^h = P^{(hij)}g_{ij}$ can be formed from $P^{(hij)}$ and $P^{(hij)}$ can be split up into two quantities

$$P^{(hij)} = \tfrac{3}{5}P^{(h}g^{ij)}+\{P^{(hij)}-\tfrac{3}{5}P^{(h}g^{ij)}\}. \tag{5.7}$$

The first one has *3* components, the components of P^h, and the second one is a tensor, without a vector part, with *7* components, called a *septor*. If the scalar part, the vector part, and the septor part are taken out of P^{hij}, we still have a quantity R^{hij} with $27-1-3-7 = 16$ components which satisfies the condition

$$R^{hij}+R^{ijh}+R^{jhi} = 0. \tag{5.8}$$

Hence
$$R^{hij} = \tfrac{2}{3}(R^{[hi]j}+R^{[hj]i})+\tfrac{2}{3}(R^{h[ij]}+R^{i[hj]}), \tag{5.9}$$

and this equation expresses the fact that R^{hij} can be split up into two parts, which only depend on $R^{[hi]j}$ and $R^{h[ij]}$ respectively. Now here a bivector is equivalent to a W-vector, hence $R^{[hi]j}$ is equivalent to a

W-affinor of valence *2* whose scalar part is zero. This means that R^{hij} can be split up into two vector parts and two W-deviator parts. Hence the decomposition of P^{hij} is

$$W\text{-scalar}+3\text{ vectors}+2\ W\text{-deviators}+\text{septor}\quad (27 = 1+9+10+7), \tag{5.10}$$

and all these parts are centro-symmetric. All other decompositions can be found in the same way.

We give a table of decompositions:

Affinor with valence		*Components*	*Scalar*	*Vector*	*Deviator*	*Septor*	*Nonor*†
1	not c.s.	*3*		*1*			
2	c.s.	*9*	*1*	*1W*	*1*		
3	not c.s.	*27*	*1W*	*3*	*2W*	*1*	
4	c.s.	*81*	*3*	*6W*	*6*	*3W*	*1*

(5.11)

W-affinor with valence		*Components*	*Scalar*	*Vector*	*Deviator*	*Septor*	*Nonor*
1	c.s.	*3*		*1W*			
2	not c.s.	*9*	*1W*	*1*	*1W*		
3	c.s.	*27*	*1*	*3W*	*2*	*1W*	
4	not c.s.	*81*	*3W*	*6*	*6W*	*3*	*1W*

(5.12)

Tensor with valence		*Components*	*Scalar*	*Vector*	*Deviator*	*Septor*	*Nonor*
1	not c.s.	*3*		*1*			
2	c.s.	*6*	*1*		*1*		
3	not c.s.	*10*		*1*		*1*	
4	c.s.	*15*	*1*		*1*		*1*

(5.13)

W-tensor with valence		*Components*	*Scalar*	*Vector*	*Deviator*	*Septor*	*Nonor*
1	c.s.	*3*		*1W*			
2	not c.s.	*6*	*1W*		*1W*		
3	c.s.	*10*		*1W*		*1W*	
4	not c.s.	*15*	*1W*		*1W*		*1W*

(5.14)

† A nonor is an affinor of valence *4* with 9 components and no scalar, vector, deviator, or septor parts. The terms deviator, septor, and nonor were introduced by the author in 1914, the last two for want of better expressions.

	Components	*Scalar*	*Vector*	*Deviator*	*Septor*	*Nonor*
Tensor-tensor $\mathfrak{c}^{\kappa\lambda\mu\nu}$; c.s.	*21*	*2*		*2*		*1*
$\mathfrak{e}^{\kappa\mu\nu} = \mathfrak{e}^{\kappa\nu\mu}$; not c.s.	*18*		*2*	*1W*	*1*	
$\tilde{\mathfrak{m}}^{\kappa\mu\nu} = \tilde{\mathfrak{m}}^{\kappa\nu\mu}$;† c.s.	*18*		*2W*	*1*	*1W*	

(5.15)

All parts of a (not) centro-symmetric quantity are (not) centro-symmetric.‡

We give a table for W-scalars, vectors and W-vectors and tensors, and W-tensors of valence *2* which are invariant for the transformations (5.3):

Transformation	*Not centro-symmetric*			*Centro-symmetric*		
	W-scalar	*Vector*	*W-tensor*	*W-vector*	*Tensor*	
$\underset{2}{x}$	$\tilde{s}$	$v_1\ 0\ 0$	$\tilde{p}_{11}\ \tilde{p}_{22}\ \tilde{p}_{33}\ \tilde{p}_{23}\ 0\ 0$	$\tilde{v}_1\ 0\ 0$	$p_{11}\ p_{22}\ p_{33}\ p_{23}\ 0\ 0$	$\Delta = +1$
$\underset{2}{y}$	$\tilde{s}$	$0\ v_2\ 0$	$\tilde{p}_{11}\ \tilde{p}_{22}\ \tilde{p}_{33}\ 0\ \tilde{p}_{31}\ 0$	$0\ \tilde{v}_2\ 0$	$p_{11}\ p_{22}\ p_{33}\ 0\ p_{31}\ 0$	
$\underset{2}{z}$	$\tilde{s}$	$0\ 0\ v_3$	$\tilde{p}_{11}\ \tilde{p}_{22}\ \tilde{p}_{33}\ 0\ 0\ \tilde{p}_{12}$	$0\ 0\ \tilde{v}_3$	$p_{11}\ p_{22}\ p_{33}\ 0\ 0\ p_{12}$	
$\underset{3}{x}, \underset{4}{x}, \underset{6}{x}$	$\tilde{s}$	$v_1\ 0\ 0$	$\tilde{p}_{11}\ \tilde{p}_{22}\ \tilde{p}_{22}\ 0\ 0\ 0$	$\tilde{v}_1\ 0\ 0$	$p_{11}\ p_{22}\ p_{22}\ 0\ 0\ 0$	
$\underset{3}{y}, \underset{4}{y}, \underset{6}{y}$	$\tilde{s}$	$0\ v_2\ 0$	$\tilde{p}_{11}\ \tilde{p}_{22}\ \tilde{p}_{11}\ 0\ 0\ 0$	$0\ \tilde{v}_2\ 0$	$p_{11}\ p_{22}\ p_{11}\ 0\ 0\ 0$	
$\underset{3}{z}, \underset{4}{z}, \underset{6}{z}$	$\tilde{s}$	$0\ 0\ v_3$	$\tilde{p}_{11}\ \tilde{p}_{11}\ \tilde{p}_{33}\ 0\ 0\ 0$	$0\ 0\ \tilde{v}_3$	$p_{11}\ p_{11}\ p_{33}\ 0\ 0\ 0$	
S	$\tilde{s}$	$v_1\ v_1\ v_1$	$\tilde{p}_{11}\ \tilde{p}_{11}\ \tilde{p}_{11}\ \tilde{p}_{23}\ \tilde{p}_{23}\ \tilde{p}_{23}$	$\tilde{v}_1\ \tilde{v}_1\ \tilde{v}_1$	$p_{11}\ p_{11}\ p_{11}\ p_{23}\ p_{23}\ p_{23}$	
C	0	$0\ 0\ 0$	$0\ 0\ 0\ 0\ 0\ 0$	$\tilde{v}_1\ \tilde{v}_2\ \tilde{v}_3$	$p_{11}\ p_{22}\ p_{33}\ p_{23}\ p_{31}\ p_{12}$	$\Delta = -1$
$\underset{x}{E} = C\underset{2}{x}$	0	$0\ v_2\ v_3$	$0\ 0\ 0\ 0\ \tilde{p}_{31}\ \tilde{p}_{12}$	$\tilde{v}_1\ 0\ 0$	$p_{11}\ p_{22}\ p_{33}\ p_{23}\ 0\ 0$	
$\underset{y}{E} = C\underset{2}{y}$	0	$v_1\ 0\ v_3$	$0\ 0\ 0\ \tilde{p}_{23}\ 0\ \tilde{p}_{12}$	$0\ \tilde{v}_2\ 0$	$p_{11}\ p_{22}\ p_{33}\ 0\ p_{31}\ 0$	
$\underset{z}{E} = C\underset{2}{z}$	0	$v_1\ v_2\ 0$	$0\ 0\ 0\ \tilde{p}_{23}\ \tilde{p}_{31}\ 0$	$0\ 0\ \tilde{v}_3$	$p_{11}\ p_{22}\ p_{33}\ 0\ 0\ p_{12}$	
$\underset{x}{S} = C\underset{4}{x}^{-1} = \underset{x4}{Ex}$	0	$0\ 0\ 0$	$0\ \tilde{p}_{22}\ -\tilde{p}_{22}\ \tilde{p}_{23}\ 0\ 0$	$\tilde{v}_1\ 0\ 0$	$p_{11}\ p_{22}\ p_{22}\ 0\ 0\ 0$	
$\underset{y}{S} = C\underset{4}{y}^{-1} = \underset{y4}{Ey}$	0	$0\ 0\ 0$	$\tilde{p}_{11}\ 0\ -\tilde{p}_{11}\ 0\ \tilde{p}_{31}\ 0$	$0\ \tilde{v}_2\ 0$	$p_{11}\ p_{22}\ p_{11}\ 0\ 0\ 0$	
$\underset{z}{S} = C\underset{4}{z}^{-1} = \underset{z4}{Ez}$	0	$0\ 0\ 0$	$\tilde{p}_{11}\ -\tilde{p}_{11}\ 0\ 0\ 0\ \tilde{p}_{12}$	$0\ 0\ \tilde{v}_3$	$p_{11}\ p_{11}\ p_{33}\ 0\ 0\ 0$	

(5.16)

By using this table we get the following quantities which exist in the thirty-two different classes:

† $\tilde{\mathfrak{m}}^{\kappa\mu\nu}$ is the quantity in the piezo-magnetic effect which corresponds to $\mathfrak{e}^{\kappa\mu\nu}$ in the piezo-electric effect.

‡ A general theory of the splitting up of quantities is developed in R.K. 1924. 1, ch. VII. A more modern treatment for the orthogonal group in 3 and 4 variables based on the theory of spin space is to be found in Littlewood 1945.1.

	Not centro-symmetric quantities				Centro-symmetric quantities		
Class	*Generating set*	*W-scalar*	*Vector*	*W-tensor*	*Generating set*	*W-vector*	*Tensor*
1. (2)	C	—	—	—	—	$\tilde{v}_1\ \tilde{v}_2\ \tilde{v}_3$	$p_{11}\ p_{22}\ p_{33}\ p_{23}\ p_{31}\ p_{12}$
2. (1)	—	$\tilde{s}$	$v_1\ v_2\ v_3$	$\tilde{p}_{11}\ \tilde{p}_{22}\ \tilde{p}_{33}\ \tilde{p}_{23}\ \tilde{p}_{31}\ \tilde{p}_{12}$			
3. (5)	C, z_2	—	—	—			
4. (4)	E_x	0	$v_1\ v_2\ 0$	$0\ \ 0\ \ 0\ \ \tilde{p}_{23}\ \tilde{p}_{31}\ 0$	z_2	$0\ 0\ \tilde{v}_3$	$p_{11}\ p_{22}\ p_{33}\ 0\ \ 0\ \ p_{12}$
5. (3)	z_2	$\tilde{s}$	$0\ 0\ v_3$	$\tilde{p}_{11}\ \tilde{p}_{22}\ \tilde{p}_{33}\ 0\ \ 0\ \ \tilde{p}_{12}$			
6. (8)	C, z_2, x_2	—	—	—			
7. (6)	z_2, x_2	$\tilde{s}$	$0\ 0\ 0$	$\tilde{p}_{11}\ \tilde{p}_{22}\ \tilde{p}_{33}\ 0\ \ 0\ \ 0$	z_2, x_2	$0\ 0\ 0$	$p_{11}\ p_{22}\ p_{33}\ 0\ \ 0\ \ 0$
8. (7)	z_2, E_x	0	$0\ 0\ v_3$	$0\ \ 0\ \ 0\ \ 0\ \ 0\ \ \tilde{p}_{12}$			
9. (20)	C, z_3, x_2	—	—				
10. (18)	z_3, x_2	$\tilde{s}$	$0\ 0\ 0$	$\tilde{p}_{11}\ \tilde{p}_{11}\ \tilde{p}_{33}\ 0\ \ 0\ \ 0$	z_3, x_2	$0\ 0\ 0$	$p_{11}\ p_{22}\ p_{33}\ 0\ \ 0\ \ 0$
11. (19)	z_3, E_x	0	$0\ 0\ v_3$	$0\ \ 0\ \ 0\ \ 0\ \ 0\ \ 0$			
12. (17)	C, z_3	—	—	—	z_3	$0\ 0\ \tilde{v}_3$	$p_{11}\ p_{11}\ p_{33}\ 0\ \ 0\ \ 0$
13. (16)	z_3	$\tilde{s}$	$0\ 0\ v_3$	$\tilde{p}_{11}\ \tilde{p}_{11}\ \tilde{p}_{33}\ 0\ \ 0\ \ 0$			
14. (15)	C, z_4, x_2	—	—	—			
15. (12)	z_4, x_2	$\tilde{s}$	$0\ 0\ 0$	$\tilde{p}_{11}\ \tilde{p}_{11}\ \tilde{p}_{33}\ 0\ \ 0\ \ 0$	z_4, x_2	$0\ 0\ 0$	$p_{11}\ p_{11}\ p_{33}\ 0\ \ 0\ \ 0$
16. (14)	z_4, E_x	0	$0\ 0\ v_3$	$0\ \ 0\ \ 0\ \ 0\ \ 0\ \ 0$			
19. (11)	S_z, x_2	0	$0\ 0\ 0$	$\tilde{p}_{11}\ -\tilde{p}_{11}\ 0\ \ 0\ \ 0\ \ 0$			
17. (13)	C, z_4	—	—	—			
18. (10)	z_4	$\tilde{s}$	$0\ 0\ v_3$	$\tilde{p}_{11}\ \tilde{p}_{11}\ \tilde{p}_{33}\ 0\ \ 0\ \ 0$	z	$0\ 0\ \tilde{v}_3$	$p_{11}\ p_{11}\ p_{33}\ 0\ \ 0\ \ 0$
20. (9)	S_z	0	$0\ 0\ 0$	$\tilde{p}_{11}\ -\tilde{p}_{11}\ 0\ \ 0\ \ 0\ \ \tilde{p}_{12}$			
21. (27)	C, z_6, x_2	—	—	—			
22. (24)	z_6, x_2	$\tilde{s}$	$0\ 0\ 0$	$\tilde{p}_{11}\ \tilde{p}_{11}\ \tilde{p}_{33}\ 0\ \ 0\ \ 0$	z_6, x_2	$0\ 0\ 0$	$p_{11}\ p_{11}\ p_{33}\ 0\ \ 0\ \ 0$
23. (26)	z_6, E_x	0	$0\ 0\ v_3$	$0\ \ 0\ \ 0\ \ 0\ \ 0\ \ 0$			
26. (22)	z_3, x_2, E_z	0	$0\ 0\ 0$	$0\ \ 0\ \ 0\ \ 0\ \ 0\ \ 0$			
24. (25)	C, z_6	—	—	—			
25. (23)	z_6	$\tilde{s}$	$0\ 0\ v_3$	$\tilde{p}_{11}\ \tilde{p}_{11}\ \tilde{p}_{33}\ 0\ \ 0\ \ 0$	z_6	$0\ 0\ \tilde{v}_3$	$p_{11}\ p_{11}\ p_{33}\ 0\ \ 0\ \ 0$
27. (21)	z_3, E_z	0	$0\ 0\ 0$	$0\ \ 0\ \ 0\ \ 0\ \ 0\ \ 0$			
28. (32)	$C, x_4, y_4, (z_4)$	—	—	—			
29. (29)	$x_4, y_4, (z_4)$	$\tilde{s}$	$0\ 0\ 0$	$\tilde{p}_{11}\ \tilde{p}_{11}\ \tilde{p}_{11}\ 0\ \ 0\ \ 0$	$x_4, y_4, (z_4)$	$0\ 0\ 0$	$p_{11}\ p_{11}\ p_{11}\ 0\ \ 0\ \ 0$
30. (31)	$S_x, S_y, (S_z)$	0	$0\ 0\ 0$	$0\ \ 0\ \ 0\ \ 0\ \ 0\ \ 0$			
31. (30)	$C, x_2, y_2, (z_2), S$	—	—	—	$x_2, y_2, (z_2), S$	$0\ 0\ 0$	$p_{11}\ p_{11}\ p_{11}\ 0\ \ 0\ \ 0$
32. (28)	$x_2, y_2, (z_2), S$	$\tilde{s}$	$0\ 0\ 0$	$\tilde{p}_{11}\ \tilde{p}_{11}\ \tilde{p}_{11}\ 0\ \ 0\ \ 0$			

(5.17)

Tables for all kinds of quantities can be constructed in the same way.

If a crystal connects two physical quantities, e.g. temperature and electric or magnetic field, electric field and strain, or stress and strain, we call this an *effect*. If the connexion is linear it is called a *linear effect*. The effect is given quantitatively by a quantity, for instance in the cases mentioned (using the group G_{or} and identifying densities and non-densities) a vector, a W-vector, an affinor of valence *3*, and an affinor of valence *4* respectively. We give a table of some linear effects:

Effect	*Connected quantities*	*Quantity connecting them*
pyro-electric	scalar, vector	vector
pyro-magnetic	scalar, W-vector	W-vector
thermo-elastic	scalar, tensor val. *2*	tensor val. *2*
dielectric	vector, vector	tensor val. *2*
magnetic	W-vector, W-vector	tensor val. *2*
piezo-electric	vector, tensor val. *2*	affinor val. *3*, symmetr. in *2* indices
piezo-magnetic	W-vector, tensor val. *2*	W-affinor val. *3*, symmetr. in *2* indices
elastic	tensor, tensor val. *2*	tensor-tensor val. *4*

(5.18)

In all these cases the effect can only exist in a certain crystal class if there is a quantity, giving the connexion, which is invariant for the group of this class. Hence it follows from table (5.17) that a pyro-electric effect only exists in the 10 classes *2*, *4*, *5*, *8*, *11*, *13*, *16*, *18*, *23*, *25* and a pyro-magnetic effect only in the 5 sub-systems containing the 13 classes *1*, *2*, *3*, *4*, *5*, *12*, *13*, *17*, *18*, *20*, *24*, *25*, *27*. The dielectric and the magnetic effect exist in all classes.

For the elastic effect the quantities giving the connexion for the adiabatic process are $\underset{\mathfrak{D}}{\mathfrak{c}}^{\kappa\lambda\mu\nu}$ and $\overset{F}{\mathfrak{s}}_{\kappa\lambda\mu\nu}$ (cf. (4.14*, 15*)) related by (4.23* *a*–*d*). Using the group G_{or} and identifying densities and non-densities we write c^{hijk} and s_{hijk} for them, dropping the affixes $\mathfrak{D}$ and F. In the same way we write T^{hi} and S_{ji} for stress and strain. Now it is not pleasant to have to use four indices in this work. So an abbreviated notation is usually introduced. In T^{hi} and c^{hijk} the combinations of indices 11, 22, 33, 23, 31, 12 are replaced by 1, 2, 3, 4, 5, 6. Then T^{hi} and c^{hijk} can be written

$$T^{A} \quad (A, B = 1,\ldots, 6), \tag{5.19}$$

$$c^{AB} = c^{BA}; \qquad 11 = 1; \qquad 23 = 4; \qquad \text{cycl. } 1, 2, 3 \text{ and } 4, 5, 6.$$

T^{A} is not a vector and c^{AB} not a tensor of valence *2* because the components transform in a more complicated way that can be derived from the transformation of T^{hi} and c^{hijk}. An abbreviated notation for S_{ji} and s_{hijk} can be introduced in the same way. But we must do this in such a way that $T^{A}S_{A}$ means the same as $T^{hi}S_{hi}$ and $T^{A}s_{AB}$ the same as $T^{hi}s_{hijk}$. These conditions are satisfied if we write

$$\begin{gathered} S_1 = S_{11}, \\ S_4 = 2S_{23}, \end{gathered}$$

$$\begin{array}{lll} & s_{1111} = s_{11}, \quad s_{2233} = s_{23}, & \\ 2s_{1123} = s_{14}, & 2s_{1131} = s_{15}, & 2s_{1112} = s_{16}, \\ 4s_{2323} = s_{44}, & 4s_{2331} = s_{45}, & 4s_{2312} = s_{46}, \end{array} \tag{5.20}$$

cycl. 1, 2, 3 and 4, 5, 6.

Then we have in fact

$$\begin{aligned} T^{ij}S_{ij} &= T^{11}S_{11}+2T^{23}S_{23}+\text{cycl. } 1, 2, 3 \\ &= T^{1}S_{1}+T^{4}S_{4}+\text{cycl. } 1, 2, 3;\ 4, 5, 6 \quad (i,j = 1,2,3). \end{aligned} \tag{5.21}$$

The transformation of indices for S_{ij} and s_{hijk} is just the same as for T^{hi} and c^{hijk}, except that a factor 2 has to be added for every set 23, 31, 12. This is a bit artificial and it is not pleasant to find that T^A, c^{AB}, S_B, and s_{BA} transform in a rather complicated way. (Never use this transformation and always go back to the transformation with 2 and 4 indices!) The abbreviated notation, however, is very satisfactory and we have to pay for that.

In order to find the possible tensor-tensors in the thirty-two classes we write out the tensor-tensors that are invariant for those transformations $\underset{2}{z}$, $\underset{3}{z}$, $\underset{4}{z}$, $\underset{6}{z}$, $\underset{2}{x}$, $\underset{4}{x}$, $\underset{2}{y}$, $\underset{4}{y}$, S which form the generating sets with respect to centro-symmetric properties.‡ Since a tensor-tensor is centro-symmetric we do not have to consider other transformations:†

$$\underset{2}{z}:\quad \begin{array}{cccccc} c_{11} & c_{12} & c_{13} & 0 & 0 & c_{16} \\ & c_{22} & c_{23} & 0 & 0 & c_{26} \\ & & c_{33} & 0 & 0 & c_{36} \\ & & & c_{44} & c_{45} & 0 \\ & & & & c_{55} & 0 \\ & & & & & c_{66} \end{array}$$

$$\underset{3}{z}:\quad \begin{array}{cccccc} c_{11} & c_{12} & c_{13} & c_{14} & -c_{25} & 0 \\ & c_{11} & c_{13} & -c_{14} & c_{25} & 0 \\ & & c_{33} & 0 & 0 & 0 \\ & & & c_{44} & 0 & c_{25} \\ & & & & c_{44} & c_{14} \\ & & & & & \frac{1}{2}(c_{11}-c_{12}) \end{array}$$

$$\underset{4}{z}:\quad \begin{array}{cccccc} c_{11} & c_{12} & c_{13} & 0 & 0 & c_{16} \\ & c_{11} & c_{13} & 0 & 0 & -c_{16} \\ & & c_{33} & 0 & 0 & 0 \\ & & & c_{44} & 0 & 0 \\ & & & & c_{44} & 0 \\ & & & & & c_{66} \end{array}$$

$$\underset{6}{z}:\quad \begin{array}{cccccc} c_{11} & c_{12} & c_{13} & 0 & 0 & 0 \\ & c_{11} & c_{13} & 0 & 0 & 0 \\ & & c_{33} & 0 & 0 & 0 \\ & & & c_{44} & 0 & 0 \\ & & & & c_{44} & 0 \\ & & & & & \frac{1}{2}(c_{11}-c_{12}) \end{array}$$

† Since the fundamental tensor is positive definite the indices can be written as upper or as lower indices. In the tables we have chosen lower indices because these are used in other publications.

‡ See Additional Note G, p. 272.

$$
\begin{array}{lcccccc}
\underset{2}{x}: & c_{11} & c_{12} & c_{13} & c_{14} & 0 & 0 \\
 & & c_{22} & c_{23} & c_{24} & 0 & 0 \\
 & & & c_{33} & c_{34} & 0 & 0 \\
 & & & & c_{44} & 0 & 0 \\
 & & & & & c_{55} & c_{56} \\
 & & & & & & c_{66}
\end{array}
\qquad
\begin{array}{lcccccc}
\underset{4}{x}: & c_{11} & c_{12} & c_{12} & 0 & 0 & 0 \\
 & & c_{22} & c_{23} & c_{24} & 0 & 0 \\
 & & & c_{22} & -c_{24} & 0 & 0 \\
 & & & & c_{44} & 0 & 0 \\
 & & & & & c_{55} & 0 \\
 & & & & & & c_{55}
\end{array}
$$

$$
\begin{array}{lcccccc}
\underset{2}{y}: & c_{11} & c_{12} & c_{13} & 0 & c_{15} & 0 \\
 & & c_{22} & c_{23} & 0 & c_{25} & 0 \\
 & & & c_{33} & 0 & c_{35} & 0 \\
 & & & & c_{44} & 0 & c_{46} \\
 & & & & & c_{55} & 0 \\
 & & & & & & c_{66}
\end{array}
\qquad
\begin{array}{lcccccc}
y: & c_{11} & c_{12} & c_{13} & 0 & c_{15} & 0 \\
 & & c_{22} & c_{12} & 0 & 0 & 0 \\
 & & & c_{11} & 0 & -c_{15} & 0 \\
 & & & & c_{44} & 0 & 0 \\
 & & & & & c_{55} & 0 \\
 & & & & & & c_{44}
\end{array}
$$

$$
\begin{array}{lcccccc}
S: & c_{11} & c_{12} & c_{12} & c_{14} & c_{15} & c_{16} \\
 & & c_{11} & c_{12} & c_{16} & c_{14} & c_{15} \\
 & & & c_{11} & c_{15} & c_{16} & c_{14} \\
 & & & & c_{44} & c_{45} & c_{45} \\
 & & & & & c_{44} & c_{45} \\
 & & & & & & c_{44}
\end{array}
\tag{5.22}
$$

From these tables we get for the eleven sub-systems:

Triclinic, *1*, *2*: Full table 21 constants†

Monoclinic, *3*, *4*, *5*; $\underset{2}{z}$:

$$
\begin{array}{cccccc}
c_{11} & c_{12} & c_{13} & 0 & 0 & c_{16} \\
 & c_{22} & c_{23} & 0 & 0 & c_{26} \\
 & & c_{33} & 0 & 0 & c_{36} \\
 & & & c_{44} & c_{45} & 0 \\
 & & & & c_{55} & 0 \\
 & & & & & c_{66}
\end{array}
$$

13 constants†

Rhombic, *6*, *7*, *8*; $\underset{2}{z}$, $\underset{2}{x}$:

$$
\begin{array}{cccccc}
c_{11} & c_{12} & c_{13} & 0 & 0 & 0 \\
 & c_{22} & c_{23} & 0 & 0 & 0 \\
 & & c_{33} & 0 & 0 & 0 \\
 & & & c_{44} & 0 & 0 \\
 & & & & c_{55} & 0 \\
 & & & & & c_{66}
\end{array}
$$

9 constants

Trigonal I, *9*, *10*, *11*; $\underset{3}{z}$, $\underset{2}{x}$:

$$
\begin{array}{cccccc}
c_{11} & c_{12} & c_{13} & c_{14} & 0 & 0 \\
 & c_{11} & c_{13} & -c_{14} & 0 & 0 \\
 & & c_{33} & 0 & 0 & 0 \\
 & & & c_{44} & 0 & 0 \\
 & & & & c_{44} & c_{14} \\
 & & & & & \tfrac{1}{2}(c_{11}-c_{12})
\end{array}
$$

6 constants

† The number of constants could be reduced to 18 and 12 respectively. No reduction is possible in the other systems.

Trigonal II, *12*, *13*; $\underset{3}{z}$:

$$\begin{array}{cccccc}
c_{11} & c_{12} & c_{13} & c_{14} & -c_{25} & 0 \\
 & c_{11} & c_{13} & -c_{14} & c_{25} & 0 \\
 & & c_{33} & 0 & 0 & 0 \\
 & & & c_{44} & 0 & c_{25} \\
 & & & & c_{44} & c_{14} \\
 & & & & & \tfrac{1}{2}(c_{11}-c_{12})
\end{array}$$

7 constants

Tetragonal I, *14*, *15*, *16*, *19*; $\underset{4}{z}$, $\underset{2}{x}$:

$$\begin{array}{cccccc}
c_{11} & c_{12} & c_{13} & 0 & 0 & 0 \\
 & c_{11} & c_{13} & 0 & 0 & 0 \\
 & & c_{33} & 0 & 0 & 0 \\
 & & & c_{44} & 0 & 0 \\
 & & & & c_{44} & 0 \\
 & & & & & c_{66}
\end{array}$$

6 constants

Tetragonal II, *17*, *18*, *20*; $\underset{4}{z}$:

$$\begin{array}{cccccc}
c_{11} & c_{12} & c_{13} & 0 & 0 & c_{16} \\
 & c_{11} & c_{13} & 0 & 0 & -c_{16} \\
 & & c_{33} & 0 & 0 & 0 \\
 & & & c_{44} & 0 & 0 \\
 & & & & c_{44} & 0 \\
 & & & & & c_{66}
\end{array}$$

7 constants

Hexagonal I, *21*, *22*, *23*, *26*; $\underset{6}{z}$, $\underset{2}{x}$ and

Hexagonal II, *24*, *25*, *27*; $\underset{6}{z}$:

$$\begin{array}{cccccc}
c_{11} & c_{12} & c_{13} & 0 & 0 & 0 \\
 & c_{11} & c_{13} & 0 & 0 & 0 \\
 & & c_{33} & 0 & 0 & 0 \\
 & & & c_{44} & 0 & 0 \\
 & & & & c_{44} & 0 \\
 & & & & & \tfrac{1}{2}(c_{11}-c_{12})
\end{array}$$

5 constants

Cubic I, *28*, *29*, *30*; $\underset{4}{x}$, $\underset{4}{y}$, $(\underset{4}{z})$ and

Cubic II, *31*, *32*; $\underset{2}{x}$, $\underset{2}{y}$, $(\underset{2}{z})$, S:

$$\begin{array}{cccccc}
c_{11} & c_{12} & c_{12} & 0 & 0 & 0 \\
 & c_{11} & c_{12} & 0 & 0 & 0 \\
 & & c_{11} & 0 & 0 & 0 \\
 & & & c_{44} & 0 & 0 \\
 & & & & c_{44} & 0 \\
 & & & & & c_{44}
\end{array}$$

3 constants

Isotropic medium; group G_{or}:

$c^{\kappa\lambda\mu\nu} = \lambda g^{\kappa\lambda}g^{\mu\nu}+2\mu g^{(\kappa|\mu|}g^{\lambda)\nu}$

$$\left.\begin{array}{l} \lambda = c_{12} \\ \mu = \frac{1}{2}(c_{11}-c_{12}) \end{array}\right\} \text{ constants of Lamé}$$

$\lambda+2\mu = c_{11}$

$$\begin{array}{cccccc} c_{11} & c_{12} & c_{12} & 0 & 0 & 0 \\ & c_{11} & c_{12} & 0 & 0 & 0 \\ & & c_{11} & 0 & 0 & 0 \\ & & & \frac{1}{2}(c_{11}-c_{12}) & 0 & 0 \\ & & & & \frac{1}{2}(c_{11}-c_{12}) & 0 \\ & & & & & \frac{1}{2}(c_{11}-c_{12}) \end{array}$$

2 constants

(5.23)

These tables are valid in the triclinic and isotropic case for all cartesian systems and in the other cases for cartesian systems which are chosen as follows with respect to the crystallographic axes:

Monoclinic: the z-axis is the one crystallographic axis that is perpendicular to the other axes and the x-axis is one of those other axes;

Rhombic: the x-, y-, and z-axis are the crystallographic axes;

Trigonal: the z-axis has threefold symmetry, the x-axis is one of the other axes;

Hexagonal: the z-axis has sixfold symmetry, the x-axis is one of the other axes;

Cubic: the x-, y-, and z-axis are the crystallographic axes.

As an example we take α-quartz (class *10*). The axes for the right-hand case are shown in Fig. 26.† According to Voigt, Mason, and Cady the values of c^{AB} and s_{AB} are in Giorgi units:

	Voigt	*Mason*	*Cady*	
$c_{11} = c_{22}$	*85·46*	*86·05*	*87·(5)*	
c_{33}	*105·62*	*107·1*	*107·(7)*	
$c_{44} = c_{55}$	*57·12*	*58·65*	*57·(3)*	
c_{12}	*7·25*	*5·05*	*7·6(2)*	$\times\, 10^{9}$ newton/m.^{2}
$c_{13} = c_{23}$	*14·35*	*10·45*	*15·(1)*	
$c_{14} = -c_{24} = c_{56}$	*16·82*	*18·25*	*17·(2)*	
$c_{66} = \frac{1}{2}(c_{11}-c_{12})$	*39·10*	*40·5*	*39·(9)*	

	Mason	*Cady*	
$s_{11} = s_{22}$	*1·279*	*1·26(9)*	
s_{33}	*0·956*	*0·97(1)*	
$s_{44} = s_{55}$	*1·978*	*2·00(5)*	
s_{12}	*−0·1535*	*−0·16(9)*	$\times\, 10^{-11}$ (m.^{2}/newton)
$s_{13} = s_{23}$	*−0·110*	*−0·15(4)*	
$s_{14} = -s_{24} = \frac{1}{2}s_{56}$	*−0·446*	*−0·43(1)*	
$s_{66} = 2(s_{11}-s_{22})$	*2·865*	*2·8(8)*	

(5.24)

† This figure is taken from Lack, Willard, and Fair 1934. 2, p. 753, with kind consent of the Bell Telephone Group.

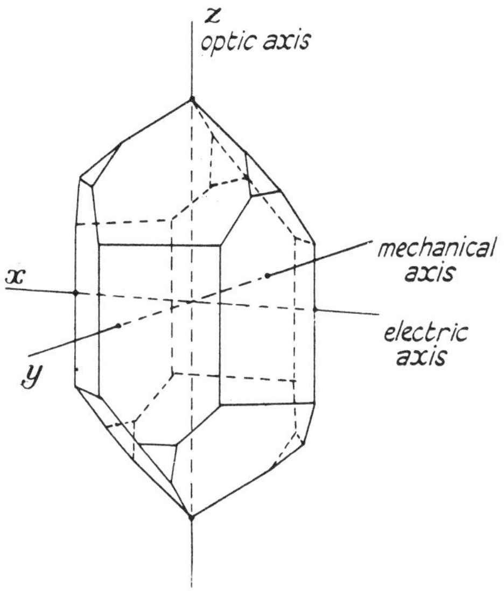

FIG. 26.

6. Piezo-electric and piezo-magnetic effect

The piezo-electric effect is expressed by the formulae (cf. (4.15*), (4.22*), (4.23*))

$$
\begin{array}{lll}
(a) & \mathfrak{D}^{\kappa} = \mathfrak{e}^{\kappa\mu\nu} S_{\mu\nu} & \multirow{2}{*}{(field zero),} \\
(b) & \mathfrak{D}^{\kappa} = d^{\kappa}_{\cdot\nu\lambda}\, \mathfrak{T}^{\nu\lambda} & \\
(c) & \mathfrak{T}^{\kappa\lambda} = -\mathfrak{e}^{\mu\kappa\lambda} F_{\mu} \quad \text{(strain zero)}, & (6.1) \\
(d) & S_{\mu\nu} = d^{\lambda}_{\cdot\mu\nu} F_{\lambda} \quad \text{(stress zero)}, & \\
(e) & \mathfrak{e}^{\mu\kappa\lambda} = \mathfrak{c}^{\kappa\lambda\rho\nu} d^{\mu}_{\cdot\rho\nu}. &
\end{array}
$$

Using orthogonal coordinate systems and identifying densities and non-densities we get

$$
\begin{array}{lll}
(a) & D^{h} = e^{hij} S_{ij} & \multirow{2}{*}{(field zero),} \\
(b) & D^{h} = d^{h}_{\cdot ij} T^{ij} & \\
(c) & T^{ij} = -e^{hij} F_{h} \quad \text{(strain zero)}, & (6.2) \\
(d) & S_{ij} = d^{h}_{\cdot ij} F_{h} \quad \text{(stress zero)}, & \\
(e) & e^{lhi} = c^{hijk} d^{l}_{\cdot jk} \quad (h, i, j, k, l = 1, 2, 3). &
\end{array}
$$

e^{hij} is an affinor which is symmetrical in the last two indices. Accordingly it has 18 independent components. As we have seen in § 5 it can be split up into two vectors (*2*×*3*), one *W*-deviator (*5*), and one septor (*7*). Since e^{hij} is not centro-symmetric we can only have the 21 classes *2*, *4*, *5*, *7*, *8*, *10*, *11*, *13*, *15*, *16*, *19*, *18*, *20*, *22*, *23*, *26*, *25*, *27*, *29*, *30*, and *32*. In these classes the vector parts exist in the 10 classes *2*, *4*, *5*,

8, 11, 13, 16, 18, 23, 25 only and the W-deviator in the 13 classes *2, 4, 5, 7, 8, 10, 13, 15, 19, 18, 20, 22,* and *25* only (cf. (5.17)). As we have here only a table of the possible tensors of the valences up to *2* and not of valence *3*, we are only able to conclude that piezo-electricity occurs for certain in the 16 classes *2, 4, 5, 7, 8, 10, 11, 13, 15, 16, 19, 18, 20, 22, 23,* and *25*. The five classes *26, 27, 29, 30,* and *32* have still to be investigated.

For the piezo-magnetic effect we have equations similar to (6.2) containing magnetic instead of electric quantities. Instead of e^{hij} they contain a W-affinor $\tilde{m}^{hij}$ which is symmetric in the last two indices. This quantity is centro-symmetric and can be split up into two W-vectors (2×3), one deviator (5), and one W-septor (7). Since central symmetry exists, we have only to look at the sub-systems. In all systems except the cubic system either the deviator or the W-vectors exist (cf. (5.17)). Hence all these systems are piezo-magnetic and we have still to investigate the cubic system.

We use the following abbreviated notations for e^{hij}, $\tilde{m}^{hij}$, and $d^h_{\cdot ij}$:

$$\left.\begin{array}{ll} e^{h1} = e^{h11}, & e^{h4} = e^{h23} \\ \tilde{m}^{h1} = \tilde{m}^{h11}, & \tilde{m}^{h4} = \tilde{m}^{h23} \\ d^h_{\cdot 1} = d^h_{\cdot 11}, & e^h_{\cdot 4} = 2e^h_{\cdot 23} \end{array}\right\} \quad \text{cycl. } 1, 2, 3;\ 4, 5, 6 \qquad (6.3)$$

To find the possible form of e^{hij} and $\tilde{m}^{hij}$ in all classes we give the values of e^{hij} which are invariant for the transformations occurring in (5.4) (the number of independent components is given in brackets):

$$\underset{2}{z}:(8)\ \begin{array}{cccccc} 0 & 0 & 0 & e_{14} & e_{15} & 0 \\ 0 & 0 & 0 & e_{24} & e_{25} & 0 \\ e_{31} & e_{32} & e_{33} & 0 & 0 & e_{36} \end{array} \qquad \underset{3}{z}:(6)\ \begin{array}{cccccc} e_{11} & -e_{11} & 0 & e_{14} & e_{15} & -e_{22} \\ -e_{22} & e_{22} & 0 & e_{15} & -e_{14} & -e_{11} \\ e_{31} & e_{31} & e_{33} & 0 & 0 & 0 \end{array}$$

$$\underset{4}{z} \text{ and } \underset{6}{z}:(4)\ \begin{array}{cccccc} 0 & 0 & 0 & e_{14} & e_{15} & 0 \\ 0 & 0 & 0 & e_{15} & -e_{14} & 0 \\ e_{31} & e_{31} & e_{33} & 0 & 0 & 0 \end{array} \qquad \underset{2}{x}:(8)\ \begin{array}{cccccc} e_{11} & e_{12} & e_{13} & e_{14} & 0 & 0 \\ 0 & 0 & 0 & 0 & e_{25} & e_{36} \\ 0 & 0 & 0 & 0 & e_{35} & e_{36} \end{array}$$

$$\underset{4}{x}:(4)\ \begin{array}{cccccc} e_{11} & e_{12} & e_{13} & 0 & 0 & 0 \\ 0 & 0 & 0 & 0 & e_{25} & e_{26} \\ 0 & 0 & 0 & 0 & e_{26} & -e_{25} \end{array} \qquad S:(6)\ \begin{array}{cccccc} e_{11} & e_{12} & e_{13} & e_{14} & e_{15} & e_{16} \\ e_{13} & e_{11} & e_{12} & e_{16} & e_{14} & e_{15} \\ e_{12} & e_{13} & e_{11} & e_{15} & e_{16} & e_{14} \end{array}$$

$$\underset{x}{E}:(10)\ \begin{array}{cccccc} 0 & 0 & 0 & 0 & e_{15} & e_{16} \\ e_{21} & e_{22} & e_{23} & e_{24} & 0 & 0 \\ e_{31} & e_{32} & e_{33} & e_{34} & 0 & 0 \end{array} \qquad \underset{z}{E}:(10)\ \begin{array}{cccccc} e_{11} & e_{12} & e_{13} & 0 & 0 & e_{16} \\ e_{21} & e_{22} & e_{23} & 0 & 0 & e_{26} \\ 0 & 0 & 0 & e_{34} & e_{35} & 0 \end{array}$$

$$\underset{x}{S}:(4)\quad \begin{matrix} 0 & e_{12} & -e_{12} & e_{14} & 0 & 0 \\ 0 & 0 & 0 & 0 & e_{25} & e_{26} \\ 0 & 0 & 0 & 0 & -e_{26} & e_{25} \end{matrix} \qquad \underset{z}{S}:(4)\quad \begin{matrix} 0 & 0 & 0 & e_{14} & e_{15} & 0 \\ 0 & 0 & 0 & -e_{15} & e_{14} & 0 \\ e_{31} & -e_{31} & 0 & 0 & 0 & e_{36} \end{matrix} \tag{6.4}$$

From these we get for the remaining 21 classes (see Note H, p. 272):

2.(1):(18) Full table

4.(4); $\underset{z}{E}$:(10)

$$\begin{matrix} e_{11} & e_{12} & e_{13} & 0 & 0 & e_{16} \\ e_{21} & e_{22} & e_{23} & 0 & 0 & e_{26} \\ 0 & 0 & 0 & e_{34} & e_{35} & 0 \end{matrix}$$

5.(3); $\underset{2}{z}$:(7)

$$\begin{matrix} 0 & 0 & 0 & e_{14} & e_{15} & 0 \\ 0 & 0 & 0 & e_{24} & e_{25} & 0 \\ e_{31} & e_{32} & e_{33} & 0 & 0 & e_{36} \end{matrix}$$

7.(6); $\underset{2}{z}$, $\underset{2}{x}$:(3)

$$\begin{matrix} 0 & 0 & 0 & e_{14} & 0 & 0 \\ 0 & 0 & 0 & 0 & e_{25} & 0 \\ 0 & 0 & 0 & 0 & 0 & e_{36} \end{matrix}$$

8.(7); $\underset{2}{z}$, $\underset{x}{E}$:(5)

$$\begin{matrix} 0 & 0 & 0 & 0 & e_{15} & 0 \\ 0 & 0 & 0 & e_{24} & 0 & 0 \\ e_{31} & e_{32} & e_{33} & 0 & 0 & 0 \end{matrix}$$

10.(18); $\underset{3}{z}$, $\underset{2}{x}$:(2)

$$\begin{matrix} e_{11} & -e_{11} & 0 & e_{14} & 0 & 0 \\ 0 & 0 & 0 & 0 & -e_{14} & -e_{11} \\ 0 & 0 & 0 & 0 & 0 & 0 \end{matrix}$$

11.(19); $\underset{3}{z}$, $\underset{x}{E}$:(4)

$$\begin{matrix} 0 & 0 & 0 & 0 & e_{15} & -e_{22} \\ -e_{22} & e_{22} & 0 & e_{15} & 0 & 0 \\ e_{31} & e_{31} & e_{33} & 0 & 0 & 0 \end{matrix}$$

13.(16); $\underset{3}{z}$:(6)

$$\begin{matrix} e_{11} & -e_{11} & 0 & e_{14} & e_{15} & -e_{22} \\ -e_{22} & e_{22} & 0 & e_{15} & -e_{14} & -e_{11} \\ e_{31} & e_{31} & e_{33} & 0 & 0 & 0 \end{matrix}$$

15.(12); $\underset{4}{z}$, $\underset{2}{x}$ }
22.(24); $\underset{6}{z}$, $\underset{2}{x}$ } :(1)

$$\begin{matrix} 0 & 0 & 0 & e_{14} & 0 & 0 \\ 0 & 0 & 0 & 0 & -e_{14} & 0 \\ 0 & 0 & 0 & 0 & 0 & 0 \end{matrix}$$

16.(14); $\underset{4}{z}$, $\underset{x}{E}$ }
23.(26); $\underset{6}{z}$, $\underset{x}{E}$ } :(3)

$$\begin{matrix} 0 & 0 & 0 & 0 & e_{15} & 0 \\ 0 & 0 & 0 & e_{15} & 0 & 0 \\ e_{31} & e_{31} & e_{33} & 0 & 0 & 0 \end{matrix}$$

18.(10); $\underset{4}{z}$ }
25.(23); $\underset{6}{z}$ } :(4)

$$\begin{matrix} 0 & 0 & 0 & e_{14} & e_{15} & 0 \\ 0 & 0 & 0 & e_{15} & -e_{14} & 0 \\ e_{31} & e_{31} & e_{33} & 0 & 0 & 0 \end{matrix}$$

19.(11); $\underset{z}{S}$, $\underset{2}{x}$:(2)

$$\begin{matrix} 0 & 0 & 0 & e_{14} & 0 & 0 \\ 0 & 0 & 0 & 0 & e_{14} & 0 \\ 0 & 0 & 0 & 0 & 0 & e_{36} \end{matrix}$$

20.(9); $\underset{z}{S}$:(4)

$$\begin{matrix} 0 & 0 & 0 & e_{14} & e_{15} & 0 \\ 0 & 0 & 0 & -e_{15} & e_{14} & 0 \\ e_{31} & -e_{31} & 0 & 0 & 0 & e_{36} \end{matrix}$$

26.(22); $\underset{3}{z}$, $\underset{2}{x}$, $\underset{z}{E}$:(1)

$$\begin{matrix} e_{11} & -e_{11} & 0 & 0 & 0 & 0 \\ 0 & 0 & 0 & 0 & 0 & -e_{11} \\ 0 & 0 & 0 & 0 & 0 & 0 \end{matrix}$$

27.(21); $\underset{3}{z}$, $\underset{z}{E}$:(2)

$$\begin{matrix} e_{11} & -e_{11} & 0 & 0 & 0 & -e_{22} \\ -e_{22} & e_{22} & 0 & 0 & 0 & -e_{11} \\ 0 & 0 & 0 & 0 & 0 & 0 \end{matrix}$$

29.(29); $\underset{4}{x}, \underset{4}{y}, (\underset{4}{z})$: All zero

$$\left.\begin{array}{l}30.(31)\ \underset{x}{S}, \underset{y}{S}, (\underset{z}{S})\\ 32.(28)\ \underset{2}{x}, \underset{2}{y}, (\underset{2}{z}), S\end{array}\right\}:(1)$$

$$\begin{array}{cccccc} 0 & 0 & 0 & e_{14} & 0 & 0\\ 0 & 0 & 0 & 0 & e_{14} & 0\\ 0 & 0 & 0 & 0 & 0 & e_{14}\end{array} \qquad (6.5)$$

e^{hij} is a septor in the classes *26*, *27*, *30*, and *32*.

The piezo-magnetic quantity $\tilde{m}_{iB}$ behaves just like e_{iB} with respect to all transformations with $\Delta = +1$. Since $\tilde{m}_{iB}$ is centro-symmetric these are the only transformations that count and consequently we have the following possibilities for $\tilde{m}_{iB}$:

Triclinic, *1*, *2*:(18) Full table.

Monoclinic, *3*, *4*, *5*; $\underset{2}{z}$:(8)

$$\begin{array}{cccccc} 0 & 0 & 0 & \tilde{m}_{14} & \tilde{m}_{15} & 0\\ 0 & 0 & 0 & \tilde{m}_{24} & \tilde{m}_{25} & 0\\ \tilde{m}_{31} & \tilde{m}_{32} & \tilde{m}_{33} & 0 & 0 & \tilde{m}_{36}\end{array}$$

Rhombic, *6*, *7*, *8*; $\underset{2}{z}, \underset{2}{x}$:(3)

$$\begin{array}{cccccc} 0 & 0 & 0 & \tilde{m}_{14} & 0 & 0\\ 0 & 0 & 0 & 0 & \tilde{m}_{25} & 0\\ 0 & 0 & 0 & 0 & 0 & \tilde{m}_{36}\end{array}$$

Trigonal I, *9*, *10*, *11*; $\underset{3}{z}, \underset{2}{x}$:(2)

$$\begin{array}{cccccc} \tilde{m}_{11} & -\tilde{m}_{11} & 0 & \tilde{m}_{14} & 0 & 0\\ 0 & 0 & 0 & 0 & -\tilde{m}_{14} & -\tilde{m}_{11}\\ 0 & 0 & 0 & 0 & 0 & 0\end{array}$$

Trigonal II, *12*, *13*; $\underset{3}{z}$:(6)

$$\begin{array}{cccccc} \tilde{m}_{11} & -\tilde{m}_{11} & 0 & \tilde{m}_{14} & \tilde{m}_{15} & -\tilde{m}_{22}\\ -\tilde{m}_{22} & \tilde{m}_{22} & 0 & \tilde{m}_{15} & -\tilde{m}_{14} & -\tilde{m}_{11}\\ \tilde{m}_{31} & \tilde{m}_{31} & \tilde{m}_{33} & 0 & 0 & 0\end{array}$$

$$\left.\begin{array}{l}\text{Tetragonal I, } \mathit{14, 15, 16, 19};\ \underset{4}{z}, \underset{2}{x}\\ \text{Hexagonal I, } \mathit{21, 22, 23, 26};\ \underset{6}{z}, \underset{2}{x}\end{array}\right\}:(1)\quad \begin{array}{cccccc} 0 & 0 & 0 & \tilde{m}_{14} & 0 & 0\\ 0 & 0 & 0 & 0 & -\tilde{m}_{14} & 0\\ 0 & 0 & 0 & 0 & 0 & 0\end{array}$$

$$\left.\begin{array}{l}\text{Tetragonal II, } \mathit{17, 18, 20};\ \underset{4}{z}\\ \text{Hexagonal II, } \mathit{24, 25, 27};\ \underset{6}{z}\end{array}\right\}:(4)\quad \begin{array}{cccccc} 0 & 0 & 0 & \tilde{m}_{14} & \tilde{m}_{15} & 0\\ 0 & 0 & 0 & \tilde{m}_{15} & -\tilde{m}_{14} & 0\\ \tilde{m}_{31} & \tilde{m}_{31} & \tilde{m}_{33} & 0 & 0 & 0\end{array}$$

Cubic I, *28*, *29*, *30*; $\underset{4}{x}, \underset{4}{y}, (\underset{4}{z})$:(0) All zero

Cubic II, *31*, *32*; $\underset{2}{x}, \underset{2}{y}, (\underset{2}{z}), S$:(1)

$$\begin{array}{cccccc} 0 & 0 & 0 & \tilde{m}_{14} & 0 & 0\\ 0 & 0 & 0 & 0 & \tilde{m}_{14} & 0\\ 0 & 0 & 0 & 0 & 0 & \tilde{m}_{14}\end{array} \qquad (6.6)$$

$\tilde{m}^{hij}$ is a *W*-septor in the classes *31* and *32*.

As an example of piezo-electric constants we take α-quartz (class *10*).

According to Cady the values of e_{iB} and d_{iB} are in electrostatic c.g.s. units and in Giorgi units:

$$\begin{aligned} e_{11} &= -e_{12} = -e_{26} & 4{\cdot}77\times 10^4 \\ e_{14} &= -e_{25} & 1{\cdot}23\times 10^4 \end{aligned}\Bigg\} \frac{\text{dyne}}{10^{-8}c \text{ volt cm.}} = \begin{matrix} 4{\cdot}77\times 10^9/c \\ 1{\cdot}23\times 10^9/c \end{matrix}\Bigg\} \frac{\text{newton}}{\text{volt m.}}$$

$$\begin{aligned} d_{11} &= -d_{12} = -\tfrac{1}{2}d_{26} & 6{\cdot}9\times 10^{-8} \\ d_{14} &= -d_{25} & -2{\cdot}0\times 10^{-8} \end{aligned}\Bigg\} \frac{\text{cm.}}{10^{-8}c \text{ volt}} = \begin{matrix} 6{\cdot}9\times 10^{-2}/c \\ -2{\cdot}0\times 10^{-2}/c \end{matrix}\Bigg\} \frac{\text{m.}}{\text{volt}}.$$

7. Waves in a homogeneous anisotropic medium†

We use cartesian coordinates for the deduction of the wave equation in crystals. The form of the theory is due to Christoffel. The general equation of motion in an anisotropic medium in orthogonal coordinates is (cf. (2.7))

$$\rho[h]\frac{\partial^2 u^k}{\partial t^2} = \partial_j \mathfrak{T}^{jk} \quad (j,k = 1,2,3), \tag{7.1}$$

where u^k is the displacement vector.

For a flat wave, the points with the same displacements lie at any fixed time in parallel planes. If the distance of such a plane from the origin is denoted by s and if the unit vector perpendicular to the planes in the direction of increasing s is n^h we have

$$\partial_j \mathfrak{T}^{jk} = n_j \frac{\partial \mathfrak{T}^{jk}}{\partial s} \quad (j,k = 1,\ 2,\ 3) \tag{7.2}$$

and accordingly

$$\rho\frac{\partial^2 u^k}{\partial t^2} = n_j \frac{\partial \mathfrak{T}^{jk}}{\partial s} \quad (j,k = 1,2,3). \tag{7.3}$$

Since the process is adiabatic we get, using (1.6) and (4.22* d)

$$\begin{aligned} \rho\frac{\partial^2 u^k}{\partial t^2} &= n_j \frac{\partial}{\partial s}(c^{hijk}S_{hi} - e^{ijk}F_i) \\ &= n_j n_h c^{hijk}\frac{\partial^2 u_i}{\partial s^2} - e^{ijk}\partial_j F_i \end{aligned} \quad (h,i,j,k = 1,2,3), \tag{7.4}$$

dropping the subscript η and writing c^{hijk}, e^{ijk} instead of $\mathfrak{c}^{hijk}$, $\mathfrak{e}^{ijk}$ as in §§ 5 and 6. Now if the crystal plate is used as a resonator for frequency control the field F_i depends only on time and is, at each moment, constant in space; hence

$$\rho\frac{\partial^2 u^k}{\partial t^2} = n_j n_h c^{hijk}\frac{\partial^2 u_i}{\partial s^2} \quad (h,i,j,k = 1,2,3). \tag{7.5}$$

† Cf. Christoffel 1877. 1; Voigt 1898. 1; Brillouin 1938. 1; Love 1944. 1; Cady 1946. 2.

Hence, if p^h is the unit vector of u^h and $u^h = up^h$ we get the equation

$$\boxed{\rho c^2 p^k = n_h n_j c^{hijk} p_i} \qquad (h,i,j,k = 1,2,3) \tag{7.6}$$

for p^h, and the linear partial differential equation of the second order

$$\frac{\partial^2 u}{\partial t^2} = c^2 \frac{\partial^2 u}{\partial s^2} \tag{7.7}$$

for u. In these equations c^2 is one of the three eigenvalues of the tensor

$$(1/\rho) n_h n_j e^{hijk}.$$

The general monochromatic solution of (7.7) with a frequency ν has the form

$$u = \underset{1}{u} e^{2\pi i(\nu t - s/\lambda)} + \underset{2}{u} e^{-2\pi i(\nu t - s/\lambda)} + \underset{3}{u} e^{2\pi i(\nu t + s/\lambda)} + \underset{4}{u} e^{-2\pi i(\nu t + s/\lambda)}, \tag{7.8}$$

where $\lambda = c/\nu$ is the wave-length and $\underset{1}{u}$, $\underset{2}{u}$, $\underset{3}{u}$, and $\underset{4}{u}$ are integration constants. This shows that c is the velocity of propagation in the direction of n^h. The first two terms represent a wave moving in the direction of increasing s and the last two a wave in the opposite direction.

If the three eigenvalues are different, a definite unit vector p^h belongs to each of them; and these vectors are mutually perpendicular.

We now consider the possibility of free vibrations in a plate which is cut from a crystal and is bounded by two parallel planes $s = 0$ and $s = a$ perpendicular to n^h. We suppose that the plate has an infinite area and that the same motion is shared by all particles which are at the same distance from the plane $s = 0$. These vibrations are often called 'thickness vibrations'. No stress occurs at the boundary planes, hence, according to (6.2 d)

$$S_{ij} = \frac{\partial u}{\partial s} n_{(j} p_{i)} = d^h_{\cdot ij} F_h \quad (h,i,j,k = 1,2,3). \tag{7.9}$$

$F_i = 0$ for free vibrations, hence $\dfrac{\partial u}{\partial s} = 0$ for $s = 0$ and $s = a$. Introducing these boundary conditions into the equation

$$\frac{\partial u}{\partial s} = \frac{2\pi i}{\lambda}\left(-\underset{1}{u} e^{2\pi i(\nu t - s/\lambda)} + \underset{2}{u} e^{-2\pi i(\nu t - s/\lambda)} + \underset{3}{u} e^{2\pi i(\nu t + s/\lambda)} - \underset{4}{u} e^{-2\pi i(\nu t + s/\lambda)}\right) \tag{7.10}$$

we get

$$\begin{aligned} 0 &= (-\underset{1}{u} + \underset{3}{u}) e^{2\pi i\nu t} + (\underset{2}{u} - \underset{4}{u}) e^{-2\pi i\nu t}, \\ 0 &= e^{2\pi i\nu t}(-\underset{1}{u} e^{-2\pi i a/\lambda} + \underset{3}{u} e^{2\pi i a/\lambda}) + e^{-2\pi i\nu t}(\underset{2}{u} e^{2\pi i a/\lambda} - \underset{4}{u} e^{-2\pi i a/\lambda}), \end{aligned} \tag{7.11}$$

or

$$\underset{1}{u} = \underset{3}{u}, \qquad \underset{2}{u} = \underset{4}{u},$$
$$\sin 2\pi \frac{a}{\lambda} = 0. \tag{7.12}$$

Hence

$$\lambda = \frac{2}{n} a, \qquad \nu = \frac{nc}{2a}, \tag{7.13}$$

where n is an integer, and the solution takes the form

$$u = 2(\underset{1}{u} e^{2\pi i \nu t} + \underset{2}{u} e^{-2\pi i \nu t}) \cos \pi \frac{n}{a} s \tag{7.14}$$

or in another form

$$u = \left(\underset{1}{C} \cos \pi \frac{nc}{a} t + \underset{2}{C} \sin \pi \frac{nc}{a} t \right) \cos \pi \frac{n}{a} s \tag{7.15}$$

with two integration constants. We see from (7.9) that it is possible to generate this free vibration by a small pulsating electric field F_i in the direction of n^h if and only if

$$n_h d^h{}_{ij} n^j p^i \neq 0 \quad (h, i, j = 1, 2, 3) \tag{7.16}$$

and if the frequency of F_i is $nc/2a$. Of course in practice the expression (7.16) must have a sufficiently large value because in our theoretical considerations we have neglected all kinds of resistance and losses of energy.

For general waves we get from (7.1), (1.6), and (4.22* d) for $\partial_j F_i = 0$

$$\rho \frac{\partial^2 u^h}{\partial t^2} = c^{hijk} \partial_i \partial_j u_k \quad (h, i, j, k = 1, 2, 3). \tag{7.17}$$

This is a linear partial differential equation of second order. We use the method developed by Lorentz† for a wave function satisfying an arbitrary linear partial differential equation of arbitrary order and write the solution in the form

$$u^h = \hat{u}^h e^{2\pi i \chi}, \qquad \hat{u}^h \overset{\text{def}}{=} |u^h| \quad (h = 1, 2, 3). \tag{7.18}$$

The factor $e^{2\pi i \chi}$ expresses the wave character. Its partial derivatives are

$$\partial_t e^{2\pi i \chi} = 2\pi i \chi_t e^{2\pi i \chi}, \qquad \chi_t \overset{\text{def}}{=} \partial_t \chi,$$
$$\partial_j e^{2\pi i \chi} = 2\pi i \chi_j e^{2\pi i \chi}, \qquad \chi_j \overset{\text{def}}{=} \partial_j \chi. \tag{7.19}$$

The wave fronts, i.e. the surfaces of equal phase at any fixed time, are given by the equation

$$\chi = \text{constant}. \tag{7.20}$$

χ^h is the vector perpendicular to the wave front and the inverse of the

† Cf. Fokker 1939. 1, 2.

length of this vector is the wave-length, i.e. the distance between two neighbouring wave-fronts which have equal phase. Hence, if n^h is the unit vector of χ^h we have

$$\chi^h = \frac{n^h}{\lambda} \quad (h = 1,2,3). \tag{7.21}$$

The wave-fronts travel in time and at some fixed point the phase is the same as for $t = 0$ if χ has suffered a change 1. Hence χ_t is the frequency ν of the wave.

By differentiation of (7.18) we get

$$\begin{aligned}
\partial_t u^h &= (\partial_t \hat{u}^h + 2\pi i \hat{u}^h \chi_t) e^{2\pi i \chi},\\
\partial_t^2 u^h &= (\partial_t^2 \hat{u}^h - 4\pi^2 \hat{u}^h \chi_t^2 + 4\pi i \chi_t \partial_t \hat{u}^h + 2\pi i \hat{u}^h \partial_t^2 \chi) e^{2\pi i \chi},\\
\partial_j u_k &= (\partial_j \hat{u}_k + 2\pi i \hat{u}_k \chi_j) e^{2\pi i \chi},\\
\partial_i \partial_j u_k &= (\partial_i \partial_j \hat{u}_k - 4\pi^2 \hat{u}_k \chi_i \chi_j + 4\pi i \chi_{(i} \partial_{j)} \hat{u}_k + 2\pi i \hat{u}_k \chi_{ij}) e^{2\pi i \chi}; \; \chi_{ij} \overset{\text{def}}{=} \partial_i \partial_j \chi
\end{aligned} \tag{7.22}$$

$$(h,i,j,k = 1,2,3)$$

and this gives by substitution in (7.17)

$$\begin{aligned}
&(a) \qquad \rho \partial_t^2 \hat{u}^h - 4\pi^2 \rho \hat{u}^h \chi_t^2 = c^{hijk} (\partial_i \partial_j \hat{u}_k - 4\pi^2 \hat{u}_k \chi_i \chi_j)\\
&(b) \qquad 2\rho \chi_t \partial_t \hat{u}^h + \rho \hat{u}^h \partial_t^2 \chi = c^{hijk} (2\chi_{(i} \partial_{j)} \hat{u}_k + \hat{u}_k \chi_{ij})\\
&\qquad\qquad (h,i,j,k = 1,2,3).
\end{aligned} \tag{7.23}$$

We suppose now that the amplitude $\hat{u}^h$, the frequency ν, the wave-length λ, and the direction of n^h undergo only small changes in space and time for displacements of the order of λ in space and for intervals of the order of $1/\nu$ in time. This can be expressed mathematically by the inequalities:

$$\begin{aligned}
\partial_i \hat{u}^h &\ll \hat{u}^h \chi_i, \dagger & \partial_i \partial_j \chi &\ll \chi_i \chi_j,\\
\partial_t \hat{u}^h &\ll \hat{u}^h \chi_t, & \partial_i \partial_t \chi &\ll \chi_i \chi_t,\\
\partial_i \partial_j \hat{u}^h &\ll \chi_i \chi_j \hat{u}^h, & \partial_t^2 \chi &\ll \chi_t \chi_t,\\
\partial_t^2 \hat{u}^h &\ll \chi_t^2 \hat{u}^h, & &\\
\partial_i \partial_t \hat{u}^h &\ll \chi_i \chi_t \hat{u}^h & (h,i,j &= 1,2,3).
\end{aligned} \tag{7.24}$$

If these inequalities hold, (7.23 a) takes the simple form

$$\boxed{\rho c^2 p^h = c^{hijk} n_i n_j p_k} \quad (h,i,j,k = 1,2,3). \tag{7.25}$$

where p^h is the unit vector of u^h and $c = \lambda\nu$. This equation gives the relation between the direction of the displacement and the normal to the wave-front. *It is the same as the relation* (7.6) *for flat waves.*

† $\ll$ means small with respect to.

We now discuss how the energy travels in a wave. The sum of the kinetic and the elastic energy in a volume τ is

$$W = \tfrac{1}{2}\int\limits_{\tau} \rho \dot{u}_i \dot{u}^i \, d\tau + \tfrac{1}{2}\int\limits_{\tau} c^{ijkl}(\partial_j u_i)(\partial_l u_k)\, d\tau. \tag{7.26}$$

Hence

$$\begin{aligned} \dot{W} &= \int\limits_{\tau} \rho \dot{u}_i \ddot{u}^i \, d\tau + \int\limits_{\tau} c^{ijkl}(\partial_j \dot{u}_i)(\partial_l u_k)\, d\tau \\ &= \int\limits_{\tau} \rho \dot{u}_i \ddot{u}^i \, d\tau + \int\limits_{\tau} c^{ijkl}\partial_j(\dot{u}_i \partial_l u_k)\, d\tau - \int\limits_{\tau} c^{ijkl}\dot{u}_i(\partial_j \partial_l u_k)\, d\tau \\ &\qquad\qquad (i,j,k,l = 1,2,3). \end{aligned} \tag{7.27}$$

The first and the third term cancel out because of the relation (7.17) and consequently, using the theorem of Stokes, we get

$$\dot{W} = \int\limits_{\sigma} c^{ijkl}\dot{u}_i(\partial_l u_k) f_j \, d\sigma \quad (i,j,k,l = 1,2,3), \tag{7.28}$$

where f^h is the unit vector perpendicular to the boundary σ. Hence the vector density of the flow of energy is

$$\mathfrak{E}^j = c^{ijkl}\dot{u}_i \partial_l u_k \quad (i,j,k,l = 1,2,3). \tag{7.29}$$

Writing the wave in the real form

$$u^h = Ap^h \cos 2\pi\chi = A^h \cos 2\pi\chi \quad (h = 1,2,3) \tag{7.30}$$

we have

$$\begin{aligned} \partial_l u_i &= (\partial_l A_i)\cos 2\pi\chi - 2\pi A_i \nu \sin 2\pi\chi, \\ \partial_l u_k &= (\partial_l A_k)\cos 2\pi\chi - \frac{2\pi}{\lambda} A_k n_l \sin 2\pi\chi \\ &\qquad\qquad (i,k,l = 1,2,3). \end{aligned} \tag{7.31}$$

From (7.24), the first terms on the right-hand side of these equations can be neglected. Hence

$$\mathfrak{E}^j = 4\pi^2 \frac{\nu}{\lambda} A^2 c^{ijkl} p_i p_k n_l \sin^2 2\pi\chi \quad (i,j,k,l = 1,2,3). \tag{7.32}$$

With the same approximation it follows from (7.26) and (7.31) that the total energy per unit volume is

$$\mathfrak{W} = 2\pi^2 \rho \nu^2 A^2 \sin^2 2\pi\chi + \frac{2\pi^2}{\lambda^2} \rho c^2 A^2 \sin^2 2\pi\chi = 4\pi^2 \rho \nu^2 A^2 \sin^2 2\pi\chi. \tag{7.33}$$

Hence the velocity of the energy flow is

$$v^h = \mathfrak{E}^h/\mathfrak{W} = \frac{1}{\rho c} c^{hijk} p_i n_j p_k \quad (h,i,j,k = 1,2,3). \tag{7.34}$$

Transvecting with n_h we get

$$n_h \underset{e}{v^h} = c \quad (h = 1, 2, 3).$$

Hence the projection of $\underset{e}{v^h}$ on n^h is the wave velocity cn^h.

For a flat wave in the direction of decreasing s we have $\chi = \frac{s}{\lambda} + \nu t$. Hence the mean value of $\mathfrak{E}^j$ and $\mathfrak{W}$ from $t = 0$ to $t = \frac{1}{\nu}$ is

$$\begin{aligned} \overline{\mathfrak{E}}^j &= \nu 4\pi^2 \frac{\nu}{\lambda} A^2 c^{ijkl} p_i p_k n_l \int_0^{1/\nu} \sin^2 2\pi \left(\frac{s}{\lambda} + \nu t\right) dt \\ &= 2\pi^2 \frac{\nu}{\lambda} A^2 c^{ijkl} p_i p_k n_l, \\ \overline{\mathfrak{W}} &= 2\pi^2 \rho \nu^2 A^2 \qquad (i, j, k, l = 1, 2, 3). \end{aligned} \tag{7.35}$$

It is interesting to work out $\mathfrak{E}^j$ and $\mathfrak{W}$ for the case of a free vibration of a plate cut perpendicular to n^h. Here we have from (7.15), taking the null point of the time conveniently

$$u^h = Cp^h \cos n\pi \frac{s}{a} \cos n\pi \frac{ct}{a} = Cp^h \cos 2\pi \frac{s}{\lambda} \cos 2\pi\nu t \qquad (h = 1, 2, 3) \tag{7.36}$$

and consequently

$$\partial_t u^h = -2\pi\nu C p^h \cos 2\pi \frac{s}{\lambda} \sin 2\pi\nu t, \tag{7.37}$$

$$\partial_j u^h = -\frac{2\pi}{\lambda} C p^h n_j \sin 2\pi \frac{s}{\lambda} \cos 2\pi\nu t \quad (h, j = 1, 2, 3). \tag{7.38}$$

Hence

$$\begin{aligned} \mathfrak{E}^j &= \frac{\pi^2 \nu}{\lambda} c^{ijkl} p_i p_k n_l C^2 \sin 4\pi \frac{s}{\lambda} \sin 4\pi\nu t \quad (i, j, k, l = 1, 2, 3), \\ \mathfrak{W} &= \pi^2 \nu^2 \rho C^2 \left(1 - \cos 4\pi \frac{s}{\lambda} \cos 4\pi\nu t\right). \end{aligned} \tag{7.39}$$

The mean value of $\mathfrak{E}^j$ between $t = 0$ and $t = 1/\nu$ is zero in this case because the standing wave is a composition of two travelling waves with the same magnitude and opposite directions. But the mean value of $\mathfrak{W}$ is $\pi^2 \rho \nu^2 C^2$.

We call the density

$$\overline{\mathfrak{E}} \overset{\text{def}}{=} c^{ijkl} p_i n_j p_k n_l \quad (i, j, k, l = 1, 2, 3) \tag{7.40}$$

the *energy function* belonging to two arbitrary directions n^h and p^h.

From (7.25) we see that this function has an extreme value for variations of p^h if n^h is left constant. The direction of p^h belonging to such an extreme value is the direction of one of the possible displacement vectors for a wave-front perpendicular to n^h and $\sqrt{(\overline{\mathfrak{E}}/\rho)}$ is the velocity of the wave belonging to this displacement.

Now we require extreme values of $\overline{\mathfrak{E}}$ for independent variations of n^h and p^h. Since these vectors are unit vectors we have $n_h\,dn^h = 0$, $p_h\,dp^h = 0$. Hence the variational equations are

$$\begin{gathered} d\overline{\mathfrak{E}} = 2c^{ijkl}\,dp_i\,n_j\,p_k\,n_l+2c^{ijkl}p_i\,dn_j\,p_k\,n_l = 0, \\ n_h\,dn^h = 0, \qquad p_h\,dp^h = 0 \\ (h,i,j,k,l = 1,2,3). \end{gathered} \tag{7.41}$$

These equations are equivalent to the equations

$$\begin{aligned} d\overline{\mathfrak{E}} = 2(c^{ijkl}n_j\,p_k\,n_l - c^{hjkl}p_h\,n_j\,p_k\,n_l\,p^i)\,dp_i + \\ +2(c^{ijkl}p_i\,p_k\,n_l - c^{ihkl}p_i\,n_h\,p_k\,p_l\,n^j)\,dn_j = 0 \\ (i,j,k,l = 1,2,3) \end{aligned} \tag{7.42}$$

without additional conditions. Hence $\overline{\mathfrak{E}}$ is extreme if and only if the two equations

$$\boxed{\begin{aligned} &(a) \quad \overline{\mathfrak{E}}p^i = c^{ijkl}n_j\,p_k\,n_l \\ &(b) \quad \overline{\mathfrak{E}}n^j = c^{ijkl}p_i\,p_k\,n_l \end{aligned}} \qquad (i,j,k,l = 1,2,3) \tag{7.43}$$

are satisfied. The first equation is identical with (7.25), hence p^h is one of the directions of the displacement vector belonging to a wave-front perpendicular to n^h, and $\overline{\mathfrak{E}} = \rho c^2$. From (7.32) the second equation expresses the fact that the flow of energy is perpendicular to the wave-front.

If two directions n^h and p^h satisfy both equations (7.43) we call them *reciprocal*. If two directions are reciprocal, then belonging to a wave-front perpendicular to one of them there is an energy flow in that same direction and a displacement vector in the other direction and vice versa. $\overline{\mathfrak{E}}$ is extreme for free variations of n^h and p^h and the extreme value is $\overline{\mathfrak{E}} = \rho c^2$.

If $n^h = p^h$ the direction of n^h is called *self-reciprocal*. The necessary and sufficient condition is that

$$c^{ijkl}n_i\,n_j\,n_k = \rho c^2 n^l \quad (i,j,k,l = 1,2,3). \tag{7.44}$$

If $n^h = \alpha \underset{1}{i^h} + \beta \underset{2}{i^h} + \gamma \underset{3}{i^h}$ is substituted into (7.44) we get three equations of the third degree in α, β, γ, and c with the coefficients (in abbreviated notation)

$$\begin{array}{rlll}
\alpha^3: & c_{11} & c_{16} & c_{15} \\
\beta^3: & c_{26} & c_{22} & c_{24} \\
\gamma^3: & c_{35} & c_{34} & c_{33} \\
\alpha^2\beta: & 3c_{16} & c_{12}+2c_{66} & c_{14}+2c_{56} \\
\beta^2\gamma: & c_{25}+2c_{64} & 3c_{24} & c_{23}+2c_{44} \\
\gamma^2\alpha: & c_{31}+2c_{55} & c_{36}+2c_{45} & 3c_{35} \\
\alpha\beta^2: & c_{12}+2c_{66} & 3c_{26} & c_{25}+2c_{64} \\
\beta\gamma^2: & c_{36}+2c_{45} & c_{23}+2c_{44} & 3c_{34} \\
\gamma\alpha^2: & 3c_{15} & c_{14}+2c_{56} & c_{31}+2c_{55} \\
\alpha\beta\gamma: & 4c_{56}+2c_{14} & 4c_{64}+2c_{25} & 4c_{45}+2c_{36} \\
\alpha: & -\rho c^2 & 0 & 0 \\
\beta: & 0 & -\rho c^2 & 0 \\
\gamma: & 0 & 0 & -\rho c^2
\end{array} \tag{7.45}$$

and one equation $\alpha^2+\beta^2+\gamma^2 = 1$. As there are four equations and four unknowns, in general, solutions exist. The necessary and sufficient conditions that all solutions of the fourth equation satisfy the other equations are

$$c_{11} = c_{22} = c_{33} = c_{23}+2c_{44} = c_{31}+2c_{55} = c_{12}+2c_{66},$$
$$c_{15} = c_{16} = c_{24} = c_{26} = c_{34} = c_{35} = 0, \quad c_{25}+2c_{64} = 0, \quad c_{36}+2c_{45} = 0,$$
$$c_{14}+2c_{56} = 0. \tag{7.46}$$

Hence the table of constants must have the form

$$\begin{array}{cccccc}
c_{11} & c_{11}-2c_{66} & c_{11}-2c_{55} & -2c_{56} & 0 & 0 \\
 & c_{11} & c_{11}-2c_{44} & 0 & -2c_{64} & 0 \\
 & & c_{11} & 0 & 0 & -2c_{45} \\
 & & & c_{44} & c_{45} & c_{46} \\
 & & & & c_{55} & c_{56} \\
 & & & & & c_{66}
\end{array} \tag{7.47}$$

If $c_{45} = c_{56} = c_{64} = 0$ the medium belongs to the rhombic system and is called a *Green medium.*† In a Green medium, for every form of the wave-front, one of the possible displacement vectors is perpendicular to the front and the other two lie in the tangent plane. $\rho c^2 = c_{11}$ for the longitudinal waves. For the two transverse waves the values of ρc^2 are the two other solutions of the equation (cf. (7.6))

$$\begin{vmatrix}
\alpha^2c_{11}+\beta^2c_{66}+\gamma^2c_{55}-\rho c^2 & (c_{11}-c_{66})\alpha\beta & (c_{11}-c_{55})\alpha\gamma \\
(c_{11}-c_{66})\beta\alpha & \alpha^2c_{66}+\beta^2c_{11}+\gamma^2c_{44}-\rho c^2 & (c_{11}-c_{44})\beta\gamma \\
(c_{11}-c_{55})\gamma\alpha & (c_{11}-c_{44})\gamma\beta & \alpha^2c_{55}+\beta^2c_{44}+\gamma^2c_{11}-\rho c^2
\end{vmatrix} = 0. \tag{7.48}$$

† Cf. Tedone 1907. 1, p. 119.

By an elementary transformation of this determinant we see that these are the solutions of the equation

$$\begin{vmatrix} c_{66}+\gamma^2(-c_{66}+c_{44})-\rho c^2 & (c_{55}-c_{44})\alpha\gamma \\ (c_{66}-c_{44})\beta\gamma & c_{55}+\beta^2(-c_{55}+c_{44})-\rho c^2 \end{vmatrix} = 0, \quad (7.49)$$

or, in another form,

$$\frac{\alpha^2}{c_{44}-\rho c^2} = \frac{\beta^2}{c_{55}-\rho c^2} = \frac{\gamma^2}{c_{66}-\rho c^2}. \quad (7.50)$$

These are the well-known equations of Fresnel. Hence in a Green medium the transverse waves satisfy the equations of Fresnel. An isotropic medium is a special case of a Green medium with $c_{44} = c_{55} = c_{66}$ (cf. (5.22)). Hence in an isotropic medium all longitudinal waves have $\rho c^2 = c_{11}$ and all transverse waves have $\rho c^2 = \frac{1}{2}(c_{11}-c_{12})$.

In a crystal of the trigonal system I, for instance α-quartz (*10*) or tourmaline (*11*), the equations (7.45) take the form (cf. (5.22))

$$c_{11}\alpha+(c_{31}+2c_{44}-c_{11})\alpha\gamma^2+6c_{14}\alpha\beta\gamma = \rho c^2\alpha;$$
$$c_{11}\beta+(c_{31}+2c_{44}-c_{11})\beta\gamma^2+3c_{14}\gamma(\alpha^2-\beta^2) = \rho c^2\beta;$$
$$-c_{14}\beta^3+(c_{33}-c_{31}-2c_{44})\gamma^3+3c_{14}\alpha^2\beta+(c_{31}+2c_{44})\gamma = \rho c^2\gamma;$$
$$\alpha^2+\beta^2+\gamma^2 = 1, \quad (7.51)$$

from which we see that the x-axis is self-reciprocal with $\rho c^2 = c_{11}$ and the z-axis self-reciprocal with $\rho c^2 = c_{33}$.† There are other solutions.

In a crystal of the cubic system the equations take the form (cf. (5.22))

$$(c_{11}-c_{12}-2c_{44})\alpha^3 = (\rho c^2-c_{12}-2c_{44})\alpha,$$
$$(c_{11}-c_{12}-2c_{44})\beta^3 = (\rho c^2-c_{12}-2c_{44})\beta, \quad (7.52)$$
$$(c_{11}-c_{12}-2c_{44})\gamma^3 = (\rho c^2-c_{12}-2c_{44})\gamma,$$

with the solutions

$$\left.\begin{aligned} &\alpha = 1, \quad \beta = 0, \quad \gamma = 0, \quad \rho c^2 = c_{11}, \\ &\alpha = \sqrt{\tfrac{1}{2}}, \quad \beta = \sqrt{\tfrac{1}{2}}, \quad \gamma = 0, \quad \rho c^2 = \tfrac{1}{2}(c_{11}+c_{12}+2c_{44}), \end{aligned}\right\} \text{cycl. } \alpha, \beta, \gamma,$$
$$\alpha = \sqrt{\tfrac{1}{3}}, \quad \beta = \sqrt{\tfrac{1}{3}}, \quad \gamma = \sqrt{\tfrac{1}{3}}, \quad \rho c^2 = \tfrac{1}{3}(c_{11}+2c_{12}+4c_{44}). \quad (7.53)$$

The conditions for the self reciprocity of the x-, y-, and z-axis or for the reciprocity of y, x; z, x; and x, y are

$$\begin{aligned} &x: \quad c_{15} = c_{16} = 0 && y, z: \quad c_{42} = c_{43} = c_{45} = c_{46} = 0 \\ &y: \quad c_{26} = c_{24} = 0 && z, x: \quad c_{53} = c_{51} = c_{56} = c_{54} = 0 \\ &z: \quad c_{34} = c_{35} = 0 && x, y: \quad c_{61} = c_{62} = c_{64} = c_{65} = 0. \end{aligned}$$

(7.54)

† Cf. Koga 1933. 2.

From the tables (5.22) we see that the following reciprocities exist:

	Self-reciprocal	*Reciprocal*	
Triclinic, *1, 2* . . .	—	—	
Monoclinic, *3, 4, 5* . .	z	—	
Rhombic, *6, 7, 8* . .	x, y, z	$y, z; z, x; x, y$	
Trigonal I, *9, 10, 11* .	x, z	z, x	(7.55)
Trigonal II, *12, 13* . .	z	—	
Tetragonal I, *14, 15, 16, 19*	x, y, z	$y, z; z, x; x, y$	
Tetragonal II, *17, 18, 20* .	z	$y, z; z, x$	
Hexagonal, *21–27* . .	x, y, z	$y, z; z, x; x, y$	
Cubic, *28–32* . . .	x, y, z	$y, z; z, x; x, y$	

If in a general homogeneous anisotropic medium we write

$$N^{ik} = (1/\rho)c^{hijk}n_h n_j,$$

then to every direction n^h there belong the values of c^2 which satisfy the equation

$$\begin{vmatrix} N_{11}-c^2 & N_{12} & N_{13} \\ N_{21} & N_{22}-c^2 & N_{23} \\ N_{31} & N_{32} & N_{33}-c^2 \end{vmatrix} = 0; \quad N_{ij} = N_{ji} \quad (i,j = 1,2,3). \tag{7.56}$$

This equation always has three positive solutions. If the solutions are different, one value of p^h belongs to each of them. If two solutions are equal, a pencil of possible values of p^h all of which are perpendicular to the p^h of the third solution, belongs to these two. If all solutions are equal, N_{ij} is equal to g_{ij} to within a scalar factor and p^h can have every direction A plane wave, with the velocity c and planes perpendicular to n^h, which starts from a plane through the origin, reaches the plane through the point cn^h or $-cn^h$ after unit time. The points cn^h and $-cn^h$ fill the *wave velocity surface*, which consists of three sheets. Hence the parametric equation in point coordinates of the wave velocity surface is

$$x^h = \pm\sqrt{\{(1/\rho)c^{lijk}n_l p_i n_j p_k\}}n^h \quad (h,i,j,k,l = 1,2,3) \tag{7.57}$$

in which p_i is one of the possible displacement unit vectors belonging to n_h. The covariant vector perpendicular to n^h which has its first plane through the origin and its second plane through the point cn^h is

$$w_h = \frac{\pm 1}{\sqrt{\{(1/\rho)c^{lijk}n_l p_i n_j p_k\}}}n_h \quad (h,i,j,k,l = 1,2,3). \tag{7.58}$$

Hence (7.58) is, in plane coordinates, the parametric equation of the surface enveloped by these planes. This surface is called the *wave surface* belonging to the origin. It also consists of three sheets. In order to find the parametric equation of the wave surface in point coordinates we require the radius vector r^h of the point of this surface where the tangent plane is perpendicular to n^h. This point is the intersection of the plane $\frac{1}{c}n_h$ and its 'neighbouring' tangent planes. Hence

$$\frac{1}{c}r^h n_h = 1, \quad r^h d\left(\frac{1}{c}n_h\right) = 0, \quad n^h\,dn_h = 0 \qquad (h = 1,2,3), \quad (7.59)$$

and consequently

$$-\frac{1}{c^2}r^i n_i\left(\frac{\partial c}{\partial n_h}+\frac{\partial c}{\partial p_j}\frac{\partial p_j}{\partial n_h}\right)+\frac{1}{c}r^h = \lambda n^h \quad (h,i,j = 1,2,3) \qquad (7.60)$$

where λ is a suitable scalar factor. Now we know that c is extreme for variations of p^h and constant n^h, hence $\frac{\partial c}{\partial p_j} = 0$ and

$$-\frac{1}{c}\frac{\partial c}{\partial n_h}+\frac{1}{c}r^h = -\frac{1}{c}\cdot\tfrac{1}{2}c^{-1}\cdot(2/\rho)c^{hijk}p_i n_j p_k+\frac{1}{c}r^h = \lambda n^h$$

$$(h,i,j,k = 1,2,3). \qquad (7.61)$$

Transvection with n_h gives $\qquad \lambda = 0, \qquad (7.62)$

and consequently

$$r^h = \frac{\partial c}{\partial n_h} = \frac{1}{\rho c}c^{hijk}p_i n_j p_k = \underset{e}{v^h} \quad (h,i,j,k = 1,2,3) \quad (\text{cf. } (7.34)). \qquad (7.63)$$

Hence r^h is equal to the velocity vector of the energy flow. This shows why the projection of $\underset{e}{v^h}$ on n^h is exactly the wave velocity vector cn^h.

We collect results in the statement:

The wave velocity surface is the locus of the end-points of the wave velocity vectors and the wave surface is the locus of the end-points of the energy velocity vectors.

In a Green medium that sheet of the wave surface which belongs to the longitudinal waves coincides with one of the sheets of the wave velocity surface. In an isotropic medium the two surfaces coincide and consist of two spheres, one for the longitudinal waves and one for the transverse ones.

8. The quartz resonator†

Crystals are used for stabilization of frequency in an electric circuit. Plates of α-quartz (trigonal *10*) are often connected with the circuit in

† Cf. Cady 1946. 2; Mason 1947. 4.

such a way that the electric field is perpendicular to the plate. A plate perpendicular to the x-axis is called an *x-cut*, and so on. We see from the table (6.5) that the only non-zero constants are $d^1_{.11}$, $d^1_{.22} = -d^1_{.11}$, $d^1_{.23}$, $d^2_{.31} = -d^1_{.23}$, and $d^2_{.12} = -d^1_{.11}$. Accordingly the piezo-electric effect is expressed by the equations (cf. (6.1 d)):

$$\begin{aligned} S_{11} &= d^1_{.11} F_1 = d_{11} F_1, \\ S_{22} &= -d^1_{.11} F_1 = -d_{11} F_1, \\ S_{33} &= 0, \\ S_{23} &= d^1_{.23} F_1 = \tfrac{1}{2} d_{14} F_1, \\ S_{31} &= -d^1_{.23} F_2 = -\tfrac{1}{2} d_{14} F_2, \\ S_{12} &= -d^1_{.11} F_2 = -d_{11} F_2. \end{aligned} \tag{8.1}$$

Hence a field parallel to the z-axis cannot produce a piezo-electric effect, a field in the x-direction gives rise to the strains S_{11}, S_{22}, and S_{23}, and a field in the y-direction to the strains S_{31} and S_{12}.

For an x-cut the equation for p^h is

$$c^{1i1k} p_i = \rho c^2 p^k \quad (i, k = 1, 2, 3) \tag{8.2}$$

where ρc^2 is one of the solutions of

$$\begin{vmatrix} c_{11} - \rho c^2 & 0 & 0 \\ 0 & c_{66} - \rho c^2 & c_{56} \\ 0 & c_{56} & c_{55} - \rho c^2 \end{vmatrix} = 0. \tag{8.3}$$

The first solution is $\underset{x1}{\rho c^2} = c_{11}$ belonging to $\underset{x1}{p^h} = \underset{1}{i^h}$. The unit vectors

$$\left. \begin{aligned} \underset{x2}{p^h} &= \underset{2}{i^h} \cos \underset{0}{\theta} - \underset{3}{i^h} \sin \underset{0}{\theta} \\ \underset{x3}{p^h} &= \underset{2}{i^h} \sin \underset{0}{\theta} + \underset{3}{i^h} \cos \underset{0}{\theta} \end{aligned} \right\} \quad (h = 1, 2, 3) \tag{8.4}$$

belong to the other solutions $\underset{x2}{\rho c^2}$ and $\underset{x3}{\rho c^2}$ (Fig. 27).

From (8.3) and (8.4) it follows that

$$\tan 2\underset{0}{\theta} = \frac{2c_{56}}{c_{55} - c_{66}}, \tag{8.5}$$

hence, according to (5.23),

$$\underset{0}{\theta} = \begin{cases} 30^\circ\ 55' \text{ (Voigt)}, \\ 31^\circ\ 33' \text{ (Cady)}, \\ 31^\circ\ 30' \text{ (Mason)}. \end{cases} \tag{8.6}$$

For a y-cut the equation is

$$\begin{vmatrix} c_{66} - \rho c^2 & 0 & 0 \\ 0 & c_{22} - \rho c^2 & c_{24} \\ 0 & c_{24} & c_{44} - \rho c^2 \end{vmatrix} = 0; \; c_{22} = c_{11}; \; c_{24} = -c_{14}. \tag{8.7}$$

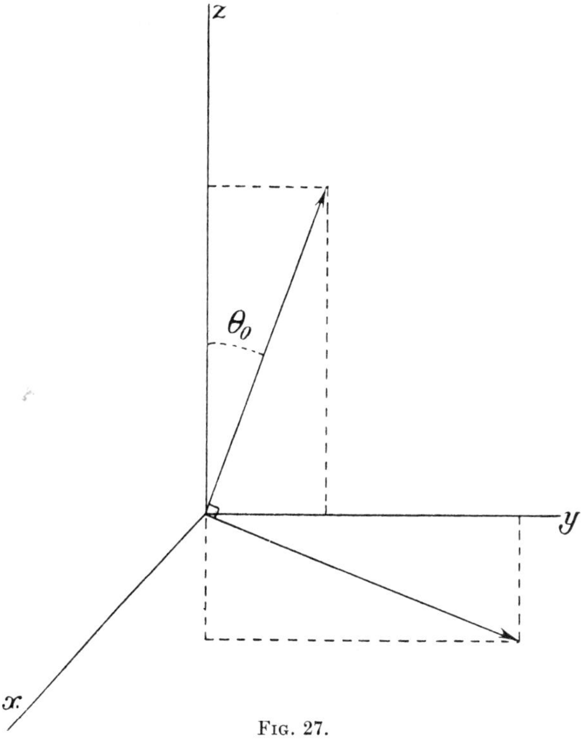

FIG. 27.

with the solutions $\rho \underset{y1}{c^2} = c_{66}$, $\rho \underset{y2}{c^2}$, $\rho \underset{y3}{c^2}$ and the unit vectors

$$\begin{aligned} \underset{y1}{p^h} &= \underset{1}{i^h}, \\ \underset{y2}{p^h} &= \underset{2}{i^h} \cos\phi - \underset{3}{i^h} \sin\phi, \qquad \tan 2\phi = \frac{2c_{24}}{c_{44}-c_{22}}. \\ \underset{y3}{p^h} &= \underset{2}{i^h} \sin\phi + \underset{3}{i^h} \cos\phi, \qquad (h = 1,2,3). \end{aligned} \tag{8.8}$$

For a z-cut the equation is

$$\begin{vmatrix} c_{55}-\rho c^2 & 0 & 0 \\ 0 & c_{44}-\rho c^2 & 0 \\ 0 & 0 & c_{33}-\rho c^2 \end{vmatrix} = 0; \; c_{55} = c_{44}, \tag{8.9}$$

with the solution $\rho \underset{z1}{c^2} = \rho \underset{z2}{c^2} = c_{44}$, $\rho \underset{z3}{c^2} = c_{33}$ and the unit vectors

$$\left.\begin{aligned} \underset{z1}{p^h} &= \underset{1}{i^h} \cos\psi - \underset{2}{i^h} \sin\psi \\ \underset{z2}{p^h} &= \underset{1}{i^h} \sin\psi + \underset{2}{i^h} \cos\psi \end{aligned}\right\} \psi \text{ arbitrary}$$

$$\underset{z3}{p^h} = \underset{3}{i^h} \qquad (h = 1,2,3). \tag{8.10}$$

The waves belonging to these directions of p^h are

x-cut

$$u^h_{x1} = i^h_1 \cos\frac{n\pi}{a}x \, . \, f(t)\dagger$$

$$u^h_{x2} = (i^h_2 \cos\theta_0 - i^h_3 \sin\theta_0)\cos\frac{n\pi}{a}x \, . \, f(t)$$

$$u^h_{x3} = (i^h_2 \sin\theta_0 + i^h_3 \cos\theta_0)\cos\frac{n\pi}{a}x \, . \, f(t)$$

y-cut

$$u^h_{y1} = i^h_1 \cos\frac{n\pi}{a}y \, . \, f(t)$$

$$u^h_{y2} = (i^h_2 \cos\phi - i^h_3 \sin\phi)\cos\frac{n\pi}{a}y \, . \, f(t)$$

$$u^h_{y3} = (i^h_2 \sin\phi + i^h_3 \cos\phi)\cos\frac{n\pi}{a}y \, . \, f(t)$$

z-cut

$$u^h_{z1} = (i^h_1 \cos\psi - i^h_2 \sin\psi)\cos\frac{n\pi}{a}z \, . \, f(t)$$

$$u^h_{z2} = (i^h_1 \sin\psi + i^h_2 \cos\psi)\cos\frac{n\pi}{a}z \, . \, f(t)$$

$$u^h_{z3} = i^h_3 \cos\frac{n\pi}{a}z \, . \, f(t) \qquad (h = 1, 2, 3). \tag{8.11}$$

By differentiation of (8.11) with respect to x, y, and z we find in the end that the only components of S_{ij} occurring in these waves are

	x-cut	*y-cut*	*z-cut*
1st wave	S_{11}	S_{12}	S_{31}, S_{33}
2nd wave	S_{12}, S_{31}	S_{22}, S_{23}	S_{31}, S_{33}
3rd wave	S_{12}, S_{31}	S_{22}, S_{23}	S_{33}

(8.12)

From (8.1) we see that in an x-cut only the first wave can be generated piezo-electrically. The lowest harmonic requires a field with the frequency $\frac{1}{2a} c_{x1}$. In a y-cut only the first wave can be generated and its lowest harmonic requires a field with the frequency $\frac{1}{2a} c_{y1}$. For the direction of the energy flow we find

x-cut: i^h_1 for all waves

y-cut: 1st wave: $c_{66}\, i^h_2 + c_{56}\, i^h_3$

2nd wave: $(c_{22}\cos^2\phi + c_{44}\sin^2\phi - 2c_{24}\sin\phi\cos\phi)i^h_2 +$
$+(c_{24}\cos^2\phi - (c_{44}+c_{33})\sin\phi\cos\phi)i^h_3$

3rd wave: $(c_{22}\sin^2\phi + c_{44}\cos^2\phi + 2c_{24}\sin\phi\cos\phi)i^h_2 +$
$+(c_{24}\sin^2\phi + (c_{44}+c_{33})\sin\phi\cos\phi)i^h_3$

† In (8.11) the periodic functions of t denoted by $f(t)$ are not all equal.

$$z\text{-cut: 1st wave: } -(c_{56}+c_{14})\sin\psi\cos\psi\,\underset{1}{i}^h+(c_{56}\cos^2\psi+c_{24}\sin^2\psi)\underset{2}{i}^h+$$
$$+(c_{55}\cos^2\psi+c_{44}\sin^2\psi)\underset{3}{i}^h = -c_{14}\underset{1}{i}^h\sin 2\psi+c_{14}\underset{2}{i}^h\cos 2\psi+c_{44}\underset{3}{i}^h$$
$$\text{2nd wave: } (c_{56}+c_{14})\sin\psi\cos\psi\,\underset{1}{i}^h+(c_{56}\sin^2\psi+c_{24}\cos^2\psi)\underset{2}{i}^h+$$
$$+(c_{55}\sin^2\psi+c_{44}\cos^2\psi)\underset{3}{i}^h = c_{14}\underset{1}{i}^h\sin 2\psi-c_{14}\underset{2}{i}^h\cos 2\psi+c_{44}\underset{3}{i}^h$$
$$\text{3rd wave: } \underset{3}{i}^h \qquad (h = 1,2,3). \tag{8.13}$$

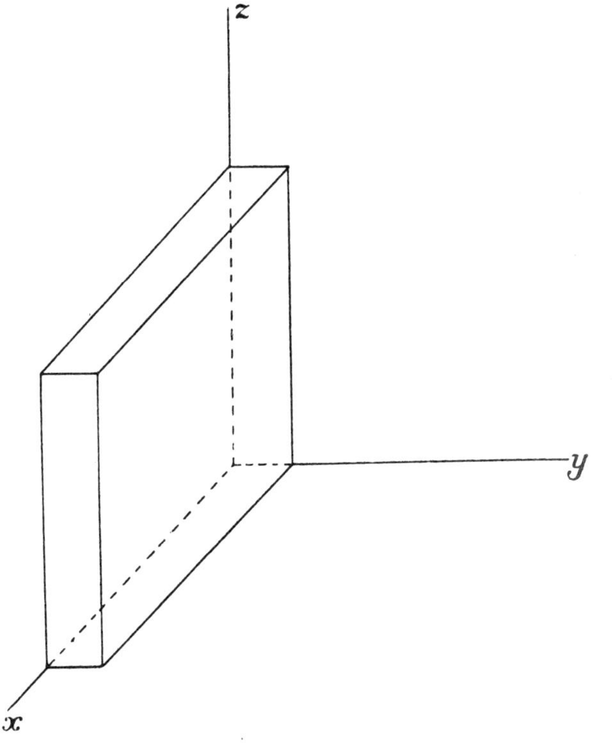

FIG. 28.

Since the energy flow for an x-cut is always perpendicular to the plate, it follows *that* $\underset{x2}{p}^h$ *and* $\underset{x3}{p}^h$ *are each reciprocal to* $\underset{1}{i}^h$. For a y-cut the energy flow is always in the yz-plane and for a z-cut it is perpendicular to the plate for the third wave.

Up till now we have been supposing that the plates have had an infinite area. But for practical use a plate has to be finite. Consider for instance, a rectangular y-cut with side faces parallel to the x- and z-axis (Fig. 28). If we try to generate the first wave by means of a field F_i parallel to the y-axis it is clear that we shall never succeed exactly, because in the first wave there is a strain S_{12} all over the plate. But if the plate is entirely free, S_{12} has to be zero in the side faces. Now a plate is never entirely free; there is the surrounding air in which sound waves can be generated, or the plate may be clamped and waves will be generated in the clamping material. This means that there will always be a possibility of stresses $\mathfrak{T}^{hi}$ in the side faces. Now we know that

$$\begin{aligned} \mathfrak{T}^{31} &= c_{55} S_{31} + c_{56} S_{12}, \\ \mathfrak{T}^{12} &= c_{56} S_{31} + c_{66} S_{12}, \end{aligned} \tag{8.14}$$

because c_{51}, c_{52}, c_{53}, c_{54}, c_{61}, c_{62}, c_{63}, and c_{64} vanish, and accordingly S_{12} induces a strain S_{31} in the side faces. This effect is called the *mechanical coupling* of S_{12} and S_{31} and it is due to the non-vanishing constant c_{56}. This has led to the name *coupling constants* for the constants which do not occur in the main diagonal. Now the whole process is highly complicated and could only be determined exactly by rather difficult integrations which take all the boundary conditions into account. But, since the plate is thin, we see without integrations that the solution will be a vibration consisting for the most part, of a first wave, and for the rest, of supplementary waves due to the boundary disturbances, adjusted in such a way that the correct values of S_{31} and S_{12} in the side faces result. It is also possible to detect the most important of these supplementary waves without integration. The plate is a y-cut, but it can also be looked upon as a rather thick x-cut or z-cut. Now the second and the third waves in an x-cut can both be generated by S_{12} and S_{31}. Since the plate is thick its main frequencies are far too small to be excited by a field with a frequency $\frac{1}{2a}\,c_{y1}$, but some very high harmonic may lie sufficiently near to this frequency to be excited. It is highly probable that this occurs because high harmonics of a low main frequency are very near to each other. The same can be said for the plate looked upon as a very thick z-cut. Here S_{12} has no effect, but the induced strain S_{31} will generate some higher harmonic wave in the xy-plane. These parasitic higher harmonics are very troublesome. If the plate has to be used as a resonator we wish to have one resonance frequency with no near neighbours. But since parasitic waves are possible, several resonance frequencies may exist near to each other, and this makes the resonator unfit for practical use because it may jump from one frequency to another while in use. There are many ways of diminishing the influence of parasitic waves, for instance, rotating the plate in its own plane, chosing length and breadth in a suitable way, rounding the edges, clamping in the right way, etc. All these things have to be found out by experiments, because the processes are too complicated for exact mathematical treatment. But there is one source of trouble that can be dealt with mathematically and that is the coupling constant c_{56}.

Lack, Willard, and Fair† gave two examples of very bad frequency curves of y-cuts which were due to parasitical waves. They proposed to use a plate that could be got from a y-cut by rotating it about the x-axis through an angle θ from z to y and they proved that for the new $x'y'z'$ system arising from that rotation the constant $c_{2'5'}$ vanishes if $\theta = 30^\circ\, 55'$ or $-59^\circ\, 5'$ (using the values of Voigt). But this gives exactly the directions of $\underset{x1}{p^h}$ and $\underset{x2}{p^h}$! Now, as van Dyl‡ pointed out, this is not accidental because it follows immediately from (8.3, 4) that $\underset{x1}{p^h}$ and $\underset{x2}{p^h}$ are the principal axes of the ellipse

$$c_{66}\, y^2 + 2c_{56}\, yz + c_{55}\, z^2 = 1, \tag{8.15}$$

and that the term with $y'z'$ vanishes if these principal axes are taken as y' and z'. For these new axes the table of constants takes the form

$$\begin{matrix} c_{1'1'} & c_{1'2'} & c_{1'3'} & c_{1'4'} & 0 & 0 \\ & c_{2'2'} & c_{2'3'} & c_{2'4'} & 0 & 0 \\ & & c_{3'3'} & c_{3'4'} & 0 & 0 \\ & & & c_{4'4'} & 0 & 0 \\ & & & & c_{5'5'} & 0 \\ & & & & & c_{6'6'} \end{matrix} \tag{8.16}$$

the constants $c_{1'5'}$, $c_{1'6'}$, $c_{2'5'}$, $c_{2'6'}$, $c_{3'5'}$, $c_{3'6'}$, $c_{4'5'}$, $c_{4'6'}$ (but *not* $c_{3'4'}$!) remaining zero. Of course the equalities existing between the c_{AB} of α-quartz do not hold for the $c_{A'B'}$. Using this coordinate system we get the equations for ρc^2

$$\begin{gathered} \text{x'-cut ($= x$-cut)} \qquad\qquad\qquad\qquad \text{y'-cut} \\ \begin{vmatrix} c_{1'1'}-\rho c^2 & 0 & 0 \\ 0 & c_{6'6'}-\rho c^2 & 0 \\ 0 & 0 & c_{5'5'}-\rho c^2 \end{vmatrix} = 0 \qquad \begin{vmatrix} c_{6'6'}-\rho c^2 & 0 & 0 \\ 0 & c_{2'2'}-\rho c^2 & c_{2'4'} \\ 0 & c_{2'4'} & c_{4'4'}-\rho c^2 \end{vmatrix} = 0 \\ \text{z'-cut} \\ \begin{vmatrix} c_{5'5'}-\rho c^2 & 0 & 0 \\ 0 & c_{4'4'}-\rho c^2 & c_{3'4'} \\ 0 & c_{3'4'} & c_{3'3'}-\rho c^2 \end{vmatrix} = 0. \end{gathered} \tag{8.17}$$

† 1934. 2. ‡ 1936. 1

The directions of u^h are

	x'-*cut*	y'-*cut*	z'-*cut*
1st wave	$i^h_{1'}$	$i^h_{1'}$	$i^h_{1'}$
2nd wave	$i^h_{2'}$	$i^h_{2'}\cos\phi' - i^h_{3'}\sin\phi'$	$i^h_{2'}\cos\psi' - i^h_{3'}\sin\psi'$
3rd wave	$i^h_{3'}$	$i^h_{2'}\sin\phi' + i^h_{3'}\cos\phi'$	$i^h_{2'}\sin\psi' + i^h_{3'}\cos\psi'$
		$\tan 2\phi' = \dfrac{2c_{2'4'}}{c_{4'4'} - c_{2'2'}}$	$\tan 2\psi' = \dfrac{2c_{3'4'}}{c_{2'2'} - c_{4'4'}}$

$(h = 1, 2, 3)$ (8.18)

and the following strains occur

	x'-*cut* ($= x$-*cut*)	y'-*cut*	z'-*cut*
1st wave	$S_{1'1'}$	$S_{1'2'}$	$S_{3'1'}$
2nd wave	$S_{1'2'}$	$S_{2'2'}, S_{2'3'}$	$S_{2'3'}, S_{3'3'}$
3rd wave	$S_{1'3'}$	$S_{2'2'}, S_{2'3'}$	$S_{2'3'}, S_{3'3'}$

(8.19)

For an x-cut the direction of the energy flow is perpendicular to the plate for all waves. The same is true for the first wave in a y'-cut and in a z'-cut. For the other waves in a y'-cut or a z'-cut the energy flow lies in the $y'z'$-plane.

A y'-cut is called an AC-cut and a z'-cut a BC-cut. For the AC-cut, only $d^{2'}_{.\,1'2'}$ is important:

$$d^{2'}_{.\,1'2'} = d^2_{.12}\cos^2\underset{0}{\theta} - d^2_{.13}\cos\underset{0}{\theta}\sin\underset{0}{\theta} = -d_{11}\cos^2\underset{0}{\theta} + d_{14}\cos\underset{0}{\theta}\sin\underset{0}{\theta}$$

$$= \begin{cases} -5{\cdot}96; & \underset{0}{\theta} = 30^\circ\ 55' \text{ (Voigt)}, \\ -5{\cdot}92; & \underset{0}{\theta} = 31^\circ\ 30' \text{ (Mason)}. \end{cases} \quad (8.20)$$

For the BC-cut, only $d^{3'}_{.\,1'3'}$ is important:

$$d^{3'}_{.\,1'3'} = d^2_{.12}\sin^2\underset{0}{\theta} + d^2_{.13}\sin\underset{0}{\theta}\cos\underset{0}{\theta} = -d_{11}\sin^2\underset{0}{\theta} - d_{14}\sin\underset{0}{\theta}\cos\underset{0}{\theta}$$

$$= \begin{cases} +0{\cdot}700; & \underset{0}{\theta} = 30^\circ\ 55' \text{ (Voigt)}, \\ +0{\cdot}703; & \underset{0}{\theta} = 31^\circ\ 30' \text{ (Mason)}. \end{cases} \quad (8.21)$$

Hence the AC-cut is preferable from the point of view of activity.

The elastic constants are functions of the temperature and from this it follows that the frequency of a resonator is a function of temperature. This function is strongly influenced by the parasitic waves. For low frequencies these waves have even been utilized to produce a low temperature coefficient for a limited temperature range. But the trouble is that the parasitic waves cause discontinuities in the characteristic frequency-temperature and frequency-thickness curves of the plate. 'In practice the plates are adjusted in such a way that there are no discontinuities in the frequency-temperature characteristic in the region where they are expected to operate, but at high frequencies it is difficult to eliminate all these discontinuities over a wide temperature range.'† This requires close temperature regulation. x-cut plates have many coupling constants and are notorious for their bad frequency curves.‡

Lack, Willard, and Fair† found out that the temperature coefficient of $c_{6'6'}$ is zero for $\theta = 35° \; 15'$ at $45°$ C. and for $\theta = -49°$ at $25°$ C. These cuts are called the AT-cut and the BT-cut respectively. They are so near the AC-cut and the BC-cut that they partake of the advantages of the latter though $c_{5'6'}$ does not vanish absolutely.

Post§ has answered the question whether the conditions for the x', y', z' axes with $\theta = 30° \; 55'$ and $c_{5'6'} = 0$ are due to accidental properties of α-quartz. In the crystal class (*10*) of α-quartz the x'-axis is self-reciprocal and reciprocal to the y'-axis and to the z'-axis. Now, taking the axes x, y, z arbitrarily in any crystal, the conditions for self-reciprocity of the x-axis are

$$c_{16} = 0, \qquad c_{15} = 0. \tag{8.22}$$

The additional conditions for the reciprocity of x- and y-axis and x- and z-axis are

$$c_{56} = 0, \quad c_{26} = 0, \quad c_{46} = 0, \quad c_{45} = 0, \quad c_{35} = 0. \tag{8.23}$$

Hence the table of constants with respect to the coordinate system chosen, which does not need to coincide with the crystallographic axes, has the form

$$\begin{matrix} c_{11} & c_{12} & c_{13} & c_{14} & 0 & 0 \\ & c_{22} & c_{23} & c_{24} & c_{25} & 0 \\ & & c_{33} & c_{34} & 0 & c_{36} \\ & & & c_{44} & 0 & 0 \\ & & & & c_{55} & 0 \\ & & & & & c_{66} \end{matrix} \tag{8.24}$$

In a y-cut with side faces parallel to the x- and z-axis there exists a wave with $p^h = \underset{1}{i^h}$ and $\rho c^2 = c_{66}$, and only the strain S_{12} occurs in this wave. But S_{12} is mechanically coupled with S_{33} and this strain can set up a parasitic wave in the plate (considered as a z-cut). In the same way

† Lack, Willard, and Fair, 1934. 2.

‡ Cf. Cady 1946. 2, where a very bad frequency curve, published by Bechmann in 1937, is reprinted.

§ 1948. 1.

in a z-cut S_{13} is mechanically coupled with S_{22} and this strain can set up a parasitic wave in the plate (considered as a y-cut). It is a special property of α-quartz that c_{25} and c_{36} vanish.

Post also considered a more special case where the three axes are all self-reciprocal and mutually reciprocal. The additional conditions are

$$c_{24} = 0, \qquad c_{34} = 0, \tag{8.25}$$

and the table becomes

$$\begin{array}{cccccc} c_{11} & c_{12} & c_{13} & c_{14} & 0 & 0 \\ & c_{22} & c_{23} & 0 & c_{25} & 0 \\ & & c_{33} & 0 & 0 & c_{36} \\ & & & c_{44} & 0 & 0 \\ & & & & c_{55} & 0 \\ & & & & & c_{66} \end{array} \tag{8.26}$$

We now have the following directions of displacement, strains, and values of ρc^2:

x-cut			y-cut			z-cut		
$p^h = i^h_1$,	S_{11},	c_{11}	$p^h = i^h_1$,	S_{21},	c_{66}	$p^h = i^h_1$,	S_{31},	c_{55}
$p^h = i^h_2$,	S_{12},	c_{66}	$p^h = i^h_2$,	S_{22},	c_{22}	$p^h = i^h_2$,	S_{32},	c_{44}
$p^h = i^h_3$,	S_{13},	c_{55}	$p^h = i^h_3$,	S_{23},	c_{44}	$p^h = i^h_3$,	S_{33},	c_{33}

(8.27)

with mechanical coupling of S_{23} and S_{11}, S_{31} and S_{22}, S_{12} and S_{33}. All flow of energy is perpendicular to the plate. The case when c_{14}, c_{25}, and c_{36} all vanish seems very advantageous. This occurs for the crystallographic axes in the rhombic system (*7*, *8*), in the tetragonal system I (*15*, *16*, *19*), in the hexagonal system (*22*, *23*, *25*, *26*, *27*), and in the cubic system (*30*, *32*). Now for generation of one of the transverse waves we need S_{23}, S_{31}, or S_{12} and this means that one of the constants

$$d_{15}, \quad d_{16}, \quad d_{24}, \quad d_{26}, \quad d_{34}, \quad d_{35} \tag{8.28}$$

must be other than zero. This precludes the use of the classes *7*, *15*, *19*, *22*, *30*, and *32*. For the generation of longitudinal waves we need S_{11}, S_{22}, or S_{33}, and hence one of the constants

$$d_{11}, \quad d_{22}, \quad d_{33}. \tag{8.29}$$

This precludes the use of the same classes. So only the rhombic class *8*, the tetragonal class *16*, the hexagonal classes *23* and *25* all with d_{15}, d_{24}, and d_{33} and the hexagonal classes *26* with d_{11} and d_{26} and *27* with d_{11}, d_{16}, d_{22}, and d_{26}, remain for plates parallel to two of the crystallographic axes.† The first four classes seem excellent for longitudinal waves in z-cuts because they have no coupling effects whatever. Of course it is possible that full reciprocity of three axes which are not crystallographic can exist in any of these or other classes. But we must remember that for practical use as a resonator there are many other properties to be considered apart from the possibility of generation and the freedom from coupling effects. For instance low temperature coefficients, mechanical strength, and chemical indifference of the material, the possibility of finding or making pure crystals of a suitable size, etc., are important. The construction of a resonator is a technical problem that cannot be solved by theoretical investigations alone. They must be supplemented by much research work in the laboratory.

EXERCISES

VII.1. Prove that

$$\underset{x}{E}\underset{\iota}{x} = C\underset{\iota}{x}^{-1}$$

and draw a figure of the transformation $\underset{x}{S}$.

VII.2. Prove the identities in the footnote on page 154.

† In a letter of 2 August 1956 Professor W. G. Cady told the author that as far as he knew the only one of the classes in which promising crystals had been discovered or grown was class *16*, pentaerythritol, and that some work was done on these crystals at the Bureau of Standards at Washington.

VIII

CLASSICAL DYNAMICS†

1. Holonomic systems

We consider a set of particles with mass $\underset{i}{m}$, cartesian coordinates $\underset{i}{x}$, $\underset{i}{y}$, $\underset{i}{z}$, and forces $\underset{i}{X}$, $\underset{i}{Y}$, $\underset{i}{Z}$ acting on these particles, such that the $\underset{i}{x}$, $\underset{i}{y}$, $\underset{i}{z}$ can be expressed in terms of n variables q^κ, the *coordinates* of the system, and the time t. Then there are two possible cases:

(*a*) *Scleronomic*: the $\underset{i}{x}$, $\underset{i}{y}$, $\underset{i}{z}$ depend only on the q^κ:

$$\underset{i}{x} = \underset{i}{x}(q^\kappa); \quad \text{cycl. } x,\, y,\, z. \tag{1.1}$$

Example: a point moving on a fixed surface; $n = 2$.

(*b*) *Rheonomic*: the $\underset{i}{x}$, $\underset{i}{y}$, $\underset{i}{z}$ depend on the q^κ and t:

$$\underset{i}{x} = \underset{i}{x}(q^\kappa, t); \quad \text{cycl. } x,\, y,\, z. \tag{1.2}$$

Example: a point moving on a surface whose transformation in time is given; $n = 2$.

In the scleronomic case the kinetic energy of the system has the form

$$T = \tfrac{1}{2} g_{\lambda\kappa}(q^\nu)\dot{q}^\kappa \dot{q}^\lambda, \tag{1.3}$$

and in the rheonomic case we have

$$T = \tfrac{1}{2} a_{\lambda\kappa}(q^\nu, t)\dot{q}^\kappa \dot{q}^\lambda + b_\kappa(q^\nu, t)\dot{q}^\kappa + c(q^\nu, t). \tag{1.4}$$

In both cases it is easy to deduce the equations

$$\frac{d}{dt}\frac{\partial T}{\partial \dot{q}^\lambda} - \frac{\partial T}{\partial q^\lambda} = X_\lambda, \tag{1.5}$$

where the

$$X_\lambda \overset{\text{def}}{=} \sum_i \left(\underset{i}{X}\,\partial_\lambda \underset{i}{x} + \underset{i}{Y}\,\partial_\lambda \underset{i}{y} + \underset{i}{Z}\,\partial_\lambda \underset{i}{z}\right) \tag{1.6}$$

are called the components of the 'generalized force'.

A displacement $d\underset{i}{x}$, $d\underset{i}{y}$, $d\underset{i}{z}$ is called *virtual* if it satisfies the conditions of the motion of the system, that is, if values of dq^κ and dt exist which give $d\underset{i}{x}$, $d\underset{i}{y}$, $d\underset{i}{z}$ exactly if substituted in (1.1) or (1.2). Now

$$X_\lambda\, dq^\lambda = \sum_i (\underset{i}{X}\, d\underset{i}{x} + \underset{i}{Y}\, d\underset{i}{y} + \underset{i}{Z}\, d\underset{i}{z}) \tag{1.7}$$

is the work done by the forces $\underset{i}{X}$, $\underset{i}{Y}$, $\underset{i}{Z}$ if all particles undergo a virtual

† General references: Whittaker 1917. 1; Birkhoff 1927. 1; Prange 1934. 1; Synge 1936. 2, this paper contains a very extensive list of literature.

displacement in a time zero. Hence X_λ depends only on those forces that actually do work in a displacement of this kind. All other forces, as for example molecular forces between particles of a rigid body or reactions of fixed or moving smooth surfaces with which parts of the system remain in contact, drop out.

In a *scleronomic* system we have (cf. V, § 4)

$$\frac{d}{dt}\frac{\partial T}{\partial \dot{q}^\lambda}-\frac{\partial T}{\partial q^\lambda} = g_{\kappa\lambda}\ddot{q}^\kappa+\dot{q}^\mu(\partial_\mu g_{\kappa\lambda})\dot{q}^\kappa-\tfrac{1}{2}(\partial_\lambda g_{\kappa\mu})\dot{q}^\kappa\dot{q}^\mu$$

$$= g_{\kappa\lambda}\frac{d\dot{q}^\kappa+\Gamma^\kappa_{\rho\mu}\dot{q}^\rho\,dq^\mu}{dt} = g_{\kappa\lambda}\frac{\delta\dot{q}^\kappa}{dt}; \quad \Gamma^\kappa_{\mu\lambda} = \left\{{}^{\kappa}_{\mu\lambda}\right\}, \tag{1.8}$$

hence

$$\frac{\delta\dot{q}^\kappa}{dt} = X^\kappa. \tag{1.9}$$

This equation expresses the fact that, for $X_\lambda = 0$, the trajectories of the force-free motion in the X_n of the q^κ is a geodesic in a V_n with the fundamental tensor $g_{\lambda\kappa}$ and that t is a natural parameter on this geodesic.

Now we know that

$$s = \int ds = \int |\sqrt{(g_{\lambda\kappa}\,dq^\lambda dq^\kappa)}| \tag{1.10}$$

is extreme on a geodesic (cf. V, § 4). But we know also that

$$\int T\,dt = \tfrac{1}{2}\int ds\frac{ds}{dt} \tag{1.11}$$

is extreme (cf. IV, § 6). How is this possible? The answer lies in the fact that in the variation of $\int T\,dt$ only the $g_{\lambda\kappa}$ suffer a variation but t is left constant (cf. IV, § 6). This means that the variation is effected in such a way that corresponding points on neighbouring trajectories are reached at the same time. Now s and t are both natural parameters on the geodesic, hence, according to (V.2.9) ds/dt is a constant for every geodesic. Consequently

$$\overset{v}{d}\int ds\frac{ds}{dt} = \overset{v}{d}\int\frac{ds}{dt}\frac{ds}{dt}\,dt = 2\int\frac{ds}{dt}\overset{v}{d}\frac{ds}{dt}\,dt = 2\frac{ds}{dt}\int\overset{v}{d}\,ds = 2\frac{ds}{dt}\overset{v}{d}\int ds, \tag{1.12}$$

and this proves that $\int T\,dt$ is extreme if $\int ds$ is extreme and vice versa.

If a trajectory is such that any other trajectory, which is sufficiently near to the first one in an $\underset{0}{\mathfrak{N}}(q^\kappa)$, will remain near to it all over, it is called *stable*. In order to find the necessary and sufficient condition for stability in the scleronomic case, we write the equation of a trajectory in the form

$$\frac{d\dot{q}^\kappa}{dt} + \Gamma^\kappa_{\mu\lambda}\dot{q}^\mu\dot{q}^\lambda = X^\kappa \tag{1.13}$$

and consider a displacement $v^\kappa\,d\epsilon$ which transforms the trajectory into another one. Now we use the same device as in V, § 4, and drag along the transformed curve and the fields $\Gamma^\kappa_{\mu\lambda}$ and X^κ over $-v^\kappa\,d\epsilon$. Then $\Gamma^\kappa_{\mu\lambda}$ and X^κ transform into $\Gamma^\kappa_{\mu\lambda} - \underset{v}{\pounds}\Gamma^\kappa_{\mu\lambda}\,d\epsilon$ and $X^\kappa - \underset{v}{\pounds}X^\kappa\,d\epsilon$. From (IV.5.16) and (V.1.27) we have

$$\underset{v}{\pounds}X^\kappa = v^\mu\nabla_\mu X^\kappa - X^\mu\nabla_\mu v^\kappa, \tag{1.14}$$

and it is easily proved that†

$$\underset{v}{\pounds}\Gamma^\kappa_{\mu\lambda} = \nabla_\mu\nabla_\lambda v^\kappa + v^\sigma K_{\sigma\mu\lambda}{}^{\cdot\cdot\cdot\kappa}. \tag{1.15}$$

This leads to the differential equation

$$\frac{d\dot{q}^\kappa}{dt} + \{\Gamma^\kappa_{\mu\lambda} - (\nabla_\mu\nabla_\lambda v^\kappa + v^\sigma K_{\sigma\mu\lambda}{}^{\cdot\cdot\cdot\kappa})\,d\epsilon\}\dot{q}^\mu\dot{q}^\lambda = X^\kappa - (v^\mu\nabla_\mu X^\kappa - X^\mu\nabla_\mu v^\kappa)\,d\epsilon. \tag{1.16}$$

From (1.13) and (1.16) it follows that

$$(\nabla_\mu\nabla_\lambda v^\kappa + v^\sigma K_{\sigma\mu\lambda}{}^{\cdot\cdot\cdot\kappa})\dot{q}^\mu\dot{q}^\lambda = v^\mu\nabla_\mu X^\kappa - X^\mu\nabla_\mu v^\kappa \tag{1.17}$$

or

$$\dot{q}^\mu\nabla_\mu\dot{q}^\lambda\nabla_\lambda v^\kappa + v^\sigma K_{\sigma\mu\lambda}{}^{\cdot\cdot\cdot\kappa}\dot{q}^\mu\dot{q}^\lambda = v^\mu\nabla_\mu X^\kappa + \dot{q}^\mu(\nabla_\mu\dot{q}^\lambda)\nabla_\lambda v^\kappa - X^\mu\nabla_\mu v^\kappa. \tag{1.18}$$

Hence, according to (1.13)

$$\frac{\delta}{dt}\frac{\delta}{dt}v^\kappa + v^\sigma\dot{q}^\mu\dot{q}^\lambda K_{\sigma\mu\lambda}{}^{\cdot\cdot\cdot\kappa} = v^\mu\nabla_\mu X^\kappa. \tag{1.19}$$

This is the differential equation which v^κ has to satisfy along the trajectory. If this equation has solutions that remain small for all values of t, the trajectory is stable.

Quite apart from any mechanical problem we may ask whether a finite displacement $v^\kappa\,d\epsilon$ can be found which transforms a given geodesic in a V_n into another geodesic. If s is used as a parameter on the first geodesic, the ds on the second geodesic transforms into (cf. V.4.16)

$$ds' = \left(1 + \frac{dq^\mu}{ds}\frac{\delta}{ds}v_\mu\,d\epsilon\right)ds \tag{1.20}$$

with the dragging along over $-v^\kappa\,d\epsilon$, and in the same way as above, but taking into account this new value ds', it can be proved that

$$\frac{\delta}{ds}\frac{\delta}{ds}v^\kappa + v^\sigma K_{\sigma\mu\lambda}{}^{\cdot\cdot\cdot\kappa}\frac{dq^\mu}{ds}\frac{dq^\lambda}{ds} - \frac{d\lambda}{ds}\frac{dq^\kappa}{ds} = 0; \quad \lambda \overset{\text{def}}{=} \frac{dq^\lambda}{ds}\frac{\delta}{ds}v_\lambda. \tag{1.21}$$

† Cf. E I. 1935. 1, pp. 142, 150.

In this equation the last term on the left-hand side reduces to zero if v^κ is always perpendicular to the geodesic, because in that case

$$\lambda = \frac{dq^\mu}{ds}\frac{\delta}{ds}v_\mu = \frac{\delta}{ds}\left(\frac{dq^\mu}{ds}v_\mu\right) - v_\mu \frac{\delta}{ds}\frac{dq^\mu}{ds} = 0 - 0 = 0. \tag{1.22}$$

(1.21) is the famous equation for 'l'écart géodésique' of Levi Civita, a generalization of an equation of Gauss which was only valid for $n = 2$. For $n = 2$ a unit vector n^κ perpendicular to a geodesic is displaced pseudo-parallel along the geodesic; hence $\delta n^\kappa = 0$. Hence, if $v^\kappa = vn^\kappa$ and $i^\kappa = \frac{dq^\kappa}{ds}$, we have

$$\frac{d^2v}{ds^2} = -vn^\nu i^\mu i^\lambda n^\kappa K_{\nu\mu\lambda\kappa} = +vKn^\nu i^\mu i^\lambda n^\kappa g_{[\nu[\lambda}g_{\mu]\kappa]} = -\frac{K}{2}v \tag{1.23}$$

for surfaces in ordinary space (signature $+++$). This is the equation of Gauss, from which it follows immediately that on surfaces with a constant positive curvature a given geodesic has neighbouring geodesics, but that if the constant curvature is negative this is not so.

The equation (1.4) of a *rheonomic* system can be written in the form

$$T = \tfrac{1}{2}g_{\phi\psi}\dot{q}^\phi\dot{q}^\psi \quad (\phi, \psi = 0, 1, ..., n), \tag{1.24}$$

where

$$q^0 \overset{\text{def}}{=} t, \qquad g_{\kappa\lambda} \overset{\text{def}}{=} a_{\kappa\lambda}, \qquad g_{0\lambda} = g_{\lambda 0} \overset{\text{def}}{=} b_\lambda, \qquad g_{00} \overset{\text{def}}{=} 2c. \tag{1.25}$$

With this notation we have

$$\begin{aligned}\frac{d}{dt}\frac{\partial T}{\partial \dot{q}^\lambda} - \frac{\partial T}{\partial q^\lambda} - X_\lambda &= g_{\lambda\phi}\ddot{q}^\phi + \dot{q}^\psi(\partial_\psi g_{\lambda\phi})\dot{q}^\phi - \tfrac{1}{2}(\partial_\lambda g_{\phi\psi})\dot{q}^\phi\dot{q}^\psi - X_\lambda \\ &= g_{\lambda\kappa}\ddot{q}^\kappa + \tfrac{1}{2}(\partial_\psi g_{\lambda\phi} + \partial_\phi g_{\lambda\psi} - \partial_\lambda g_{\phi\psi})\dot{q}^\phi\dot{q}^\psi - X_\lambda \dot{q}^0\dot{q}^0 = 0.\end{aligned} \tag{1.26}$$

Now if $g^{\kappa\lambda}$ is the inverse of $g_{\lambda\kappa}$ and if we write

$$\Gamma^0_{\chi\psi} \overset{\text{def}}{=} 0, \qquad \Gamma^\kappa_{\chi\psi} = \tfrac{1}{2}g^{\kappa\lambda}(\partial_\chi g_{\lambda\psi} + \partial_\psi g_{\lambda\chi} - \partial_\lambda g_{\psi\chi}) - \delta^0_\chi \delta^0_\psi X^\kappa \tag{1.27}$$

the last n equations of

$$\frac{\delta\dot{q}^\phi}{dt} \overset{\text{def}}{=} \frac{d\dot{q}^\phi}{dt} + \Gamma^\phi_{\chi\psi}\dot{q}^\chi\dot{q}^\psi = 0\dagger \tag{1.28}$$

are equivalent to (1.26) and the first equation is an identity. Hence in the X_{n+1} of the q^κ and t (*film-space*) we have found a symmetric displacement depending on $a_{\lambda\kappa}$, b_κ, c, and X_λ which makes from this X_{n+1} an A_{n+1}. Those geodesics in this A_{n+1} on which q^0 is a natural parameter are the *world-lines* of the system in film-space. If the A_{n+1} is

† Wundheiler 1932. 2. Cf. for other displacements Synge 1936. 2.

reduced with respect to the curves with the equations $q^\kappa =$ constant, that is if we consider each of these curves as a point of an X_n (cf. Ch. I, § 3), we get the X_n of the q^κ and the world-lines become the trajectories. If X_λ is fixed, there are ∞^{2n} world-lines, but if

$$q^\kappa = f^\kappa(t), \qquad q^0 = t \tag{1.29}$$

is a world-line, then according to (1.28)

$$q^\kappa = f^\kappa(t), \qquad q^0 = \underset{1}{c}t+\underset{2}{c}, \qquad \underset{1}{c}, \underset{2}{c} = \text{constants} \tag{1.30}$$

is also a world line, belonging to the same trajectory. This agrees with the number of trajectories ($\infty^{2(n-1)}$).

2. Anholonomic coordinates and anholonomic mechanical systems†

If a *holonomic* mechanical system is given, in the X_{n+1} of the q^κ and $q^0 = t$, anholonomic coordinates (cf. IV, § 7) can be introduced by means of the formulae

$$\begin{gathered}(dq)^p = A^p_\phi\, dq^\phi = A^p_\kappa\, dq^\kappa + A^p_0\, dq^0 \\ dq^\phi = A^\phi_p\, (dq)^p = A^\phi_h\, (dq)^h + A^\phi_0\, (dq)^0; \quad \mathrm{Det}(A^p_\phi) \neq 0 \\ (p = 0, 1,..., \mathrm{n};\ h = 1,..., \mathrm{n};\ \phi = 0, 1,..., n).\end{gathered} \tag{2.1}$$

Now it is convenient to do this in such a way that $(dq)^0 = dq^0$, i.e. so that

$$\begin{aligned} A^0_\kappa &= 0, & A^0_0 &= 1, \\ A^0_h &= 0, & A^0_0 &= 1. \end{aligned} \tag{2.2}$$

Then q^0 remains holonomic so that we may write $(dq)^0 = dq^0 = dq^0$ and the transformation formulae take the form

$$\begin{gathered}(dq)^h = A^h_\kappa\, dq^\kappa + A^h_0\, dt \\ dq^\kappa = A^\kappa_h (dq)^h + A^\kappa_0\, dt; \quad \mathrm{Det}\,(A^h_\kappa) \neq 0 \quad (h = 1,..., \mathrm{n}).\end{gathered} \tag{2.3}$$

The object of anholonomity (cf. IV, § 7)

$$\Omega^p_{rq} = A^{\chi\psi}_{rq}\, \partial_{[\chi} A^p_{\psi]} \quad (p, q, r = 0, 1,..., \mathrm{n};\ \psi, \chi = 0, 1,..., n) \tag{2.4}$$

satisfies the conditions

$$\Omega^0_{rq} = 0 \quad (q, r = 0, 1,..., \mathrm{n}). \tag{2.5}$$

If we write

$$v^p \overset{\text{def}}{=} \frac{(dq)^p}{dt} = A^p_\phi\, \dot{q}^\phi = A^p_\kappa\, \dot{q}^\kappa + A^p_0; \quad v^0 = 1$$
$$(p = 0, 1,..., \mathrm{n};\ \phi = 0, 1,..., n) \tag{2.6}$$

† Cf. Whittaker 1917. 1; Vranceanu 1926. 3; Hořák 1928. 1; Prange 1934. 1; E. I., 1935. 1; Synge 1936. 2, this paper contains a number of titles of other papers of Hořák and Vranceanu.

the v^p are the anholonomic components of the generalized velocity vector, which depends on $\dot{q}^\kappa$ and 1 in the same way as the $(dq)^p$ depend on dq^κ and dt. T is a function of q^κ, t, $\dot{q}^\kappa$ and can now be considered as a function of q^κ, t, and v^h also:

$$T^*(q^\kappa, t, v^h) = T(q^\kappa, t, \dot{q}^\kappa). \tag{2.7}$$

Then the left-hand side of (1.5) takes the form

$$\frac{d}{dt}\frac{\partial T}{\partial \dot{q}^\lambda} - \frac{\partial T}{\partial q^\lambda} = \frac{d}{dt}\left(\frac{\partial T^*}{\partial v^i} A^i_\lambda\right) - \frac{\partial T^*}{\partial q^\lambda} - \frac{\partial T^*}{\partial v^i}\frac{\partial v^i}{\partial q^\lambda}$$

$$= A^i_\lambda \frac{d}{dt}\frac{\partial T^*}{\partial v^i} - A^i_\lambda \frac{\partial T^*}{(\partial q)^i} + \frac{\partial T^*}{\partial v^i}(\partial_t A^i_\lambda + \dot{q}^\mu \partial_\mu A^i_\lambda - \dot{q}^\mu \partial_\lambda A^i_\mu - \partial_\lambda A^i_0),$$

$$\frac{\partial T^*}{(\partial q)^i} \overset{\text{def}}{=} A^\lambda_i \frac{\partial T^*}{\partial q^\lambda}; \quad \partial_t = \frac{\partial}{\partial t} \quad (i = 1,..., \mathrm{n}). \tag{2.8}$$

If this expression is transvected with the inverse A^λ_j of A^i_λ we get the equation†

$$\frac{d}{dt}\frac{\partial T^*}{\partial v^j} - \frac{\partial T^*}{(\partial q)^j} + 2v^h \frac{\partial T^*}{\partial v^i}\Omega^{\cdot\, i}_{hj} + 2\frac{\partial T^*}{\partial v^i}\Omega^{\cdot\, i}_{0j} = X_j \quad (h, i, j = 1,..., \mathrm{n}), \tag{2.9}$$

and all the non-vanishing components of $\Omega^{\cdot\, p}_{rq}$ occur in this equation.

In the special case when the A^i_λ and A^λ_j are independent of t and when $A^h_0 = 0$, $A^\kappa_0 = 0$:

$$\begin{aligned}(dq)^h &= A^h_\kappa dq^\kappa, \\ dq^\kappa &= A^\kappa_h (dq)^h \quad (h = 1,..., \mathrm{n}),\end{aligned} \tag{2.10}$$

all $\Omega^{\cdot\, i}_{0j}$ vanish and (2.9) takes the simpler form

$$\frac{d}{dt}\frac{\partial T^*}{\partial v^j} - \frac{\partial T^*}{(\partial q)^j} + 2v^h \frac{\partial T^*}{\partial v^i}\Omega^{\cdot\, i}_{hj} = X_j \quad (h, i, j = 1,..., \mathrm{n}). \tag{2.11}$$

By means of anholonomic coordinates it is possible to treat an *anholonomic* mechanical system very satisfactorily. If the coordinates q^κ have to satisfy certain kinematical conditions expressed by $n-m$ ordinary independent equations between the q^κ and t, $n-m$ of the q^κ can be eliminated and a system with m degrees of freedom remains. If the system was scleronomic, it may happen that the new system is rheonomic, but there are no further difficulties. Another case arises, however, if the kinematical conditions can only be expressed by means of $n-m$ linearly independent linear relations between the differentials dq^κ and dt

$$C^x_\kappa dq^\kappa + C^x_0 dt = 0 \quad (x = \mathrm{m}+1,..., \mathrm{n}) \tag{2.12}$$

† Hamel 1904. 1.

with coefficients C^x_κ and C^x_0 depending on q^κ and t. In this case we call the system *rheonomic anholonomic*. The geometrical meaning is that an E_{m+1} is fixed at every point of the X_{n+1} of the q^κ and t, and that the direction of each world-line must at every one of its points lie in the local E_{m+1}. As an example we mention the case of a sphere moving without sliding along a surface whose motion is prescribed ($n = 5$, $m = 3$).

If we start from a scleronomic system and if the conditions are such that the C^x_κ do not depend on t and that $C^x_0 = 0$, we get a *scleronomic anholonomic* system. In this case an E_m is fixed at every point of the X_n of the q^κ, and the direction of each of the trajectories must at every one of its points lie in the local E_m. As an example we mention the case of a sphere moving without sliding along a fixed surface ($n = 5$, $m = 3$).

Anholonomic coordinates can now be used in a very elegant way to get rid of superfluous equations. To do this the coordinates q^κ are rearranged in such a way that $dq^1,..., dq^m$ are linearly independent of the $n-m$ expressions in (2.12).† Then the anholonomic coordinate system is defined as follows:

$$\begin{gathered} q^0 \overset{\text{def}}{=} t, \qquad (dq)^0 = dq^0 = dt, \\ q^a \overset{\text{def}}{=} \delta^a_\alpha q^\alpha, \qquad (dq)^a = dq^a = \delta^a_\alpha dq^\alpha, \\ (dq)^x \overset{\text{def}}{=} C^x_\kappa dq^\kappa + C^x_0 dt \\ (a = 1,..., \mathrm{m};\ \alpha = 1,..., m;\ x = \mathrm{m}+1,..., \mathrm{n}) \end{gathered} \tag{2.13}$$

from which we see that the $(dq)^\phi$ ($\phi = 0, 1,..., \mathrm{n}$) are linearly independent. The coordinates q^0, q^a; $a = 1,..., \mathrm{m}$, remain holonomic because they are numerically equal to $q^0 = t$, q^α ($\alpha = 1,..., m$). The components v^ϕ of the generalized velocity

$$\begin{gathered} v^0 = 1, \qquad v^a = \delta^a_\alpha \dot{q}^\alpha, \qquad v^x = C^x_\kappa \dot{q}^\kappa + C^x_0 \\ (a = 1,..., \mathrm{m};\ x = \mathrm{m}+1,..., \mathrm{n}) \end{gathered} \tag{2.14}$$

belong to the anholonomic coordinate system, and we have for $\Omega^{\ \phi}_{\chi\psi}$

$$\begin{gathered} \Omega^0_{ji} = 0, \qquad \Omega^0_{j0} = 0, \\ \Omega^a_{ji} = 0, \qquad \Omega^a_{j0} = 0, \\ (a = 1,..., \mathrm{m};\ i, j = 1,..., \mathrm{n}). \end{gathered} \tag{2.15}$$

The conditions (2.12) take the form

$$(dq)^x = 0 \quad (x = \mathrm{m}+1,..., \mathrm{n}). \tag{2.16}$$

Since the system has to satisfy these conditions it is necessary that besides X_i there should be still another generalized force X'_i and that

† It is clear that dt is always linearly independent of them.

this force (often called *moving constraint*) should not do any work in displacements which satisfy (2.16). Hence

$$X'_a = 0 \quad (a = 1,..., \mathrm{m}). \tag{2.17}$$

The equations (2.9) now take the form

$$\frac{d}{dt}\frac{\partial T^*}{\partial v^j} - \frac{\partial T^*}{(\partial q)^j} + 2v^h \frac{\partial T^*}{\partial v^i}\Omega_{hj}^{i} + 2\frac{\partial T^*}{\partial v^i}\Omega_{0j}^{i} = X_j + X'_j \tag{2.18}$$

$$(h, i, j = 1,..., \mathrm{n}).$$

From (2.15) and (2.16) and $v^x = 0$, the first m equations of (2.18) can be written†

$$\frac{d}{dt}\frac{\partial T^*}{\partial v^b} - \frac{\partial T^*}{(\partial q)^b} + 2v^c \frac{\partial T^*}{\partial v^x}\Omega_{cb}^{x} + 2\frac{\partial T^*}{\partial v^x}\Omega_{0b}^{x} = X_b \tag{2.19}$$

$$(b, c = 1,..., \mathrm{m};\ x = \mathrm{m}+1,..., \mathrm{n}).$$

In these equations we have to put $v^a = \delta_\alpha^a \dot{q}^\alpha$ and $v^x = 0$ after the first differentiation. Then exactly m differential equations of second order in the q^κ arise, and from these m equations and the $n-m$ differential equations (2.12) of first order in the q^κ we must determine the q^κ as functions of t.

The other $n-m$ equations (2.18) are not necessary for the process of integration, it is just the merit of the method described above that they drop out automatically. One can, however, use them after the integration if it is desirable to know the components X'_x of the moving constraint.

3. Homogenization of the equations of Lagrange and of Hamilton‡

If the forces $\underset{i}{X}$, $\underset{i}{Y}$, $\underset{i}{Z}$ depend on a potential, a function V of q^κ and t exists such that

$$X_\lambda = -\partial_\lambda V. \tag{3.1}$$

A more general case is when X_λ depends in the following way on a potential V which itself depends on q^κ, t, and $\dot{q}^\kappa$ (kinetic potential)

$$X_\lambda = \frac{d}{dt}\frac{\partial V}{\partial \dot{q}^\lambda} - \frac{\partial V}{\partial q^\lambda}. \tag{3.2}$$

In both cases we may introduce the function L of Lagrange, defined by

$$L \stackrel{\text{def}}{=} T - V. \tag{3.3}$$

† Hamel 1904. 1.

‡ Cf. Dirac 1933. 4; v. Dantzig 1934. 5.

Then the equations of motion take the form

$$\boxed{\frac{d}{dt}\frac{\partial L}{\partial \dot{q}^\lambda}-\frac{\partial L}{\partial q^\lambda} = 0,} \tag{3.4}$$

which are known as the *Lagrangian equations* of the system.

If we write

$$p_\lambda \stackrel{\text{def}}{=} \frac{\partial L}{\partial \dot{q}^\lambda} \tag{3.5}$$

the p_λ are called the *momenta* belonging to the coordinates q^κ. We always assume that the $\dot{q}^\kappa$ can be found from (3.5) as functions of p_λ, q^κ, and t. If we write

$$H \stackrel{\text{def}}{=} p_\lambda \dot{q}^\lambda - L, \tag{3.6}$$

the *Hamiltonian function*, H, can be considered as a function of p_λ, q^κ, and t.

Now, according to (3.4, 5), we have

$$\begin{aligned} dL &= \frac{\partial L}{\partial q^\lambda}dq^\lambda + \frac{\partial L}{\partial \dot{q}^\lambda}d\dot{q}^\lambda + \frac{\partial L}{\partial t}dt \\ &= \dot{p}_\lambda dq^\lambda + p_\lambda d\dot{q}^\lambda + \frac{\partial L}{\partial t}dt, \end{aligned} \tag{3.7}$$

and consequently

$$dH = \dot{q}^\lambda dp_\lambda + p_\lambda d\dot{q}^\lambda - \dot{p}_\lambda dq^\lambda - p_\lambda d\dot{q}^\lambda - \frac{\partial L}{\partial t}dt. \tag{3.8}$$

But, on the other hand, we also have

$$dH = \frac{\partial H}{\partial p_\lambda}dp_\lambda + \frac{\partial H}{\partial q^\lambda}dq^\lambda + \frac{\partial H}{\partial t}dt; \tag{3.9}$$

hence

$$\begin{aligned} &(a)\quad \frac{\partial H}{\partial p_\lambda} = \dot{q}^\lambda; \quad \frac{\partial H}{\partial q^\lambda} = -\dot{p}_\lambda, \\ &(b)\quad \frac{\partial H}{\partial t} = \frac{dH}{dt}, \qquad (c)\quad \frac{\partial H}{\partial t} = -\frac{\partial L}{\partial t}. \end{aligned} \tag{3.10}$$

The *Hamiltonian equations* (3.10 *a*) of the system are $2n$ differential equations of first order in p_λ, q^κ, equivalent to the n Lagrangian equations of second order. (3.10 *b*) is a consequence of (3.10 *a*). Notice that the function H can never be homogeneous of first degree in the p_λ because in that case the problem would vanish since

$$L = p_\lambda \dot{q}^\lambda - p_\lambda \frac{\partial H}{\partial p_\lambda} = 0.$$

The Lagrangian equations (3.4) are a special case of the more general Lagrangian equations of IV, § 6. Here the one variable t plays the role of the ξ^κ in IV, § 6, and the q^κ must be considered as *scalars* in t-space because they do not undergo a transformation which depends on a transformation of t. Up till now t has not been transformed at all. But if we transform t in the same way as the ξ^κ in IV, § 6, that is, if we introduce a new variable τ, depending only on t,

$$\tau = \tau(t), \qquad t = t(\tau), \tag{3.11}$$

we know that L transforms with Δ^{-1}, $\Delta = \dfrac{d\tau}{dt}$. Writing $\mathfrak{L}$ for the transformed L belonging to τ, we have

$$\mathfrak{L} = L\overset{*}{t}, \qquad \overset{*}{t} \overset{\text{def}}{=} \frac{dt}{d\tau}. \tag{3.12}$$

$\mathfrak{L}$ is really a function of q^κ, t, $\overset{*}{q}{}^\kappa$, and $\overset{*}{t}$, if we denote the differentiation with respect to τ by $*$. Because

$$\mathfrak{L} = \overset{*}{t} L(q^\kappa, \overset{*}{q}{}^\kappa \overset{*}{t}{}^{-1}, t) \tag{3.13}$$

this function is homogeneous of first degree in $\overset{*}{q}{}^\kappa$, $\overset{*}{t}$. As we have seen in IV, § 6, the Lagrangian equation is invariant for the transformation $t \to \tau$, $L \to \mathfrak{L}$, hence

$$\frac{d}{d\tau}\frac{\partial \mathfrak{L}}{\partial \overset{*}{q}{}^\lambda} - \frac{\partial \mathfrak{L}}{\partial q^\lambda} = 0. \tag{3.14}$$

But this can also be demonstrated directly.† If the Lagrangian equation in L and t is written in the form

$$d\frac{\partial L\,dt}{\partial\, dq^\lambda} - \frac{\partial L\,dt}{\partial q^\lambda} = 0, \tag{3.15}$$

from $L\,dt = \mathfrak{L}\,d\tau$ it follows that

$$d\frac{\partial \mathfrak{L}\,d\tau}{\partial\, dq^\lambda} - \frac{\partial \mathfrak{L}\,d\tau}{\partial q^\lambda} = 0, \tag{3.16}$$

and this is equivalent to (3.14). Notice that the invariance of the equations (3.4) and (3.14) for transformations of the q^κ and fixed t has nothing to do with the invariance dealt with in IV, § 6.

We will prove now that there exists yet another equation

$$\frac{d}{d\tau}\frac{\partial \mathfrak{L}}{\partial \overset{*}{t}} - \frac{\partial \mathfrak{L}}{\partial t} = 0, \tag{3.17}$$

† Cf. v. Dantzig 1934. 5.

besides (3.14). If we write

$$\begin{aligned} p_0 &\overset{\text{def}}{=} -H = L - p_\lambda \dot{q}^\lambda, \\ q^0 &\overset{\text{def}}{=} t, \end{aligned} \tag{3.18}$$

we have

$$\mathfrak{L}\, d\tau = L\, dt = p_\lambda dq^\lambda + p_0 dq^0 = p_\phi dq^\phi \quad (\phi = 0, 1,..., n), \tag{3.19}$$

and consequently

$$\mathfrak{L} = p_\phi \overset{*}{q}{}^\phi \quad (\phi = 0, 1,..., n). \tag{3.20}$$

Now we have seen that $\mathfrak{L}$ is homogeneous of degree *1* in $\overset{*}{q}{}^\phi$. This does not follow from (3.20) because in that equation $\mathfrak{L}$ is not expressed in q^ϕ and $\overset{*}{q}{}^\phi$. From (3.20) it only follows that p_ϕ, expressed in q^ϕ and $\overset{*}{q}{}^\phi$ has to be homogeneous of degree zero in $\overset{*}{q}{}^\phi$. But from (3.13) by differentiation we get immediately

$$\frac{\partial \mathfrak{L}}{\partial \overset{*}{q}{}^\phi} = p_\phi \quad (\phi = 0, 1,..., n), \tag{3.21}$$

hence for $\phi = 0$, according to (3.10 *b*, *c*) and (3.13),

$$\frac{d}{d\tau}\frac{\partial \mathfrak{L}}{\partial \overset{*}{q}{}^0} = \frac{dp_0}{d\tau} = -\frac{dH}{d\tau} = -\overset{*}{t}\frac{dH}{dt} = \overset{*}{t}\frac{\partial L}{\partial t} = \frac{\partial \mathfrak{L}}{\partial t}. \tag{3.22}$$

Taking (3.14) and (3.17) together we get the *homogeneous Lagrangian equations*

$$\boxed{\frac{d}{d\tau}\frac{\partial \mathfrak{L}}{\partial \overset{*}{q}{}^\phi} - \frac{\partial \mathfrak{L}}{\partial q^\phi} = 0} \quad (\phi = 0, 1,..., n), \tag{3.23}$$

and these equations are invariant for transformations of the $n+1$ coordinates q^0, q^1,..., q^n and for transformations of the parameter τ. With these latter transformations, $\mathfrak{L}$ transforms as a scalar density of weight $+1$ in one-dimensional τ-space.

In order to make the Hamiltonian equations homogeneous we introduce the function

$$\mathfrak{H} \overset{\text{def}}{=} p_\phi \overset{*}{q}{}^\phi - \mathfrak{L} = \overset{*}{t}(p_\lambda \dot{q}^\lambda + p_0 - L) = \overset{*}{t}\{p_0 + H(p_\lambda, q^\kappa, t)\} \quad (\phi = 0, 1,..., n). \tag{3.24}$$

Then instead of a Hamiltonian function H (which is now called $-p_0$) we get a *Hamiltonian relation*

$$\mathfrak{H}(p_\psi, q^\phi) = 0 \quad (\phi, \psi = 0, 1,..., n). \tag{3.25}$$

In fact $\mathfrak{H}$ may be considered as a function of the p_ψ and q^ϕ because the $\overset{*}{q}{}^\phi$ can be expressed in terms of p_ψ and q^ϕ by means of (3.21).

By differentiating (3.24) and taking (3.25) into account we get

$$(a)\quad \frac{\partial \mathfrak{H}}{\partial p_\phi} = \overset{*}{q}{}^\phi, \qquad (b)\quad \frac{\partial \mathfrak{H}}{\partial q^\psi} = -\overset{*}{p}_\psi \qquad (\phi, \psi = 0, 1, \ldots, n), \tag{3.26}$$

the *homogeneous Hamiltonian equations*. Notice that $\mathfrak{H}$ can never be homogeneous in the p_ϕ because then, from (3.25), $\mathfrak{L}$ would vanish:

$$\mathfrak{L} = -\mathfrak{H}+p_\phi \overset{*}{q}{}^\phi = -\mathfrak{H}+p_\phi \frac{\partial \mathfrak{H}}{\partial p_\phi} = 0+0 = 0. \tag{3.27}$$

$\mathfrak{H}$ is not the only function which satisfies (3.25) and (3.26). Instead of $\mathfrak{H}$ we may take any function $F(\mathfrak{H})$, provided that $F(\mathfrak{H}) = 0$ and $\dfrac{\partial F}{\partial \mathfrak{H}} = 1$ are consequences of $\mathfrak{H} = 0$ (see Note I, p. 272).

The q^ϕ are arbitrary coordinates in film-space. If a world-line is given through a point q^ϕ we know $\overset{*}{q}{}^\phi$ at that point and consequently also $\mathfrak{L}\,d\tau = p_\phi\, dq^\phi$ and the covariant vector p_ψ. This means that to every point of a world-line there belongs not only a direction but also a covariant vector and its n-direction. Conversely, if an n-direction is given at any point of film-space, that is, if we have a covariant vector ρp_ψ defined to within an arbitrary scalar factor ρ, this factor can be determined by means of (3.25), and a world-line through the point belongs to each value of ρ. A point with an n-direction in this point is called an *element*. Hence a world-line is a row of elements, and it can be determined by one point and one direction at this point but not by one of its elements because a finite number of world-lines may belong to one element. If an element is given and one of the possible values of ρ is chosen, the first Hamiltonian equation (3.10 a) gives $\dot{q}^\lambda$, that is, the direction of the world-line. Then the second Hamiltonian equation (3.10 b) gives the value of p_λ in the 'succeeding' point of the world-line. Hence the Hamiltonian equations represent a process by which the world-line could be approximated step by step if the steps were chosen sufficiently small.

Through each point of film-space there are ∞^n world-lines. Hence the total is ∞^{2n}. Accordingly the solution of the problem must depend on $2n$ parameters, for instance, the values of q^κ and $\dot{q}^\kappa$ for $t = 0$. In the non-homogeneous treatment the solution gives q^κ as functions of t

and the $2n$ parameters. But, using the homogeneous method, we get n equations between the q^ϕ and the $2n$ parameters, and n of the q^ϕ can be solved from these as functions of the one remaining q^ϕ and the parameters.

4. Theory of integration†

If we have two points $\underset{0}{q}^\phi$ and q^ϕ of film-space there is always a finite number of world-lines going through them. On each of these world-lines the integral

$$\overset{0}{R}(q^\phi, \underset{0}{q}^\phi) = \int\limits_{\underset{0}{\tau}}^{\tau} \mathfrak{L}\, d\tau \quad (\phi = 0,1,\ldots,n) \tag{4.1}$$

has an extreme value in accordance with the results of IV, § 6. $\overset{0}{R}$ is a function of two points and may be called the *special eiconal function.*

We consider a variation $\overset{v}{d}q^\phi$ of a world-line between $\underset{0}{q}^\phi$ and q^ϕ. Then from (3.21, 23) we have

$$\begin{aligned}
\overset{v}{d}\overset{0}{R} = \overset{v}{d}\int\limits_{\underset{0}{\tau}}^{\tau} \mathfrak{L}\, d\tau &= \int\limits_{\underset{0}{\tau}}^{\tau} \left(\frac{\partial\mathfrak{L}}{\partial q^\phi}\overset{v}{d}q^\phi + \frac{\partial\mathfrak{L}}{\partial \overset{*}{q}^\phi}\overset{v*}{d}q^\phi\right) d\tau \\
&= \int\limits_{\underset{0}{\tau}}^{\tau} \left\{\left(\frac{d}{d\tau}\frac{\partial\mathfrak{L}}{\partial \overset{*}{q}^\phi}\right)\overset{v}{d}q^\phi + p_\phi\overset{v*}{d}q^\phi\right\} d\tau \\
&= \int\limits_{\underset{0}{\tau}}^{\tau} (\overset{*}{p}_\phi\overset{v}{d}q^\phi + p_\phi\overset{v*}{d}q^\phi)\, d\tau = \int\limits_{\underset{0}{\tau}}^{\tau} d(p_\phi\overset{v}{d}q^\phi) \\
&= (p_\phi)_{\tau=\tau}\overset{v}{d}q^\phi - (p_\phi)_{\tau=\tau_0}\overset{v}{d}\underset{0}{q}^\phi \quad (\phi = 0,1,\ldots,n),
\end{aligned} \tag{4.2}$$

from which we see once more that $\overset{0}{R}$ takes an extreme value. Now, on the other hand, we have

$$\overset{v}{d}\overset{0}{R} = \frac{\partial\overset{0}{R}}{\partial q^\phi}\overset{v}{d}q^\phi + \frac{\partial\overset{0}{R}}{\partial \underset{0}{q}^\phi}\overset{v}{d}\underset{0}{q}^\phi \quad (\phi = 0,1,\ldots,n), \tag{4.3}$$

hence

$$(a)\quad \frac{\partial\overset{0}{R}}{\partial q^\phi} = p_\phi; \qquad (b)\quad \frac{\partial\overset{0}{R}}{\partial \underset{0}{q}^\phi} = -\underset{0}{p}_\phi \quad (\phi = 0,1,\ldots,n). \tag{4.4}$$

† Cf. Whittaker 1917. 1.

If $\underset{0}{q}^{\phi}$ is now fixed, $\overset{0}{R}$ is a function of the q^{ϕ} only and consequently, from the Hamiltonian relation (3.25) we get

$$\mathfrak{H}\left(\frac{\partial \overset{0}{R}}{\partial q^{\psi}}, q^{\phi}\right) = 0 \quad (\phi, \psi = 0, 1, ..., n). \tag{4.5}$$

The equation

$$\mathfrak{H}\left(\frac{\partial R}{\partial q^{\psi}}, q^{\phi}\right) = 0 \quad (\phi, \psi = 0, 1, ..., n), \tag{4.6}$$

where R denotes some unknown function of the q^{ϕ}, is called the *equation of Hamilton–Jacobi* in film-space. The *general solution* of such an equation depends on arbitrary functions. A *special solution* is one of the solutions contained in the general one. Now $\overset{0}{R}$ is neither the general nor a special solution, it is a solution depending on $n+1$ parameters $\underset{0}{q}^{\phi}$ (*complete solution*). But we will prove that *every* solution of (4.6) can be obtained if $\overset{0}{R}$ is known. Let $R(q^{\phi})$ be a special solution of (4.6) and $\underset{0}{q}^{\phi}$ a point chosen on the X_n with the equation

$$R(q^{\phi}) = c \quad (\phi = 0, 1, ..., n) \tag{4.7}$$

in film-space. Then $\partial_{\psi} R$ has the n-direction tangent to this X_n. Because of the relation (4.6), $\partial_{\psi} R$ in $\underset{0}{q}^{\phi}$ is a possible value of p_{ψ}. The world-line through $\underset{0}{q}^{\phi}$ which belongs to this value has the equation

$$\overset{*}{q}^{\chi} = \frac{\partial \mathfrak{H}(\partial_{\psi} R, q^{\phi})}{\partial(\partial_{\chi} R)} \quad (\phi, \psi, \chi = 0, 1, ..., n). \tag{4.8}$$

By moving $\underset{0}{q}^{\phi}$ on the X_n, we get in this way a congruence of ∞^n world-lines in film-space, one through every point of film-space sufficiently near to X_n. From (4.8) and (4.6) it follows that

$$\begin{aligned} \frac{d}{d\tau} \partial_{\chi} R = \overset{*}{q}^{\omega} \partial_{\omega} \partial_{\chi} R &= \frac{\partial \mathfrak{H}(\partial_{\psi} R, q^{\phi})}{\partial(\partial_{\omega} R)} \partial_{\omega} \partial_{\chi} R \\ &= -\frac{\partial \mathfrak{H}(\partial_{\psi} R, q^{\phi})}{\partial q^{\chi}} \quad (\phi, \psi, \chi, \omega = 0, 1, ..., n), \end{aligned} \tag{4.9}$$

hence, along the world-line, $\partial_{\chi} R$ satisfies the second Hamiltonian equation, and this means that the relation $p_{\psi} = \partial_{\psi} R$ holds not only in $\underset{0}{q}^{\kappa}$ but also in every point of the world-line. Accordingly we have

$$dR = \frac{\partial R}{\partial q^{\phi}} dq^{\phi} = p_{\phi} dq^{\phi} = \mathfrak{L}\, d\tau \tag{4.10}$$

along this line, and consequently

$$\overset{0}{R}(q^\phi, \underset{0}{q}^\phi) = \int_{\underset{0}{\tau}}^{\tau} \mathfrak{L}\, dt = R(q^\phi) - R(\underset{0}{q}^\phi). \tag{4.11}$$

This proves that $R(q^\phi)$ is known to within a constant if the X_n (4.7) is given and if the special eiconal function is known. Hence every solution of (4.6) can be constructed by chosing an X_n, constructing the world-lines belonging to all elements of X_n, and finally adjoining to every point q^κ, lying on a world-line through a point $\underset{0}{q}^\kappa$ of X_n, the value $\overset{0}{R}(q^\phi, \underset{0}{q}^\phi)$ plus an arbitrary constant.

The general solution of (4.6) is an X_n-point function, that is, it adjoins a value to every combination of an X_n and a point of film-space. This solution may be called the *general eiconal function*. If the X_n degenerates to a point it degenerates into the special eiconal function. Hence the general eiconal function contains the special one, but integrations are necessary if we wish to obtain the general one from the special one.

If, starting from the X_n (4.7), the world-lines are constructed, the elements of these world-lines are at each of their points tangent to one of the X_n's with an equation (4.7) where c has now an arbitrary value. This proves the proposition:

If we take an arbitrary congruence of ∞^n rows from the ∞^{2n} rows of elements in film-space and if one X_n exists whose tangent E_n at every point coincides with an element of one of these rows, all ∞^{n+1} elements of these ∞^n rows form a set of ∞^1 X_n's.

If we know the special eiconal function, the problem can be solved without integrations. Two of the $2n+2$ parameters $\underset{0}{p}_\psi$, $\underset{0}{q}^\phi$ can be eliminated from the equations (4.4 b) and (3.25) for $p_\psi = \underset{0}{p}_\psi$, $q^\phi = \underset{0}{q}^\phi$. Then we get n equations between the q^ϕ and $2n$ parameters from which n of the q^ϕ can be solved as functions of the one remaining q^ϕ and the parameters. It is not, however, easy to find $\overset{0}{R}$ because $\mathfrak{L}\, d\tau$ is not a complete differential and the integration in (4.1) cannot be effected.

It is not necessary to know $\overset{0}{R}$ for our present purpose. We only need a solution R of (4.6), provided that this solution contains n parameters c_λ. If the c_λ take all possible values the equation $R =$ const. represents for every set of values ∞^1 X_n's in film-space to which one congruence of ∞^n world-lines belongs. Hence we get all ∞^{2n} world-lines in the end.

In order to obtain the solution of the problem we prove that the $\partial R/\partial c_\lambda$ are constant. In the first place we have from (3.26 *a*)

$$\frac{d}{d\tau}\frac{\partial R}{\partial c_\lambda} = \frac{\partial^2 R}{\partial q^\phi \partial c_\lambda}\overset{*}{q}{}^\phi = \frac{\partial^2 R}{\partial q^\phi \partial c_\lambda}\frac{\partial \mathfrak{H}}{\partial p_\phi} = \frac{\partial \mathfrak{H}}{\partial p_\phi}\frac{\partial p_\phi}{\partial c_\lambda} \quad (\phi = 0, 1, ..., n), \tag{4.12}$$

where p_ψ is to be considered as a function of q^ϕ and c_λ. Now $\mathfrak{H} = 0$ for every point q^ϕ and every possible value of p_ψ in that point. Hence

$$\mathfrak{H}'(q^\phi, c_\lambda) \overset{\text{def}}{=} \mathfrak{H}\{p_\psi(q^\phi, c_\lambda), q^\phi\} = 0 \tag{4.13}$$

for every choice of c_λ and

$$\frac{d}{d\tau}\frac{\partial R}{\partial c_\lambda} = \frac{\partial \mathfrak{H}'}{\partial c_\lambda} = 0. \tag{4.14}$$

Consequently, if we put

$$\frac{\partial R(q^\phi, c_\mu)}{\partial c_\lambda} = b^\lambda \text{ (= constants)} \quad (\phi = 0, 1, ..., n) \tag{4.15}$$

we have n equations in the q^ϕ and the $2n$ parameters b^κ, c_λ, from which n of the q^ϕ can be solved as functions of the remaining q^ϕ and the parameters. This solves the problem.

The equation (4.10) can only be used for the determination of R if $\mathfrak{L}\,d\tau = p_\phi\,dq^\phi$ is a complete differential. A *first integral* of the equations (3.23) or (3.26) is a relation between p_ψ and q^ϕ

$$F(p_\psi, q^\phi) = c \text{ (constant)} \quad (\phi, \psi = 0, 1, ..., n) \tag{4.16}$$

from which we know that it will be satisfied by all solutions of (3.23) or (3.26). If n independent first integrals are known from which the p_ψ can be solved (having regard to (3.25)) as functions of the q^ϕ and n integration constants c_λ, we get equations of the form

$$p_\psi = f_\psi(q^\phi, c_\lambda) \quad (\phi, \psi = 0, 1, ..., n), \tag{4.17}$$

and these equations represent $n+1$ first integrals of a special form. One of these integrals depends on the others because of (3.25). If the p_ψ from (4.17) are substituted in (4.10) we get

$$dR = f_\psi(q^\phi, c_\lambda)\,dq^\psi \quad (\phi, \psi = 0, 1, ..., n) \tag{4.18}$$

and this expression is a complete differential if and only if

$$\partial_{[\chi} f_{\psi]} = 0 \quad (\chi, \psi = 0, 1, ..., n). \tag{4.19}$$

This condition is over-severe because it can be proved that

$$\partial_{[0} f_{\lambda]} = 0 \tag{4.20}$$

is a consequence of

$$\partial_{[\mu} f_{\lambda]} = 0. \tag{4.21}$$

In fact, from (3.26) and (4.17) we have

$$\frac{\partial f_\chi}{\partial q^\omega}\overset{*}{q}^\omega = \overset{*}{f}_\chi = -\frac{\partial \mathfrak{H}(f_\psi, q^\phi)}{\partial q^\chi} = \frac{\partial \mathfrak{H}(f_\psi, q^\phi)}{\partial f_\psi}\frac{\partial f_\psi}{\partial q^\chi} = \overset{*}{q}^\psi \frac{\partial f_\psi}{\partial q^\chi}$$

$$(\phi, \psi, \chi, \omega = 0, 1,..., n) \qquad (4.22)$$

or

$$\overset{*}{q}^\kappa \partial_{[\kappa}\, f_{\chi]} + \overset{*}{q}^0 \partial_{[0}\, f_{\chi]} = 0 \quad (\chi = 0, 1,..., n), \qquad (4.23)$$

which proves the proposition.

If we write $$\phi_\psi \overset{\text{def}}{=} p_\psi - f_\psi(q^\phi, c_\lambda) = 0 \quad (\phi, \psi = 0, 1,..., n) \qquad (4.24)$$

the equations (4.19) can be written in the form

$$(\phi_\chi, \phi_\psi) \overset{\text{def}}{=} \frac{\partial \phi_\chi}{\partial p_\phi}\frac{\partial \phi_\psi}{\partial q^\phi} - \frac{\partial \phi_\chi}{\partial q^\phi}\frac{\partial \phi_\psi}{\partial p_\phi} = 0 \quad (\phi, \psi, \chi = 0, 1,..., n), \qquad (4.25)$$

and in the same way as above it can be proved, for instance, that $(\phi_1, \phi_\psi) = 0$ is a consequence of those other equations that do not contain ϕ_1.

(ϕ_χ, ϕ_ψ) is called the *Poisson bracket* of the two functions ϕ_χ and ϕ_ψ. Two functions are said to be in *involution* if their Poisson bracket vanishes. If a system of equations $\phi_\psi = 0$ is equivalent to another system $\psi_\psi = 0$ and if the ϕ_ψ are in involution, it can be proved easily that the ψ_ψ are also in involution. The equations $\phi_\psi = 0$ are also said to be in involution.

The condition of integrability of (4.18) can now be expressed in a more convenient way.

In order that dR should be a complete differential it is necessary and sufficient that there should be n first integrals with n integration constants

$$F_\lambda(p_\psi, q^\phi) = c_\lambda \quad (\phi, \psi = 0, 1,..., n) \qquad (4.26)$$

in involution and this is the case if and only if the functions F_λ are in involution.

If this condition is satisfied the function $R(q^\phi, c_\lambda)$ can be determined by means of one quadrature.

The method of integration explained here is an application of the following theorem from the theory of partial differential equations.

If a system of equations of the first order with $n+1$ independent variables and one unknown function

$$F_\chi\left(\frac{\partial R}{\partial q^\psi}, q^\phi\right) = c_\chi \text{ (constants)} \quad (\phi, \psi, \chi = 0, 1,..., n) \qquad (4.27)$$

satisfies the condition that the functions F_χ are in involution, and if the p_ψ are solved from the auxiliary equations

$$F_\chi(p_\psi, q^\phi) = c_\chi \quad (\phi, \psi, \chi = 0, 1,..., n) \qquad (4.28)$$

as functions of the q^ϕ and c_χ and substituted in the differential form $p_\psi dq^\psi$, this expression is a complete differential and

$$R = \int p_\psi dq^\psi \quad (\psi = 0, 1, ..., n) \tag{4.29}$$

is a solution of (4.27) *which depends on* $n+1$ *constants* c_χ.

In fact in the case of classical mechanics we have the equation (4.6)

$$\mathfrak{H}\left(\frac{\partial R}{\partial q^\psi}, q^\phi\right) = 0 \quad (\phi, \psi = 0, 1, ..., n) \tag{4.30}$$

and in the most favourable case also n first integrals

$$F_\lambda(p_\psi, q^\phi) = c_\lambda \quad (\phi, \psi = 0, 1, ..., n) \tag{4.31}$$

from which we know that the F_λ are in involution. Hence it is only necessary to prove that the F_λ are in involution with $\mathfrak{H}(p_\psi, q^\phi)$. In fact we have

$$\frac{\partial \mathfrak{H}}{\partial p_\phi}\frac{\partial F_\lambda}{\partial q^\phi} - \frac{\partial \mathfrak{H}}{\partial q^\phi}\frac{\partial F_\lambda}{\partial p_\phi} = \overset{*}{q}{}^\phi \frac{\partial F_\lambda}{\partial q^\phi} + \overset{*}{p}_\phi \frac{\partial F_\lambda}{\partial p^\phi} = \frac{dF_\lambda}{d\tau} = \frac{dc_\lambda}{d\tau} = 0$$
$$(\phi = 0, 1, ..., n). \tag{4.32}$$

But now we also see the way that can be followed in order to get a solution of (4.30) which depends on n integration constants if no first integrals are known. First we have to look for a solution of the auxiliary equation

$$(\mathfrak{H}(p_\psi, q^\phi), F(p_\psi, q^\phi)) = 0 \quad (\phi, \psi = 0, 1, ..., n). \tag{4.33}$$

Then, if such a solution F_1 has been chosen, a solution of the system

$$(\mathfrak{H}, F) = 0, \qquad (F_1, F) = 0 \tag{4.34}$$

has to be determined, etc. After n steps we arrive at the system in involution

$$\mathfrak{H} = 0, \qquad F_\lambda = c_\lambda, \tag{4.35}$$

and from this system a solution of (4.30) depending on the c_λ can be found by means of a quadrature. If m first integrals in involution were known from the beginning, the first m steps would drop out. Also, during the process described it may happen that (of one of the systems to be solved) more solutions than one are known. Then some very important short cuts are possible. A complete theory is given by S. Lie.†

5. Special cases of first integrals‡

It may happen that the function $\mathfrak{L}$ does not depend on one of the coordinates, for instance q^n (it may, however, contain $\overset{*}{q}{}^n$!). Then,

† See Schouten and v. d. Kulk 1949. 1 for an elaborate treatment and references to literature.

‡ Cf. Whittaker 1917. 1.

according to (3.24), $\mathfrak{H}$ is independent of q^n too. In this case we get

$$\frac{d}{d\tau}\frac{\partial \mathfrak{L}}{\partial \overset{*}{q}{}^n} = \overset{*}{p}_n = 0 \tag{5.1}$$

from the homogeneous Lagrangian equations, and this also follows from the second set of homogeneous Hamiltonian equations

$$\overset{*}{p}_n = -\frac{\partial \mathfrak{H}}{\partial q^n} = 0. \tag{5.2}$$

Hence we have a first integral

$$p_n = c_n, \tag{5.3}$$

and if this is substituted in the other Hamiltonian equations we get

$$\frac{\partial \mathfrak{H}'(p_i, q^h)}{\partial p_k} = \overset{*}{q}{}^k, \qquad \frac{\partial \mathfrak{H}'(p_i, q^h)}{\partial q^k} = -\overset{*}{p}_k,$$
$$\mathfrak{H}'(p_i, q^h) \overset{\text{def}}{=} \mathfrak{H}(p_i, c_n, q^k) \quad (h, i, k = 0, 1, ..., n-1), \tag{5.4}$$

and
$$\frac{\partial \mathfrak{H}(p_i, c_n, q^k)}{\partial c_n} = \overset{*}{q}{}^n \quad (i, k = 0, 1, ..., n-1). \tag{5.5}$$

(5.4) is a set of Hamiltonian equations in the X_n that arises if film-space is reduced with respect to the congruence of the coordinate curves of q^n, that is, if we consider these curves as points of an X_n by ignoring the coordinate q^n (cf. Ch. I, § 3). This is the reason why in dynamics a coordinate which does not occur in $\mathfrak{L}$ (though its derivative may occur) is often called 'ignorable'.

The equations (5.4) are the homogeneous Hamiltonian equations of a dynamical problem in this X_n and the corresponding Lagrangian equations are

$$\frac{d}{d\tau}\frac{\partial \mathfrak{L}'}{\partial \overset{*}{q}{}^k} - \frac{\partial \mathfrak{L}'}{\partial q^k} = 0,$$
$$\mathfrak{L}' \overset{\text{def}}{=} p_i \overset{*}{q}{}^i - \mathfrak{H}' = \mathfrak{L}(q^h, \overset{*}{q}{}^h, \overset{*}{q}{}^n) - c_n \overset{*}{q}{}^n \tag{5.6}$$
$$(h, i, k = 0, 1, ..., n-1)$$

if $\overset{*}{q}{}^n$ is expressed as a function of q^h, $\overset{*}{q}{}^h$, and c_n by means of (5.5) and (3.21).

For each value of c_n we now have exactly $\infty^{2(n-1)}$ world-lines in the X_n, hence altogether we have ∞^{2n-1}. If

$$q^\phi = f^\phi(\tau) \quad (\phi = 0, 1, ..., n) \tag{5.7}$$

is one of the world-lines in X_{n+1}, the equations

$$q^h = f^h(\tau) \quad (h = 0, 1, ..., n-1) \tag{5.8}$$

represent a world-line in the X_n and it follows from (5.5) that the equations

$$\begin{aligned} &(a) \qquad q^h = f^h(\tau) \quad (h = 0, 1, ..., n-1), \\ &(b) \qquad q^n = f^n(\tau) + c, \end{aligned} \tag{5.9}$$

where c is an arbitrary constant, represent another world-line in X_{n+1}. Hence to every world-line in X_n there belong ∞^1 world-lines in X_{n+1} and this agrees with the total number ∞^{2n} of world-lines in X_{n+1}. More-

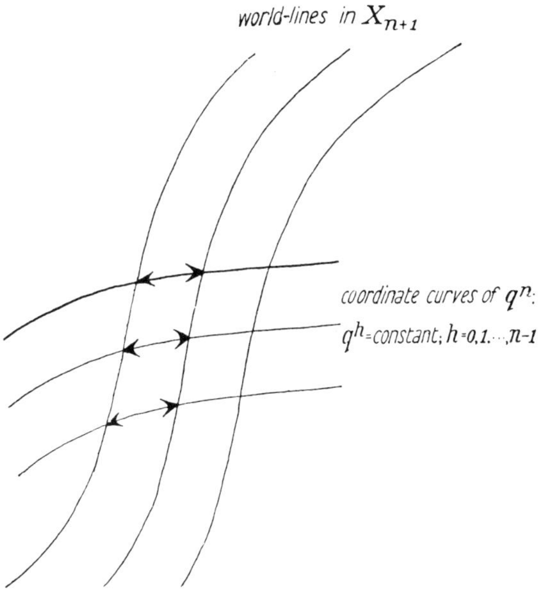

FIG. 29.

over we see that two world-lines in X_{n+1} which belong to the same world-line in X_n cut off segments from the intersecting parametric curves of q^n, that have the same 'length' if measured with q^n.

The process of integration is now simplified by reducing from $n+1$ to n. If the world-lines in X_n are known, the world-lines in X_{n+1} can be found by means of a quadrature. To do this the p_i have to be expressed in q^k and $\overset{*}{q}{}^k$ by means of the equation

$$p_i = \frac{\partial \mathfrak{L}'}{\partial \overset{*}{q}{}^i} \quad (i = 0, 1, ..., n-1). \tag{5.10}$$

If then these values and the values of q^h from (5.9a) are substituted in (5.5) we get an equation of the form

$$\overset{*}{q}^{n} = F(\tau), \tag{5.11}$$

from which q^n can be found by means of a quadrature.

If there are more ignorable coordinates, e.g. $q^1,\ldots, q^m$ the problem can be reduced in the same way to a problem in an X_{n-m+1}. The transformation of $\mathfrak{L}$ into $\mathfrak{L}'$ is called the transformation of *Routh–Helmholtz*.

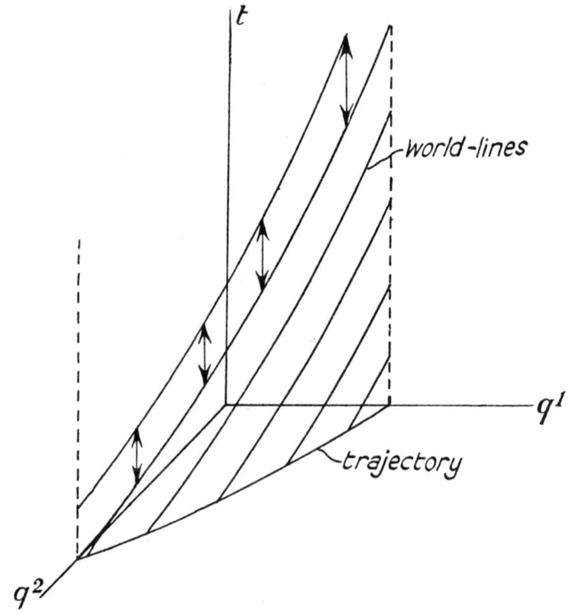

Fig. 30.

It may happen that $\overset{0}{q} = t$ is an ignorable coordinate. In that case we may also return to the non-homogeneous treatment. Then $\tau = t$ and $\mathfrak{L} = L$ and L only depends on q^κ and $\dot{q}^\kappa$. It follows from (3.10 *b*, *c*) that $dH/dt = 0$ or $H =$ constant. Because

$$L\,dt = -H\,dt + p_\lambda\,dq^\lambda \tag{5.12}$$

the variational equation of the trajectories now takes the form

$$\delta \int L\,dt = \delta \int p_\lambda\,dq^\lambda = 0. \tag{5.13}$$

This equation is known as the *variational equation of Maupertuis* (established by Euler and Lagrange) or the *principle of least action*. In fact $p_\lambda\,dq^\lambda$ has the dimension $[ml^2t^{-1}]$ of an action (quantum). To every trajectory there now belong ∞^1 world-lines and these transform into each other by a translation in the t-direction.

EXERCISES

VIII.1. The position of a homogeneous sphere is given by the coordinates x, y, z of the centre and the Eulerian angles ϕ, θ, ψ chosen in the way shown in Fig. 31. If w_1, w_2, and w_3 are the components of the angular velocity about axes

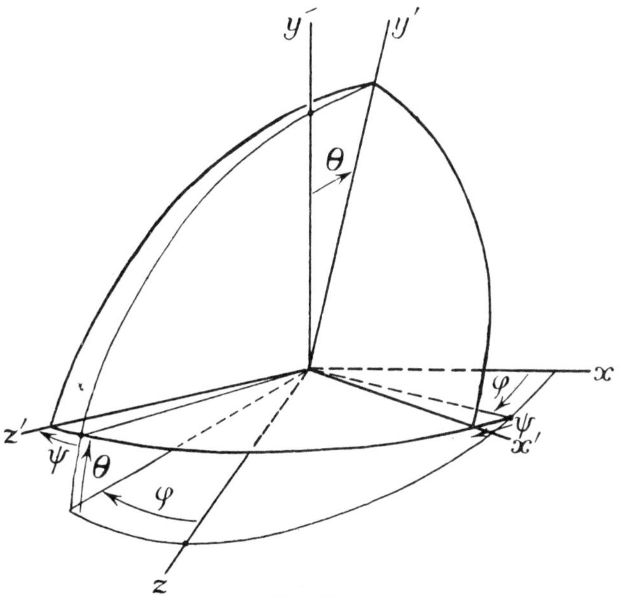

FIG. 31.

through the centre parallel to the x-, y-, and z-axes, prove that

$$\begin{aligned} w_1 &= \qquad -\dot\theta\cos\phi - \dot\psi\sin\theta\sin\phi \\ w_2 &= -\dot\phi \qquad -\dot\psi\cos\theta \\ w_3 &= \qquad -\dot\theta\sin\phi + \dot\psi\sin\theta\cos\phi. \end{aligned} \tag{1 α}$$

Now let $\quad q^1 = x,\quad q^2 = y,\quad q^3 = z,\quad q^4 = \phi,\quad q^5 = \theta,\quad q^6 = \psi,$ (1 β)

and introduce the non-holonomic system (h) $(h = 1,..., 6)$

$$q^1 = x,\quad q^2 = y,\quad q^3 = z,\quad \dot q^4 = w_1,\quad \dot q^5 = w_2,\quad \dot q^6 = w_3. \tag{1 γ}$$

Prove that the only non-vanishing components of Ω^h_{ji} are

$$\Omega^4_{56} = \Omega^5_{64} = \Omega^6_{45} = -\tfrac{1}{2} \tag{1 δ}$$

and write out the Lagrangian equations for the non-holonomic system.†

VIII.2. If we impose on the sphere of Ex. 1 the condition that it moves on the perfectly rough plane $y = -R$, then there are three equations

$$y = 0,\qquad \dot x + w_3 R = 0,\qquad \dot z - w_1 R = 0. \tag{2 α}$$

Holonomic coordinates are now

$$q^1 = x,\quad q^2 = z,\quad q^3 = \phi,\quad q^4 = \theta,\quad q^5 = \psi. \tag{2 β}$$

Introduce the non-holonomic system q^h $(h = 1,..., 5)$

$$q^1 = x,\quad q^2 = z,\quad q^3 = \phi,\quad \dot q^4 = \dot x + w_3 R,\quad \dot q^5 = \dot z - w_1 R, \tag{2 γ}$$

† Cf. Whittaker 1917. 1, p. 44.

and prove that the only non-vanishing components of Ω^h_{ji} are

$$
\begin{aligned}
\Omega^4_{12} &= \frac{-\cos\theta\cos\phi}{R\sin\theta} & \Omega^4_{23} &= -\frac{1}{2} \\
\Omega^4_{15} &= \frac{\cos\theta\cos\phi}{R\sin\theta} & \Omega^4_{24} &= -\frac{\cos\theta\cos\phi}{R\sin\theta} \\
\Omega^4_{35} &= -\frac{1}{2} & \Omega^4_{45} &= -\frac{\cos\theta\cos\phi}{R\sin\theta} \\
\Omega^5_{12} &= -\frac{\cos\theta\sin\phi}{R\sin\theta} & \Omega^5_{24} &= -\frac{\cos\theta\sin\phi}{R\sin\theta} \\
\Omega^5_{13} &= \frac{1}{2} & \Omega^5_{34} &= \frac{1}{2} \\
\Omega^5_{15} &= \frac{\cos\theta\sin\phi}{R\sin\theta} & \Omega^5_{45} &= -\frac{\cos\theta\sin\phi}{R\sin\theta}.
\end{aligned}
\tag{2δ}
$$

Write out the Lagrangian equations for the non-holonomic system.†

VIII.3. The equations of classical electrodynamics (cf. VI, § 2)

$$m\dot{\mathbf{v}} = e\mathbf{F}+e\mathbf{v}\times\mathbf{B} \tag{3α}$$

or

$$m\ddot{x} = -eF_1-e\dot{y}B_{21}-e\dot{z}B_{31};\quad \text{cycl. } 1,2,3 \tag{3β}$$

can be derived from the Lagrangian function

$$L = \tfrac{1}{2}m(\dot{x}^2+\dot{y}^2+\dot{z}^2)+\frac{e}{c}(\phi_1\dot{x}+\phi_2\dot{y}+\phi_3\dot{z})+e\phi_4 \tag{3γ}$$

provided that

$$
\begin{aligned}
F_1 &= \partial_4\phi_1-\partial_1\phi_4, \qquad \partial_4 = \frac{1}{c}\frac{\partial}{\partial t} \\
B_{23} &= \frac{1}{c}(\partial_2\phi_3-\partial_3\phi_2), \quad \text{cycl. } 1,2,3
\end{aligned}
\tag{3δ}
$$

(cf. the notations cF_{41} and F_{23} in IX, § 2). The corresponding Hamiltonian function is

$$H = \frac{1}{2m}\Big(p_1-\frac{e}{c}\phi_1\Big)^2+\text{cycl.}-e\phi_4. \tag{3ε}$$

By homogenization according to VIII, § 3, and taking

$$d\tau = \frac{1}{c}\sqrt{(c^2\,dt^2-dx^2-dy^2-dz^2)} \tag{3ζ}$$

we get

$$\mathfrak{L} = L\overset{*}{t} = \tfrac{1}{2}m(\overset{*}{x}{}^2+\overset{*}{y}{}^2+\overset{*}{z}{}^2)/\overset{*}{t}+\frac{e}{c}(\phi_1\overset{*}{x}+\phi_2\overset{*}{y}+\phi_3\overset{*}{z}+\phi_4 c\overset{*}{t}) \tag{3η}$$

(* denotes differentiation with respect to τ)

for the homogeneous Lagrangian function.

From a relativistic point of view (cf. Ch. IX, § 4) the form (3 η) is rather disappointing, since the first term on the right-hand side does not have four-dimensional invariance. But this is quite natural because we started from classical mechanics and considered m as a constant and not as $m_0\overset{*}{t}$, that is, as a function of the velocity. So it is clear that it is impossible to get a relativistic form of $\mathfrak{L}$ merely by the formal process of homogenization described in VIII, § 3.

† Cf. for more difficult exercises Whittaker 1917. 1, pp. 214 ff.

Prove that the four-dimensional form of $\mathfrak{L}$

$$\mathfrak{L} = -Qmc^2+\frac{e}{c}(\phi_1\overset{\bullet}{x}+\phi_2\overset{\bullet}{y}+\phi_3\overset{\bullet}{z}+\phi_4 c\overset{\bullet}{t}), \tag{3θ}$$

where

$$Q \overset{\text{def}}{=} \frac{1}{c}\sqrt{(c^2\overset{\bullet}{t}{}^2-\overset{\bullet}{x}{}^2-\overset{\bullet}{y}{}^2-\overset{\bullet}{z}{}^2)} \quad (=1) \tag{3ι}$$

gives the right equations and that the fourth of these equations expresses the conservation of energy (cf. IX. 4.19 b). From this $\mathfrak{L}$ we get

$$\begin{aligned} p_1 &= \frac{\partial\mathfrak{L}}{\partial\overset{\bullet}{x}} = m\dot{x}+\frac{e}{c}\phi_1; \quad \text{cycl. } 1,2,3, \\ p_4 &= \frac{1}{c}\frac{\partial\mathfrak{L}}{\partial\overset{\bullet}{t}} = mc+\frac{e}{c}\phi_4. \end{aligned} \tag{3κ}$$

Derive the Hamiltonian relation

$$\mathfrak{H} = -\mathfrak{L}+p_1\overset{\bullet}{x}+p_2\overset{\bullet}{y}+p_3\overset{\bullet}{z}+p_4 c\overset{\bullet}{t} = 0 \tag{3λ}$$

and prove in addition that

$$\overset{0}{\mathfrak{H}} = -\tfrac{1}{2}Q\underset{0}{m}c^2+\tfrac{1}{2}\underset{0}{m}c^2 = \tfrac{1}{2}\underset{0}{m}c^2-\frac{1}{2\underset{0}{m}}\left\{-\left(p_1-\frac{e}{c}\phi_1\right)^2-\text{cycl.}+\left(p_4-\frac{e}{c}\phi_4\right)^2\right\} = 0$$

is also a suitable form of the Hamiltonian relation.

VIII.4. If the $\overset{\bullet}{q}{}^\phi$ are considered as functions of the q^ψ, the p_λ are also functions of the q^ψ. If now

$$\bar{\partial}_\psi \overset{\text{def}}{=} \frac{\partial}{\partial q^\psi}+\frac{\partial\overset{\bullet}{q}{}^\chi}{\partial q^\psi}\frac{\partial}{\partial\overset{\bullet}{q}{}^\chi} \quad (\phi,\psi,\chi = 0,1,\ldots,n) \tag{4α}$$

prove that the Lagrangian equations (3.23) are equivalent to

$$2\overset{\bullet}{q}{}^\psi\bar{\partial}_{[\psi}p_{\phi]} = 0 \quad (\phi,\psi = 0,1,\ldots,n) \tag{4β}$$

or

$$\pounds p_\phi = 0 \quad (\phi = 0,1,\ldots,n) \tag{4γ}$$

where $\pounds$ is the symbol of the Lie derivative with respect to the field $\overset{\bullet}{q}{}^\psi/\mathfrak{L}$.† If the $\overset{\bullet}{q}{}^\phi$ are unknown functions of the q^ψ, (4 γ) represents a system of n independent differential equations for the n ratios of the $\overset{\bullet}{q}{}^\phi$.

† Cf. v. Dantzig 1934. 5, p. 645 f.

IX

RELATIVITY†

1. Introduction

TENSOR calculus would not exist in its modern form if there had never been a theory of relativity. The ties between these two branches of mathematics and physics are so many that they would fill a big textbook. So we have had to make a selection. After some necessary preliminaries we give, in the last section, a short introduction to relativistic thermo-hydrodynamics, since this is a subject which demands a certain amount of ability in tensor calculations. This treatment is based wholly on the work of van Dantzig. Although the affine invariant form of the electromagnetic equations was not unknown to preceding authors,‡ van Dantzig was the first to develop in a long series of publications a consistent theory of relativity which was independent of metrical geometry. In the later publications of this series he dealt with thermo-hydrodynamics. It is not possible to give a survey of the whole theory here. We have therefore restricted ourselves to the phenomenological point of view, and have simplified the assumptions in such a way that an elementary introduction could be set out. This method has had the disadvantage that nothing has remained of van Dantzig's original idea of the independence of metrical geometry, and that most of his results have not been mentioned at all. But we hope that our last section, short and insufficient though it may be, will induce many readers to study the very interesting, though rather difficult, original papers of van Dantzig, Synge, and other authors.§

2. Four-dimensional invariant electromagnetic equations||

In order to get the four-dimensional form of the electromagnetic equations (VI. 2.5 *a–f*) *in vacuo* in cartesian coordinates, we first introduce the following change of notation.

† General references: Einstein 1916. 1; v. Laue 1920. 1; Pauli 1921. 1; Eddington 1923. 1; Fokker 1929. 1; v. Dantzig 1934. 3, 4, 5, 6.

‡ Cf. e.g. Weyl 1921. 2, p. 118.

§ Cf. for a list of literature v. Dantzig 1940. 2, p. 401.

|| In this chapter we use $h, i, j, k, l = 1,..., 4$ for cartesian coordinates and $\kappa, \lambda, \mu, \nu, \rho, \sigma, \tau = 1, 2, 3, 4$ for general rectilinear or curvilinear coordinates in space time. In space we use $a, b, c, d = 1,..., 3$ for cartesian coordinates and $\alpha, \beta, \gamma = 1, 2, 3$ for general rectilinear or curvilinear coordinates.

$$
\begin{aligned}
&(a) && x^4 \stackrel{\text{def}}{=} ct,\\
&(b) && F_{b4} = -F_{4b} \stackrel{\text{def}}{=} -\frac{1}{c} F_b, \qquad F_{cb} \stackrel{\text{def}}{=} B_{cb},\\
&(c) && \mathfrak{F}^{4a} = -\mathfrak{F}^{a4} \stackrel{\text{def}}{=} \mathfrak{D}^a, \qquad \mathfrak{F}^{ab} \stackrel{\text{def}}{=} \frac{1}{c} \mathfrak{H}^{ab},\\
&(d) && \mathfrak{s}^a \stackrel{\text{def}}{=} \frac{\rho}{c} s^a, \qquad s^a \stackrel{\text{def}}{=} u^a, \qquad \mathfrak{s}^4 \stackrel{\text{def}}{=} \rho
\end{aligned}
$$

$$(a, b, c = 1, 2, 3). \tag{2.1}$$

Then we get

$$
\begin{aligned}
&(a) && \partial_2 F_{34} + \partial_3 F_{42} + \partial_4 F_{23} = 0,\\
&(b) && \partial_1 F_{23} + \partial_2 F_{31} + \partial_3 F_{12} = 0,\\
&(c) && \partial_4 \mathfrak{F}^{41} + \partial_2 \mathfrak{F}^{21} + \partial_3 \mathfrak{F}^{31} = -\mathfrak{s}^1,\\
&(d) && \partial_1 \mathfrak{F}^{14} + \partial_2 \mathfrak{F}^{24} + \partial_3 \mathfrak{F}^{34} = -\mathfrak{s}^4,\\
&(e) && \mathfrak{F}^{41} = -\underset{0}{\epsilon} c F_{41},\\
&(f) && \mathfrak{F}^{23} = \frac{1}{\underset{0}{\overset{*}{\mu}} c} F_{23}; \quad \underset{0}{\epsilon}\underset{0}{\overset{*}{\mu}} = 1/c^2; \text{ cycl. } 1, 2, 3.
\end{aligned}
\tag{2.2}
$$

Let the fundamental tensor g_{ih} $(i, h = 1,..., 4)$ whose space component is the fundamental tensor g_{ab} $(a, b = 1, 2, 3)$:

$$g_{11} = -1, \quad g_{22} = -1, \quad g_{33} = -1, \quad g_{44} = +1, \tag{2.3}$$

be introduced in the E_4 of $x^1 = x$, $x^2 = y$, $x^3 = z$, $x^4 = ct$. Then (2.2) takes the form

$$
\begin{aligned}
&(a) && \partial_{[j} F_{ih]} = 0,\\
&(b) && \partial_j \mathfrak{F}^{jh} = -\mathfrak{s}^h,\\
&(c) && \underset{0}{\overset{*}{\mu}} c \mathfrak{F}^{hi} = \mathfrak{g}^{\frac{1}{2}} g^{hj} g^{ik} F_{jk}; \quad \mathfrak{g} \stackrel{\text{def}}{=} |\det(g_{\lambda\kappa})|
\end{aligned}
\tag{2.4}
$$

$$(h, i, j, k = 1,..., 4).$$

These equations have the four-dimensional invariant form if F_{ih} is considered as a covariant bivector in R_4, $\mathfrak{F}^{jh}$ as a contravariant bivector density and $\mathfrak{s}^h$ as a contravariant vector density, both of weight $+1$, and accordingly they can be written in general rectilinear or curvilinear coordinates:

$$
\begin{aligned}
&(a) && \partial_{[\mu} F_{\lambda\kappa]} = 0,\\
&(b) && \partial_\mu \mathfrak{F}^{\mu\kappa} = -\mathfrak{s}^\kappa,\\
&(c) && \underset{0}{\overset{*}{\mu}} c \mathfrak{F}^{\kappa\lambda} = \mathfrak{g}^{\frac{1}{2}} g^{\kappa\mu} g^{\lambda\nu} F_{\mu\nu}
\end{aligned}
\tag{2.5}
$$

$$(\kappa, \lambda, \mu, \nu = 1,..., 4).$$

$F_{\mu\lambda}$ is composed of **F** and **B**, $\mathfrak{F}^{\kappa\lambda}$ of **D** and **H**, and $\mathfrak{s}^\kappa$ of ρ and **u**.

3. Relativistic kinematics

The equations (2.4) are invariant for all orthogonal transformations in the R_4 of x, y, z, and ct. Such a transformation leaves the null cone

$$-x^2-y^2-z^2+c^2t^2 = 0 \tag{3.1}$$

invariant. If the two blades of this cone are not interchanged it is called a *Lorentz transformation.* A Lorentz transformation leaves the relation past-future invariant. Now it can be proved that every Lorentz transformation can be obtained by an ordinary rotation or rotation with reflection in space followed by a so-called *special Lorentz transformation*, that is a transformation of the form

$$\begin{aligned} x' &= x\cosh\phi - ct\sinh\phi, & x &= x'\cosh\phi + ct'\sinh\phi, \\ y' &= y, & y &= y', \\ z' &= z, & z &= z', \\ ct' &= -x\sinh\phi + ct\cosh\phi, & ct &= x'\sinh\phi + ct'\cosh\phi, \end{aligned}$$
$$\cosh^2\phi - \sinh^2\phi = 1. \tag{3.2}$$

The first transformation is not important because we know quite well that we are allowed to pass from one cartesian coordinate system at rest in space to another one equally at rest. So we have only to consider the special Lorentz transformation (3.2).

From Fig. 32 we see that a point that is at rest with respect to the new coordinate system, $x' = 0$, has a velocity $v = c\dfrac{\sinh\phi}{\cosh\phi}$ with respect to the original system. Hence

$$\sinh\phi = \frac{\beta}{\sqrt{(1-\beta^2)}}; \qquad \cosh\phi = \frac{1}{\sqrt{(1-\beta^2)}}; \qquad \beta \overset{\text{def}}{=} v/c. \tag{3.3}$$

From the reciprocity of the formulae (3.2) it follows that a point that is at rest with respect to the original coordinate system, $x = 0$, has a velocity $-c\dfrac{\sinh\phi}{\cosh\phi}$ with respect to the new system. But in this new system the units of length and $c\times$time are the segments OA and OB cut out by the two unit hyperbolas. The geometry in the (x, ct)-plane is not an ordinary one but a *Minkowski geometry* with a fundamental tensor $g_{11} = -1$, $g_{44} = +1$. The linear element in this geometry is

$$c\,ds = \sqrt{(c^2dt^2-dx^2)} = c\,dt\sqrt{(1-\beta^2)}. \tag{3.4}$$

The transformations (3.2) can now be written in the form

$$x' = \frac{x}{\sqrt{(1-\beta^2)}} - \frac{\beta ct}{\sqrt{(1-\beta^2)}}; \qquad x = \frac{x'}{\sqrt{(1-\beta^2)}} + \frac{\beta ct'}{\sqrt{(1-\beta^2)}},$$

$$t' = -\frac{x\beta}{c\sqrt{(1-\beta^2)}} + \frac{t}{\sqrt{(1-\beta^2)}}; \qquad t = \frac{x'\beta}{c\sqrt{(1-\beta^2)}} + \frac{t'}{\sqrt{(1-\beta^2)}}. \tag{3.5}$$

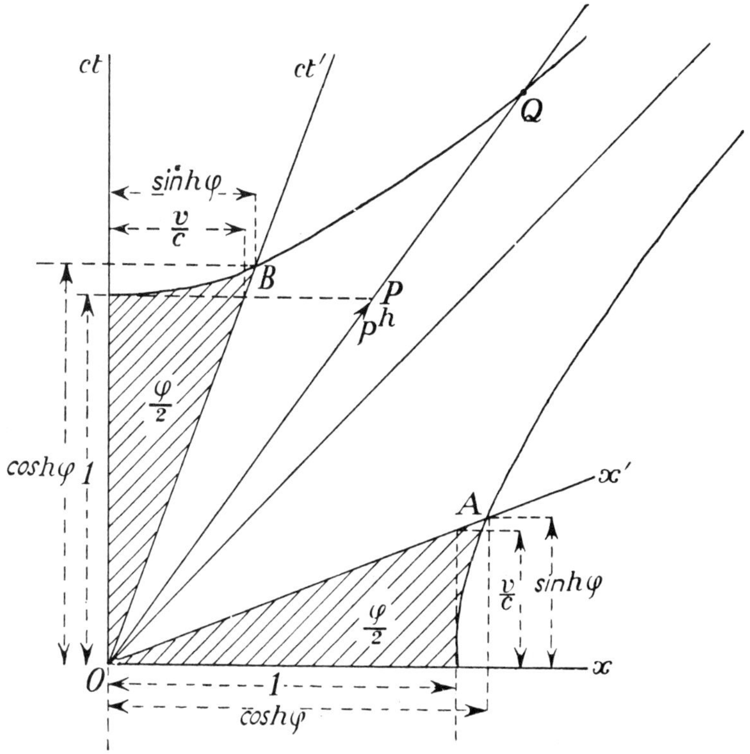

FIG. 32.

Length and time are measured with a ruler and a clock. Since the electromagnetic equations are invariant for Lorentz transformations, a ruler and a clock, based on electromagnetic principles only, would mark x and t if at rest in the original system but x' and t' if at rest in the new system. Hence, measuring only in an electromagnetic way, x' and t' would be the real distance and the real time for the moving observer and not only auxiliary variables. From this point of view consider now a point with a velocity u with respect to the original system. Then this point describes a straight world-line OP (Fig. 32) and at the time 1 for the (x, t)-observer it is at the point P with the coordinates $p^1 = u/c$, $p^4 = 1$ with respect to (x, ct). The coordinates of P with respect to (x', t') are

$$p^{1'} = \frac{1}{\sqrt{(1-\beta^2)}}p^1 - \frac{\beta}{\sqrt{(1-\beta^2)}}p^4 = \frac{1}{\sqrt{(1-\beta^2)}}\frac{u}{c} - \frac{\beta}{\sqrt{(1-\beta^2)}},$$
$$p^{4'} = -\frac{\beta}{\sqrt{(1-\beta^2)}}p^1 + \frac{1}{\sqrt{(1-\beta^2)}}p^4 = -\frac{\beta}{\sqrt{(1-\beta^2)}}\frac{u}{c} + \frac{1}{\sqrt{(1-\beta^2)}}. \tag{3.6}$$

Hence for the (x', t')-observer (using electromagnetic measuring instruments only) the moving point has a velocity

$$u' = c\frac{u/c-\beta}{-\beta(u/c)+1} = \frac{u-v}{1-uv/c^2}. \tag{3.7}$$

This is the relativistic composition of velocities. If $uv \ll c^2$ the relative velocity with respect to (x', t') tends to the classical value $u-v$. But if $u = c$ we find $u' = c$ and this means that light has the same velocity with respect to both observers (Michelson–Morley experiment).

A velocity v^a is a three-dimensional vector, but if $v < c$ it is uniquely determined by the four-dimensional unit vector in its world-line $\underset{I}{v}{}^h = \dfrac{dx^h}{c\,ds}$ with the components

$$\underset{I}{v}{}^1 = \frac{v^1}{c\sqrt{(1-\beta^2)}}, \quad \underset{I}{v}{}^2 = \frac{v^2}{c\sqrt{(1-\beta^2)}}, \quad \underset{I}{v}{}^3 = \frac{v^3}{c\sqrt{(1-\beta^2)}}, \quad \underset{I}{v}{}^4 = \frac{1}{\sqrt{(1-\beta^2)}}. \tag{3.8}$$

This vector is called the *four-dimensional velocity vector* ('Vierergeschwindigkeit').

If a bar with length 1 is at rest with respect to (x, t) the world-lines of its end-points cut out a segment OB on the x'-axis (Fig. 33). But since the unit of length on the x'-axis is OA', the length of the bar from the point of view of the (x', t')-observer is $\dfrac{1}{\cosh\phi} = \sqrt{(1-\beta^2)}$. In the same way, if a bar with length 1 is at rest with respect to (x', t') the segment cut out from the x-axis is OB' and consequently the length from the point of view of the (x, t)-observer is

$$\frac{OB'}{OA} = \frac{\cosh\phi - \sinh\phi(\sinh\phi/\cosh\phi)}{1} = \frac{1}{\cosh\phi} = \sqrt{(1-\beta^2)}. \tag{3.9}$$

From this it follows that a parallelepiped between the planes $x = 0$, $x = a$, $y = 0$, $y = b$, $z = 0$, $z = c$ at rest with respect to (x, t), is seen by the (x', t')-observer as a parallelepiped with the dimensions $a\sqrt{(1-\beta^2)}$, b, c and the volume $abc\sqrt{(1-\beta^2)}$. If this parallelepiped is filled with electricity which has a density ρ for the (x, t)-observer, the density for the (x', t')-observer is $\rho/\sqrt{(1-\beta^2)}$. Hence a charge that has a density ρ_0 (*proper density*) at some point of it, with respect to an observer moving with the same velocity, has a density

$$\rho = \rho_0 \Big/ \sqrt{(1-\beta^2)} \tag{3.10}$$

at that same point for an observer who has a velocity v with respect to the charge.

Let the (x, t)-observer put a row of (electromagnetic) clocks, all at

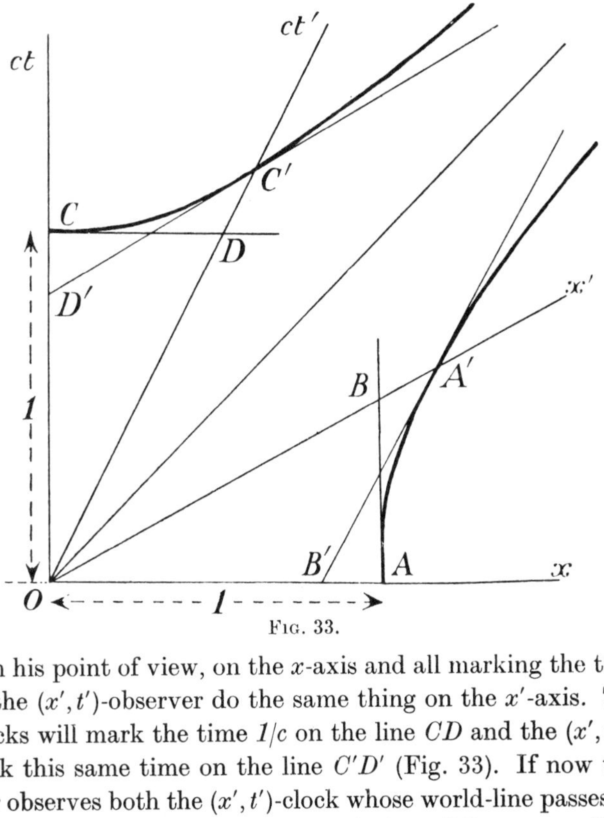

FIG. 33.

rest from his point of view, on the x-axis and all marking the time zero, and let the (x', t')-observer do the same thing on the x'-axis. Then the (x, t)-clocks will mark the time $1/c$ on the line CD and the (x', t')-clocks will mark this same time on the line $C'D'$ (Fig. 33). If now the (x, t)-observer observes both the (x', t')-clock whose world-line passes through O and also his own (x, t)-clocks, he will find no difference at O, because both clocks in that point will mark the time zero. But at D one of the (x, t)-clocks marks the time $1/c$ and the (x', t')-clock marks

$$\frac{1}{c}\frac{OD}{OC'} = \frac{1}{c}\frac{1}{\cosh\phi} = \frac{1}{c}\sqrt{(1-\beta^2)}.$$

In the same way the (x', t')-observer observing the (x, t)-clock whose world-line passes through O and his own (x', t')-clocks, finds no difference at O, but at D' the (x, t)-clock marks

$$\frac{1}{c}\frac{OD'}{OC} = \frac{1}{c}\sqrt{(1-\beta^2)}$$

and the (x', t')-clock marks $1/c$. Hence both observers find that the clocks of the other observer are going too slow in the ratio $\sqrt{(1-\beta^2)}$.

Now in the experiments with electromagnetic rulers and clocks just described there is still no theory of relativity. The statement that the electromagnetic equations *in vacuo* are invariant for Lorentz transformations is not a theory, it expresses only a well-established physical fact that was discovered by Lorentz long before Einstein published his special relativity postulate. But from the point of view of Lorentz x' and t' were artificial auxiliary variables, not a real distance or a real time. Mechanical rulers and clocks were supposed to satisfy the laws of classical mechanics, and classical mechanics are not invariant for Lorentz transformations but for the Galilei group, consisting of rotations of a coordinate system at rest and translations without acceleration. According to classical mechanics rulers and clocks will single out the one real distance x and the one real time t from all artificial distances and times that are equivalent for electromagnetic phenomena. Hence there are two possibilities. Either there are two different kinds of physical phenomena, one invariant for the Lorentz group and one for the Galilei group and there is no general invariance at all, or else all phenomena are invariant for the Lorentz group and classical mechanics is only valid approximately for velocities that are small with respect to c. Now this latter assumption is the special relativity postulate of Einstein. Experiments in all branches of physics have since made this postulate one of the best established physical laws.

4. Relativistic dynamics

If we accept this postulate we have now to look for another kind of dynamics, different from classical dynamics and converging to classical dynamics if v/c tends to zero. We must of course start from the electrodynamical equations

$$\mathbf{K} = e\mathbf{F} + e\mathbf{v} \times \mathbf{B}, \tag{4.1}$$

where e is the charge, $\mathbf{v}$ the velocity of the charge, and $\mathbf{K}$ the force exerted by the field on the charge. If we introduce the notation (2.1 b) in the equations of the components

$$K^1 = eF^1 + e(v^2B^3 - v^3B^2); \text{ cycl. } 1, 2, 3 \tag{4.2}$$

we get the relations

$$K_1 = -ecF_{14}-e(v^2F_{12}-v^3F_{31});\ \text{cycl. } 1,2,3, \tag{4.3}$$

which must be the constituent parts of a four-dimensional invariant relation. It is clear that, in this invariant relation, $\underset{I}{v^h}$ has to appear instead of v^a. Eliminating v^a by means of (3.8) we get

$$K_1\frac{1}{\sqrt{(1-\beta^2)}} = ec\underset{I}{v^2}F_{21}+ec\underset{I}{v^3}F_{31}+ec\underset{I}{v^4}F_{41};\ \text{cycl. } 1,2,3. \tag{4.4}$$

Now, if we introduce K_4 by the equation

$$K_4\frac{1}{\sqrt{(1-\beta^2)}} = ec\underset{I}{v^1}F_{14}+ec\underset{I}{v^2}F_{24}+ec\underset{I}{v^3}F_{34} \tag{4.5}$$

we get the invariant equation

$$\overset{4}{K}_i = ec\underset{I}{v^h}F_{hi} \quad (h,i = 1,2,3,4), \tag{4.6}$$

where $\overset{4}{K}_i$ is a four-dimensional vector with the components

$$\overset{4}{K}_b = K_b\frac{1}{\sqrt{(1-\beta^2)}},$$

$$\overset{4}{K}_4 = ec\underset{I}{v^a}F_{a4} = -\frac{1}{c\sqrt{(1-\beta^2)}}K_a v^a = \frac{1}{c\sqrt{(1-\beta^2)}}\sum_a K^a v^a \tag{4.7}$$

$$(a,b = 1,2,3).$$

$\sum_a K^a v^a$ is the energy lost by the fields **F** and **B** and gained by the charge e in unit time. Hence the force K_b and $1/c\times$ this energy are not, as they stand, components of a four-dimensional vector but can be made so if we multiply them by $1/\sqrt{(1-\beta^2)}$.

Since (4.6) has the invariant form it can be written in general rectilinear or curvilinear coordinates

$$\overset{4}{K}_\lambda = ec\underset{I}{v^\kappa}F_{\kappa\lambda}. \tag{4.8}$$

Now suppose that we have not just one charge but a distribution of charges with the density ρ for an observer with respect to whom the charge has a velocity v. The force per measuring parallelepiped $\mathfrak{k}^a$ is a vector density of weight $+1$ and it satisfies the equations

$$\mathfrak{k}_1 = -\rho cF_{14}-\rho(v^2F_{12}-v^3F_{31});\ \text{cycl. } 1,2,3 \tag{4.9}$$

or, according to (3.8) and (3.10)

$$\mathfrak{k}_1 = \underset{0}{\rho}c\underset{I}{v^h}F_{h1};\ \text{cycl. } 1,2,3 \quad (h = 1,2,3,4). \tag{4.10}$$

Here we have to introduce a fourth component

$$\mathfrak{k}_4 \overset{\text{def}}{=} \underset{0}{\rho}c\underset{I}{v^a}F_{a4} = -\frac{1}{c}\mathfrak{k}_a v^a = \frac{1}{c}\sum \mathfrak{k}^a v^a \quad (a = 1,2,3) \tag{4.11}$$

in order to get the invariant equations

$$\mathfrak{f}_i = \underset{0}{\rho} c \underset{I}{v}^h F_{hi} \quad (h, i = 1, 2, 3, 4) \tag{4.12}$$

or, in general rectilinear or curvilinear coordinates,

$$\mathfrak{f}_\lambda = \underset{0}{\rho}\, c \underset{I}{v}^\kappa F_{\kappa\lambda}. \tag{4.12a}$$

$\mathfrak{f}_\lambda$ is a four-dimensional vector density of weight $+1$ and $c\mathfrak{f}_4$ is the energy lost by the field and gained by the charges per unit volume and per unit time. Hence, unlike the three-dimensional force K_b, the three-dimensional force density $\mathfrak{f}_b$ is itself a component of a four-dimensional quantity. $\underset{0}{\rho}$ is a four-dimensional scalar density of weight $+1$. This also appears from (2.1 d) if we write this equation in the invariant form

$$\mathfrak{s}^h = \underset{0}{\rho} \underset{I}{s}^h \quad (h = 1, 2, 3, 4) \tag{4.13}$$

or, in general rectilinear or curvilinear coordinates,

$$\mathfrak{s}^\kappa = \underset{0}{\rho} \underset{I}{s}^\kappa, \tag{4.14}$$

where $\underset{I}{s}^\kappa$ is the four-dimensional velocity vector of the charge.

We are now able to find the relativistic equation that has to replace the equation

$$K^a = m\frac{d^2x^a}{dt^2} \quad (a = 1, 2, 3) \tag{4.15}$$

of classical mechanics. Instead of K^a this equation must contain $\overset{4}{K}{}^h$, and dt has to be replaced by an invariant differential. Now the only invariant differential is ds. Further, the new equation has to reduce to (4.15) for $\beta = 0$, that is for a particle at rest with respect to x, y, z. Let us try the four-dimensional invariant equation

$$\overset{4}{K}{}^h = \underset{0}{m}\frac{d^2x^h}{ds^2} \quad (h = 1, 2, 3, 4), \tag{4.16}$$

where $\underset{0}{m}$ is a suitable factor that must be connected in some way with the mass. This equation splits up into

$$\begin{aligned} \overset{4}{K}{}^a &= K^a\frac{1}{\sqrt{(1-\beta^2)}} = \underset{0}{m}\frac{1}{\sqrt{(1-\beta^2)}}\frac{d}{dt}\frac{dx^a}{dt\sqrt{(1-\beta^2)}}, \\ \overset{4}{K}{}^4 &= -\frac{1}{c\sqrt{(1-\beta^2)}}K_a\frac{dx^a}{dt} = \underset{0}{m}\frac{1}{\sqrt{(1-\beta^2)}}\frac{d}{dt}\frac{c}{\sqrt{(1-\beta^2)}} \quad (a = 1, 2, 3). \end{aligned} \tag{4.17}$$

The first equation reduces to (4.15) for $\beta = 0$ and $\underset{0}{m} = m$ and the second equation reduces then to the identity $0 = 0$. If we now write

$$m = \frac{\underset{0}{m}}{\sqrt{(1-\beta^2)}}, \tag{4.18}$$

(4.17) can be written in the form

$$\begin{aligned} &(a) \qquad K^a = \frac{d}{dt} m \frac{dx^a}{dt}, \\ &(b) \qquad \sum K^a \frac{dx^a}{dt} = \frac{d}{dt} mc^2 \end{aligned} \qquad (a = 1, 2, 3). \tag{4.19}$$

Since these equations are valid for a coordinate system which is at rest with respect to the moving particle, they must be valid, because of their four-dimensional invariance, for all coordinate systems with constant velocities. This proves that (4.19) are the equations of relativistic mechanics required. $\underset{0}{m}$ is the *rest mass* or *proper mass*, that is the mass with respect to an observer moving with the same velocity, and m is no longer a constant but depends on the rest mass and the velocity with respect to the observer. As in classical mechanics the force is the first time derivative of the momentum vector mv^a, but we are no longer allowed to bring m to the left side of d/dt. The second equation brings a surprise; the left-hand side is the work done by the applied force, but the right-hand side is no longer $\frac{1}{2}mv^2$ as in classical mechanics but mc^2. If $v \ll c$ we have

$$mc^2 = \frac{\underset{0}{m}c^2}{\sqrt{\{1-(v^2/c^2)\}}} \cong \underset{0}{m}c^2\{1+(v^2/c^2)\}^{\frac{1}{2}} = \underset{0}{m}c^2 + \tfrac{1}{2}mv^2 + \ldots, \tag{4.20}$$

and this is in accordance with the classical formula except for the presence of the term $\underset{0}{m}c^2$. Hence it appears that in a mass $\underset{0}{m}$ an energy $\underset{0}{m}c^2$ is always stored up. In these days of atomic bombs this is common knowledge, but in the early days of relativity it was a very important discovery. From the fact that $m \to \infty$ for $v \to c$, it follows immediately that no mass can ever move with the velocity of light with respect to any observer.

The four-dimensional vector $\underset{0}{m}\dfrac{dx^h}{ds}$ is called the *momentum-energy* vector. Its cartesian components are

$$\begin{aligned} \underset{0}{m}\frac{dx^a}{ds} &= \underset{0}{m}c\underset{I}{v^a} = \underset{0}{m}c\frac{v^a}{c\sqrt{(1-\beta^2)}} = mv^a = \text{momentum}, \\ \underset{0}{m}\frac{dx^4}{ds} &= \underset{0}{m}c\underset{I}{v^4} = \underset{0}{m}c\frac{1}{\sqrt{(1-\beta^2)}} = mc = \frac{\text{energy}}{c} \end{aligned}$$

$$(a = 1, 2, 3). \tag{4.21}$$

Hence

$$\text{momentum} = mv^a = \frac{1}{c^2}mc^2v^a = \frac{1}{c^2}\times \text{energy flow}.$$

If the matter is distributed continuously and $\mathfrak{k}^a$ is the force per measuring parallelepiped, and if μ is the mass density and $d\tau$ the volume element, both for an observer with respect to whom the mass in $d\tau$ has the velocity v^a, we have according to (4.19)

$$\begin{aligned} &(a) \qquad \mathfrak{k}^a\, d\tau = \frac{d}{dt}\mu v^a\, d\tau, \\ &(b) \qquad \sum_a \mathfrak{k}^a v^a\, d\tau = \frac{d}{dt}\mu c^2\, d\tau \end{aligned} \qquad (a = 1,2,3). \tag{4.22}$$

If now $\underset{0}{\mu}$ is the mass density and $\underset{0}{\tau}$ the volume, both with respect to the observer moving with the same velocity, we have

$$d\tau = d\underset{0}{\tau}\sqrt{(1-\beta^2)},$$

$$\underset{0}{\mu}\, d\underset{0}{\tau} = \mu\, d\tau\sqrt{(1-\beta^2)} \tag{4.23}$$

and consequently

$$\underset{0}{\mu} = \mu(1-\beta^2). \tag{4.24}$$

This transformation differs from the transformation of the charge density (3.10) because in the case of a mass density there are two different causes for a change, firstly the change of the mass and secondly the change of volume. Introducing $\underset{0}{\mu}$ and ds in (4.22) we get

$$\begin{aligned} \mathfrak{k}^a &= \underset{0}{\mu}\frac{d}{ds}\frac{dx^a}{ds} \\ \mathfrak{k}^4 &= \frac{1}{c}\sum \mathfrak{k}^a v^a = \underset{0}{\mu}\frac{d}{ds}\frac{dx^4}{ds} \end{aligned} \qquad (a = 1,2,3). \tag{4.25}$$

or

$$\mathfrak{k}^h = \underset{0}{\mu}\frac{d}{ds}\frac{dx^h}{ds} \quad (h = 1,2,3,4). \tag{4.26}$$

This equation can be written in another form if we introduce the tensor density

$$\underset{m}{\mathfrak{P}}^{hi} = \underset{0}{\mu}\frac{dx^h}{ds}\frac{dx^i}{ds} \quad (h,i = 1,2,3,4). \tag{4.27}$$

The divergence of this tensor density is

$$\nabla_i \underset{m}{\mathfrak{P}}^{hi} = \partial_i \underset{m}{\mathfrak{P}}^{hi} = \underset{0}{\mu}\frac{dx^i}{ds}\partial_i\frac{dx^h}{ds} + \frac{dx^h}{ds}\partial_i \underset{0}{\mu}\frac{dx^i}{ds} = \mathfrak{k}^h + \frac{dx^h}{ds}\partial_i \underset{0}{\mu}\frac{dx^i}{ds}$$

$$(h,i = 1,2,3,4). \tag{4.28}$$

Now it can be proved that the second term on the right-hand side of

(4.28) vanishes. The number of particles in the volume element is proportional to $\underset{0}{\mu}\,d\tau$, hence, for the observer moving with the same velocity, $\underset{0}{\mu}$ not only gives the mass density but also the *matter density*. But, for the observer at rest, $d\tau = \underset{0}{d\tau}\sqrt{(1-\beta^2)}$, hence for him the matter density is $\underset{*}{\mu} \overset{\text{def}}{=} \underset{0}{\mu}/\sqrt{(1-\beta^2)} = \mu\sqrt{(1-\beta^2)}$, and this is different from the mass density μ. As the matter does not vanish, the equation of continuity must hold for this matter density $\underset{*}{\mu}$. This equation is

$$\begin{aligned} 0 &= \partial_a \underset{*}{\mu} v^a + \partial_t \underset{*}{\mu} = \partial_a \underset{0}{\mu}\sqrt{(1-\beta^2)}\frac{dx^a}{dt} + \partial_t \underset{0}{\mu}\sqrt{(1-\beta^2)} \\ &= \partial_a \underset{0}{\mu}\frac{dx^a}{ds} + \partial_4 \underset{0}{\mu}\frac{dx^4}{ds} = \partial_i \underset{0}{\mu}\frac{dx^i}{ds}, \\ \partial_t &\overset{\text{def}}{=} \frac{\partial}{\partial t} \qquad (a = 1,2,3;\ i = 1,2,3,4), \end{aligned} \tag{4.29}$$

and this proves that in fact

$$\mathfrak{k}^h = \partial_i \underset{m}{\mathfrak{P}}^{hi} = \nabla_i \underset{m}{\mathfrak{P}}^{hi} \quad (h, i = 1,2,3,4). \tag{4.30}$$

Since this equation has the invariant form it can be written in general rectilinear or curvilinear coordinates

$$\mathfrak{k}^\kappa = \nabla_\lambda \underset{m}{\mathfrak{P}}^{\kappa\lambda}. \tag{4.31}$$

$\underset{m}{\mathfrak{P}}^{\kappa\lambda}$ is called the *momentum-energy tensor density* of *continuous* matter. Its cartesian components are

$$\begin{aligned} &(a) \qquad \underset{m}{\mathfrak{P}}^{ab} = \underset{0}{\mu}\frac{dx^a}{ds}\frac{dx^b}{ds} = \mu v^a v^b \quad (a, b = 1,2,3), \\ &(b) \qquad \underset{m}{\mathfrak{P}}^{4a} = \underset{m}{\mathfrak{P}}^{a4} = \underset{0}{\mu}\frac{dx^4}{ds}\frac{dx^a}{ds} = c\mu v^a = c \times \text{momentum per unit volume}, \\ &\qquad\qquad = \frac{1}{c}\mu c^2 v^a = \frac{1}{c} \times \text{energy flow}, \\ &(c) \qquad \underset{m}{\mathfrak{P}}^{44} = \underset{0}{\mu}\frac{dx^4}{ds}\frac{dx^4}{ds} = \mu c^2 = \text{energy per unit volume}. \end{aligned} \tag{4.32}$$

We see from (4.32 *b*) that in relativistic mechanics not only a moving mass but also moving energy has a momentum.

In an electromagnetic field we have, according to (2.4 *a*),

$$F^{jh}\partial_i F_{jh} = -F^{jh}\partial_j F_{hi} - F^{jh}\partial_h F_{ij} = -2F^{jh}\partial_j F_{hi} \quad (h, i, j = 1,2,3,4). \tag{4.33}$$

Hence, since $\mathfrak{f}^h$ is the force per measuring parallelepiped exerted on the charges, we get according to (2.4, 4.12, 13)

$$
\begin{aligned}
\mathfrak{f}_i = c\mathfrak{s}^h F_{hi} &= -c(\partial_j \mathfrak{F}^{jh})F_{hi} \\
&= -c^2 \underset{0}{\epsilon}\, \mathfrak{g}^{\frac{1}{2}}(\partial_j F^{jh})F_{hi} \\
&= -\partial_j\, c^2 \underset{0}{\epsilon}\, \mathfrak{g}^{\frac{1}{2}} F^{jh}F_{hi} + c^2 \underset{0}{\epsilon}\, \mathfrak{g}^{\frac{1}{2}} F^{jh}\partial_j F_{hi} \\
&= -\partial_j\{c^2 \underset{0}{\epsilon}\, \mathfrak{g}^{\frac{1}{2}} F^{jh}F_{hi} + \tfrac{1}{4}c^2 \underset{0}{\epsilon}\, \mathfrak{g}^{\frac{1}{2}} A_i^j\, F^{kh}F_{kh}\} \\
&= -\partial_j\, c\{\mathfrak{F}^{jh}\, F_{hi} + \tfrac{1}{4}A_i^j\, \mathfrak{F}^{kh}F_{kh}\} \\
&= -\partial_j \underset{e}{\mathfrak{P}}{}^{j}_{.i} = -\nabla_j \underset{e}{\mathfrak{P}}{}^{j}_{.i},
\end{aligned} \tag{4.34}
$$

or, in general rectilinear or curvilinear coordinates,

$$
\mathfrak{f}_\lambda = -\nabla_\mu \underset{e}{\mathfrak{P}}{}^{\mu}_{.\lambda}, \tag{4.35}
$$

where

$$
\underset{e}{\mathfrak{P}}{}^{\mu\kappa} \overset{\text{def}}{=} c(\mathfrak{F}^{\mu\lambda}F_{\lambda\nu}g^{\nu\kappa} + \tfrac{1}{4}\mathfrak{F}^{\nu\lambda}F_{\nu\lambda}g^{\mu\kappa}) \tag{4.36}
$$

is called the *momentum-energy tensor density* of the electromagnetic field. Its cartesian components are (cf. 2.1 *b*, *c*)

$$
\begin{aligned}
\underset{e}{\mathfrak{P}}{}^{11} &= -c\mathfrak{F}^{12}F_{21} - c\mathfrak{F}^{13}F_{31} - c\mathfrak{F}^{14}F_{41} - \\
&\quad -\tfrac{1}{2}c\mathfrak{F}^{12}F_{12} - \tfrac{1}{2}c\mathfrak{F}^{23}F_{23} - \tfrac{1}{2}c\mathfrak{F}^{31}F_{31} - \\
&\quad -\tfrac{1}{2}c\mathfrak{F}^{14}F_{14} - \tfrac{1}{2}c\mathfrak{F}^{24}F_{24} - \tfrac{1}{2}c\mathfrak{F}^{34}F_{34} \\
&= \tfrac{1}{2}(-H^1B^1 + H^2B^2 + H^3B^3) + \tfrac{1}{2}(-D^1F^1 + D^2F^2 + D^3F^3), \\
\underset{e}{\mathfrak{P}}{}^{23} = \underset{e}{\mathfrak{P}}{}^{32} &= -c\mathfrak{F}^{21}F_{13} - c\mathfrak{F}^{24}F_{43} = -H^3B^2 - D^2F^3 \\
&= -\frac{1}{\underset{0}{\overset{*}{\mu}}}B^3B^2 - \underset{0}{\epsilon}F^2F^3, \\
\underset{e}{\mathfrak{P}}{}^{14} &= -c\mathfrak{F}^{12}F_{24} - c\mathfrak{F}^{13}F_{34} = -\frac{1}{c}H^3F^2 + \frac{1}{c}H^2F^3 \\
&= \frac{1}{c\underset{0}{\overset{*}{\mu}}}(B^2F^3 - B^3F^2), \\
\underset{e}{\mathfrak{P}}{}^{41} &= -c\mathfrak{F}^{42}F_{21} - c\mathfrak{F}^{43}F_{31} = -cD^2B^3 + cD^3B^2 \\
&= c\underset{0}{\epsilon}(B^2F^3 - B^3F^2) = \underset{e}{\mathfrak{P}}{}^{14},
\end{aligned}
$$

$$\underset{e}{\mathfrak{P}}_{44} = c\mathfrak{F}^{41}F_{14}+c\mathfrak{F}^{42}F_{24}+c\mathfrak{F}^{43}F_{34}$$
$$+\tfrac{1}{2}c\mathfrak{F}^{12}F_{12}+\tfrac{1}{2}c\mathfrak{F}^{23}F_{23}+\tfrac{1}{2}c\mathfrak{F}^{31}F_{31}$$
$$+\tfrac{1}{2}c\mathfrak{F}^{14}F_{14}+\tfrac{1}{2}c\mathfrak{F}^{24}F_{24}+\tfrac{1}{2}c\mathfrak{F}^{34}F_{34}$$
$$=\tfrac{1}{2}(D^1F^1+D^2F^2+D^3F^3)+\tfrac{1}{2}(B^1H^1+B^2H^2+B^3H^3). \tag{4.37}$$

$c\underset{e}{\mathfrak{P}}^{14}$ is the Poynting vector of the field and $\underset{e}{\mathfrak{P}}^{44}$ the energy of the field per unit volume. The fourth component of

$$\mathfrak{k}^h = -\partial_i \underset{e}{\mathfrak{P}}^{ih} \quad (h, i = 1, 2, 3, 4) \tag{4.38}$$

multiplied by $-c$ is

$$\partial_b\, c\underset{e}{\mathfrak{P}}^{b4}+\partial_t \underset{e}{\mathfrak{P}}^{44} = -c\mathfrak{k}^4 \quad (b = 1, 2, 3), \tag{4.39}$$

and this equation has the form of the equation of continuity of a fluid (cf. VI.2.2)

$$\partial_b\, \mathfrak{v}^b+\partial_t \mu = \mathfrak{q} \quad (b = 1, 2, 3), \tag{4.40}$$

where $\mathfrak{v}^b$ is the current vector density, μ the density, and $\mathfrak{q}$ the mass put in from the outside per unit volume and per unit time. Now, as we have seen, $-c\mathfrak{k}^4$ is the energy per unit volume and per unit time gained by the field, and $\underset{e}{\mathfrak{P}}^{44}$ is the energy density. Hence $c\underset{e}{\mathfrak{P}}^{b4}$ is the vector density of the energy current. Now, taking one of the other components,

$$\partial_b \underset{e}{\mathfrak{P}}^{b1}+\partial_t \frac{1}{c} \underset{c}{\mathfrak{P}}^{41} = -\mathfrak{k}^1 \quad (b = 1, 2, 3), \tag{4.41}$$

we see that this equation has the same form. A force is always the first time derivative of a momentum, hence $-\mathfrak{k}^1$ represents the x-component of the momentum per unit volume and per unit time gained by the field. Accordingly $\frac{1}{c}\underset{c}{\mathfrak{P}}^{41} = \frac{1}{c}\underset{e}{\mathfrak{P}}^{14}$ is the x-component of the momentum of the field per unit volume and $\underset{c}{\mathfrak{P}}^{b1}$ is the vector density of the flow of the x-component of the momentum. From this it follows once more that

$$\text{momentum} = 1/c^2 \times \text{flow of energy}.$$

From (4.30) and (4.38) it follows that, for continuous matter,

$$\nabla_i(\underset{m}{\mathfrak{P}}^{hi}+\underset{e}{\mathfrak{P}}^{hi}) = 0 \quad (h, i = 1, 2, 3, 4), \tag{4.42}$$

and this means that the divergence of the total momentum-energy tensor density of matter and field vanishes. This equation expresses the *conservation of momentum and energy* for the special case of continuously distributed matter. In fact, according to (4.30) and (4.38) the momentum and energy lost by the matter is gained by the field and vice versa.

(4.42) can be written in the invariant form

$$\nabla_\mu \mathfrak{P}^{\mu\kappa} = 0, \tag{4.43}$$

where $\mathfrak{P}^{\kappa\lambda}$ is the total momentum-energy tensor density of matter and electromagnetic field. (4.43) is valid in general rectilinear or curvilinear coordinates. This equation is the bridge to general relativity.

5. Gravitation

In classical mechanics the gravitational force acting on a point with mass *1* can be derived from a potential Φ (cf. VIII.3.1)

$$\frac{d^2x^a}{dt^2} = g^{ab}\partial_b \Phi \quad (a, b = 1, 2, 3), \tag{5.1}$$

and this potential satisfies the differential equations

$$\Delta\Phi \overset{\text{def}}{=} -g^{cb}\partial_c \partial_b \Phi = 4\pi f\mu \quad (b, c = 1, 2, 3), \tag{5.2}$$

where μ is the mass density and

$$f = 6{\cdot}7 \times 10^{-8} \text{ gm.}^{-1} \text{ cm.}^3 \text{ sec.}^{-2} \tag{5.3}$$

is Newton's gravitational constant. Now the idea of general relativity is that space-time is not an R_4 but a V_4 and that a mass, influenced only by gravitational forces, always moves in such a way that its world-line is a geodesic in V_4. Since gravitational forces depend on the distribution of matter, this is only possible if the fundamental tensor $g_{\lambda\kappa}$ of V_4 is wholly dependent on this distribution.

In order to see if we get the equation (5.1) of classical dynamics for small gravitational forces and small velocities, we assume that for some coordinate system (κ) the $g_{\lambda\kappa}$ have values which are slightly different from *1* and *0*:

$$g_{\lambda\kappa} \overset{*}{=} \overset{0}{g}_{\lambda\kappa} + \epsilon_{\lambda\kappa}, \qquad \overset{0}{g}_{\lambda\kappa} \overset{*}{=} \delta^\kappa_\lambda, \qquad \epsilon_{\lambda\kappa} \ll 1. \tag{5.4}$$

We also suppose that the velocity of the mass is small with respect to c and that the $\partial_4 \epsilon_{\lambda\kappa}$ are small with respect to the $\partial_\gamma \epsilon_{\lambda\kappa}$ $(\gamma = 1, 2, 3)$. Then we have, neglecting all terms that are small of second order,

$$\Gamma^\kappa_{\mu\lambda} \overset{*}{=} \tfrac{1}{2}\overset{0}{g}{}^{\kappa\rho}(\partial_\mu \epsilon_{\rho\lambda} + \partial_\lambda \epsilon_{\rho\mu} - \partial_\rho \epsilon_{\mu\lambda}) \tag{5.5}$$

and

$$\Gamma^\kappa_{\mu\lambda} \frac{d\xi^\mu}{ds}\frac{d\xi^\lambda}{ds} \overset{*}{=} \Gamma^\kappa_{44} \frac{d\xi^4}{ds}\frac{d\xi^4}{ds} = c^2 \Gamma^\kappa_{44} = -\tfrac{1}{2}c^2 \overset{0}{g}{}^{\kappa\gamma}\partial_\gamma g_{44} \quad (\gamma = 1, 2, 3). \tag{5.6}$$

Substituting these values in the equations of the geodesics

$$\frac{d^2\xi^\kappa}{ds^2} = -\Gamma^\kappa_{\mu\lambda}\frac{d\xi^\mu}{ds}\frac{d\xi^\lambda}{ds} \tag{5.7}$$

we get ($\xi^4 = ct$):

$$\frac{d^2\xi^\alpha}{dt^2} = \tfrac{1}{2}c^2 g^{\alpha\beta}\partial_\beta g_{44} \quad (\alpha, \beta = 1, 2, 3). \tag{5.8}$$

But this is in fact an equation of the form (5.1) with $\frac{1}{2}c^2 g_{44}$ playing the role of the potential.

We have now to look for a relation between the $g_{\lambda\kappa}$ and the distribution of masses, which will replace the equation (5.2) of classical mechanics. Since mass is a form of energy and since energy and momentum are components of the momentum-energy tensor density $\mathfrak{P}^{\kappa\lambda}$, this tensor density, whose divergence vanishes, must come in instead of the mass density μ. Hence we have to look for a tensor density of weight $+1$, whose divergence vanishes, and which is a differential concomitant of the field $g_{\lambda\kappa}$ containing derivatives of the $g_{\lambda\kappa}$ up to the second order. According to (V.5.44) the simplest tensor density with these properties is

$$\mathfrak{G}^{\kappa\lambda} + \alpha \mathfrak{g}^{\frac{1}{2}} g^{\kappa\lambda}, \tag{5.9}$$

where α is an arbitrary constant. If $\mathfrak{G}^{\kappa\lambda}$ equals $\mathfrak{g}^{\frac{1}{2}} g^{\kappa\lambda}$ to within a scalar factor the V_n is an *Einstein space* (cf. V, § 5). It can be proved that the scalar curvature is constant in such a space. The constant α in (5.9) only plays a part in investigations concerning the properties of the V_4 of space-time in the large. It proves to be very small, and it can be put equal to zero if movements of bodies of our solar system only are concerned. Then we get the equation

$$\mathfrak{G}^{\kappa\lambda} = \frac{\kappa}{c^2}\mathfrak{P}^{\kappa\lambda}, \tag{5.10}$$

where κ is a new gravitational constant.

In order to compute this constant we assume that there is no electromagnetic field and that $\underset{m}{\mathfrak{P}}^{\kappa\lambda}$ is given by the equation (4.27). Further, we return to the case where $g_{\lambda\kappa}$ differs only slightly from $\overset{0}{g}_{\lambda\kappa}$ (5.4) and where we have small velocities and small derivatives of $\epsilon_{\lambda\kappa}$ in the time direction. Then all components of $\mathfrak{P}^{\kappa\lambda}$ can be neglected when compared with $\mathfrak{P}^{44} = \mu c^2$. Now it follows from (5.10) and (V.5.36) that

$$K_{\mu\lambda} = \frac{\kappa}{c^2}(P_{\mu\lambda} - \tfrac{1}{2}P g_{\mu\lambda}), \qquad P \overset{\text{def}}{=} P_{\lambda\kappa} g^{\lambda\kappa}, \tag{5.11}$$

hence

$$K_{44} = \frac{\kappa}{c^2}(P_{44} - \tfrac{1}{2}P g_{44}) = \frac{1}{2}\frac{\kappa}{c^2}P_{44} = \tfrac{1}{2}\kappa\mu. \tag{5.12}$$

But we have on the other hand according to (V.5.28),

$$K_{44} = 2\partial_{[\kappa}\Gamma^{\kappa}_{4]4} + 2\Gamma^{\kappa}_{[\kappa|\rho|}\Gamma^{\rho}_{4]4} = \partial_{\kappa}\Gamma^{\kappa}_{44} = \partial_{\gamma}\Gamma^{\gamma}_{44}$$
$$= -\tfrac{1}{2}\underset{0}{g}{}^{\gamma\beta}\partial_{\gamma}\partial_{\beta}g_{44} = \tfrac{1}{2}\Delta g_{44} \quad (\beta,\gamma = 1,2,3), \qquad (5.13)$$

hence
$$\Delta\tfrac{1}{2}c^2 g_{44} = \tfrac{1}{2}\kappa c^2\mu, \qquad (5.14)$$

and, according to (5.2)

$$\kappa = \frac{8\pi f}{c^2} = \frac{8\pi}{c^2}\, 6{\cdot}7\times 10^{-8} = 1{\cdot}87\times 10^{-27}\ \text{gm.}^{-1}\ \text{cm.} \qquad (5.15)$$

By integration of (5.10) it is possible to find the linear element of space-time in the neighbourhood of the world-line of a central mass, e.g. the sun. As there is no matter except on this world-line, $G_{\lambda\kappa}$ and $K_{\lambda\kappa}$ vanish at all points outside the world-line. Hence space-time is not a general V_4 but a V_4 with a vanishing Ricci tensor (cf. V, § 5).

Since space-time is not euclidean in the neighbourhood of the world-line of the sun, slight corrections arise in the movements of planets and light rays. As far as these corrections are measurable, they agree fairly well with astronomical observations.

6. Relativistic hydrodynamics†

We consider matter consisting of particles of the same size and mass and neglect effects of radiation. Let $d\tau$ be a part of space and $d\sigma$ an infinitesimal part of its boundary surface. $d\tau$ is supposed to be small but to contain a great number of particles. If, at any moment, this number is $N^{d\tau}$ and if $N^{d\tau} = \mathfrak{N}\,d\tau$, $\mathfrak{N}$ may be considered, from a macroscopic point of view, as the *particle density* of matter. If the velocity v^{α} of each particle is known we may compute the mean velocity $\dot{u}^{\alpha}$ of the particles in $d\tau$.‡ Then $\mathfrak{N}\dot{u}^{\alpha}$ is the *mean particle current density*. We write $\mathfrak{E}\,d\tau$ for the total energy, kinetic and potential, of all particles in $d\tau$. We suppose that the potential energy does not depend on the velocities but only on the coordinates of the particles in space. $\mathfrak{E}$ is the *mean energy per measuring parallelepiped*. In the same way we write $\mathfrak{M}^{\alpha}\,d\tau$ for the total momentum of all particles in $d\tau$. Then $\mathfrak{M}^{\alpha}$ is the *mean momentum per measuring parallelepiped*. Obviously $\mathfrak{E}$ and $\mathfrak{M}^{\alpha}$ are a scalar density and a vector density both of weight $+1$.

† Cf. also for literature Synge 1934. 7; 1937. 3; v. Dantzig 1934. 3, 4, 5, 6; 1939. 5; 1940. 2.

‡ We write $\dot{u}^{\alpha} = \dfrac{du^{\alpha}}{dt}$ here in order to have the same notation as in Chapter VII. Then u^{α} may be considered as a small displacement.

Now consider an infinitesimal tetrahedron which is at rest and is bounded by $\mathfrak{f}_\beta d\sigma$, $\overset{1}{\mathfrak{f}}_\beta d\overset{1}{\sigma}$, $\overset{2}{\mathfrak{f}}_\beta d\overset{2}{\sigma}$, and $\overset{3}{\mathfrak{f}}_\beta d\overset{3}{\sigma}$ with an orientation from inside to outside. Then we have the identity (cf. VII.2.1)

$$-\mathfrak{f}_\beta d\sigma = \overset{1}{\mathfrak{f}}_\beta d\overset{1}{\sigma}+\overset{2}{\mathfrak{f}}_\beta d\overset{2}{\sigma}+\overset{3}{\mathfrak{f}}_\beta d\overset{3}{\sigma} \quad (\beta = 1,2,3). \tag{6.1}$$

If the energy and momentum passing from the outside through the boundary in unit time is denoted by $E\,d\sigma$, $\underset{1}{E}\,d\overset{1}{\sigma}$, $\underset{2}{E}\,d\overset{2}{\sigma}$, $\underset{3}{E}\,d\overset{3}{\sigma}$ and $p^\alpha d\sigma$, $\underset{1}{p}{}^\alpha d\overset{1}{\sigma}$, $\underset{2}{p}{}^\alpha d\overset{2}{\sigma}$, $\underset{3}{p}{}^\alpha d\overset{3}{\sigma}$ respectively, the changes of energy and of momentum of the matter in $d\tau$ can be neglected because they are of third order in the differentials of the coordinates. Hence it follows from the law of conservation of energy and momentum that (cf. VII.2.2)

$$\begin{aligned} -E\,d\sigma &= \underset{1}{E}\,d\overset{1}{\sigma}+\underset{2}{E}\,d\overset{2}{\sigma}+\underset{3}{E}\,d\overset{3}{\sigma}, \\ -p^\alpha d\sigma &= \underset{1}{p}{}^\alpha d\overset{1}{\sigma}+\underset{2}{p}{}^\alpha d\overset{2}{\sigma}+\underset{3}{p}{}^\alpha d\overset{3}{\sigma} \quad (\alpha = 1,2,3) \end{aligned} \tag{6.2}$$

for *every* choice of the boundary. Hence (cf. VII.2.3) a vector density $\mathfrak{E}^\alpha$ exists such that

$$E = -\mathfrak{E}^\alpha \mathfrak{f}_\alpha \tag{6.3}$$

and there is an affinor density $\mathfrak{T}^{\alpha\beta}$ such that

$$p^\alpha = \mathfrak{T}^{\alpha\beta}\mathfrak{f}_\beta \quad (\alpha,\beta = 1,2,3). \tag{6.4}$$

$\mathfrak{E}^\alpha$ is the mean energy current density. It is the sum of energy $\times$ velocity of all particles in a small volume per unit volume and accordingly is not equal to $\mathfrak{E}\dot{u}^\alpha$, which is a product of two mean values. $-\mathfrak{T}^{\alpha\beta}$ is the mean momentum current density. It is not equal to $\mathfrak{M}^\alpha \dot{u}^\beta$ which is a product of two mean values. $p^\alpha d\sigma$ is the force that the outside matter exerts on the inside matter on $\mathfrak{f}_\beta d\sigma$. Here is an essential difference from continuous matter. If the matter is distributed continuously we can go to the limit $d\tau \to 0$. Then $\dot{u}^\alpha \to v^\alpha$, $\mathfrak{E} \to \mu c^2$, $\mathfrak{E}^\alpha \to \mu c^2 v^\alpha = \mathfrak{E} v^\alpha$, $\mathfrak{M}^\alpha \to \mu v^\alpha$, $\mathfrak{T}^{\alpha\beta} \to \mu v^\alpha v^\beta = \mathfrak{M}^\alpha v^\beta$ (cf. § 2). But in our case $d\tau$ must always have a reasonable size and contain a sufficiently great number of particles.

Up till now we have considered a boundary at rest. Now take a boundary whose points have a velocity w^α, that may depend on x, y, and z. The flow of energy and momentum through the moving element $\mathfrak{f}_\beta d\sigma$ in a sense opposite to its orientation per unit of time is

$$\begin{aligned} p^\alpha d\sigma &= (\mathfrak{T}^{\alpha\beta}+\mathfrak{M}^\alpha w^\beta)\mathfrak{f}_\beta d\sigma, \\ E\,d\sigma &= (-\mathfrak{E}^\beta+\mathfrak{E} w^\beta)\mathfrak{f}_\beta d\sigma \quad (\alpha,\beta = 1,2,3). \end{aligned} \tag{6.5}$$

As before $p^{\alpha} d\sigma$ is the force exerted in $\mathfrak{f}_{\beta} d\sigma$ on the matter at the back of $\mathfrak{f}_{\beta} d\sigma$ by the matter at the front. Instead of the stress tensor $\mathfrak{T}^{\alpha\beta}$ we get a stress affinor $\mathfrak{T}^{\alpha\beta}+\mathfrak{M}^{\alpha}w^{\beta}$ which, in general, is not symmetric.

In order to write (6.5) in a four-dimensional invariant form we take cartesian coordinates in space time. We remember that $p^{\alpha} d\sigma dt$ is a momentum and $\frac{1}{c} E\, d\sigma dt$ an energy divided by c, and that consequently these quantities may be considered as components of a momentum-energy vector. The covariant vector density $\mathfrak{f}_b d\sigma$, which is a part of a plane with an outer orientation, describes in the time dt in space-time a part of an R_3, and its orientation induces an outer orientation in this R_3. Hence the figure described is a four-dimensional covariant vector density $\mathfrak{B}_i d\omega$ in R_4. Now $\mathfrak{f}_b d\sigma$ is the section of $\mathfrak{B}_i d\omega$ with space, hence

$$\mathfrak{B}_b d\omega = \mathfrak{f}_b d\sigma dt \quad (b = 1, 2, 3). \tag{6.6}$$

The direction of the four-dimensional velocity vector $\underset{1}{w}^h$ lies in $\mathfrak{B}_i d\omega$, hence

$$(w^a/c)\mathfrak{B}_a d\omega+\mathfrak{B}_4 d\omega = (w^a/c)\,\mathfrak{f}_a d\sigma dt+\mathfrak{B}_4 d\omega = 0 \quad (a = 1, 2, 3), \tag{6.7}$$

and consequently

$$\mathfrak{B}_4 d\omega = -(w^a/c)\,\mathfrak{f}_a d\sigma dt \quad (a = 1, 2, 3). \tag{6.8}$$

We have used the notation of Chapter VII here for the surface element $\mathfrak{f}_b d\sigma$ where $d\sigma$ is the area measured in m.^{2} But in relativity the length of a bar or the area of a part of an R_2 are not the same for observers with different velocities. Hence, if $\mathfrak{f}_b$ and $\mathfrak{B}_i$ are to have any sense there must be some agreement about how $d\sigma$ and $d\omega$ are to be measured. We agree that $d\sigma$ shall be the proper area, that is the area measured by an observer at rest with respect to the surface element and that $d\omega$ shall be measured by an observer whose world-line lies in $\mathfrak{B}_4$. The duration of the world-line described by a point of $\mathfrak{f}_b d\sigma$ in the time dt is

$$ds = dt\sqrt{(1-w^2/c^2)}. \tag{6.9}$$

This is the time measured by the observer at rest with respect to the surface element. It is called the proper time, and we have

$$d\omega = d\sigma ds = d\sigma dt\sqrt{(1-w^2/c^2)}. \tag{6.10}$$

The difficulty can be avoided by always using the expressions $\mathfrak{f}_b d\sigma$ and $\mathfrak{B}_i d\omega$ and never $\mathfrak{f}_b$ or $\mathfrak{B}_i$ by themselves. Then no agreement is necessary, but only the full expressions $\mathfrak{f}_b d\sigma$ and $\mathfrak{B}_i d\omega$ have any sense and $\mathfrak{f}_b$, $\mathfrak{B}_i$, $d\sigma$ or $d\omega$ alone have no sense at all. Van Dantzig follows this path and marks the fact that only $\mathfrak{f}_b d\sigma$ and $\mathfrak{B}_i d\omega$ have a sense by writing

$\mathfrak{f}_b^{d\sigma}$ and $\mathfrak{V}_i^{d\omega}$. This notation is certainly very good because for many purposes it does not matter how $d\sigma$ and $d\omega$ are measured, but we will not use it here because we wish to have the same notation in this chapter as in the others.

Introducing the values (6.6, 8) into (6.5) we get, for cartesian coordinates,

$$p^a\, d\sigma dt = \mathfrak{T}^{ab}\mathfrak{V}_b\, d\omega - c\mathfrak{M}^a\mathfrak{V}_4\, d\omega,$$

$$\frac{1}{c}E\, d\sigma dt = -\frac{1}{c}\mathfrak{E}^b\mathfrak{V}_b\, d\omega - \mathfrak{E}\mathfrak{V}_4\, d\omega \quad (a,b = 1,2,3). \tag{6.11}$$

Now if we write

$$\begin{aligned} &\mathfrak{P}^{ab} \overset{\text{def}}{=} -\mathfrak{T}^{ab}, && M^a\, d\omega \overset{\text{def}}{=} p^a\, d\sigma dt,\\ &\mathfrak{P}^{a4} \overset{\text{def}}{=} c\mathfrak{M}^a, && M^4\, d\omega \overset{\text{def}}{=} \frac{1}{c}E\, d\sigma dt,\\ &\mathfrak{P}^{4b} \overset{\text{def}}{=} \frac{1}{c}\mathfrak{E}^b, && \\ &\mathfrak{P}^{44} \overset{\text{def}}{=} \mathfrak{E}, && (a,b = 1,2,3) \end{aligned} \tag{6.12}$$

the equations (6.11) take the form

$$M^h\, d\omega = -\mathfrak{P}^{hi}\mathfrak{V}_i\, d\omega \quad (h,i = 1,2,3,4). \tag{6.13}$$

Here $M^h\, d\omega$ is the momentum-energy vector whose orthogonal components represent the momentum and the energy flowing through $\mathfrak{f}_b\, d\sigma$ from the outside to the inside in the time dt.

In order to determine the flow of heat by conduction through a surface element we take $w^b = \dot{u}^b$. Then the *total* amount of matter flowing through $\mathfrak{f}_b\, d\sigma$ is zero and there is no flow of heat by *convection*, only by *conduction*. The energy passing in the time dt through $\mathfrak{f}_b\, d\sigma$ in a sense opposite to its orientation is

$$(-\mathfrak{E}^b + \mathfrak{E}\dot{u}^b)\mathfrak{f}_b\, d\sigma dt \quad (b = 1,2,3). \tag{6.14}$$

But part of this energy is the work done by the force $p^a\, d\sigma$ in the time dt. This part is

$$-p^a\dot{u}_a\, d\sigma dt = (-\mathfrak{T}^{ab}\dot{u}_a - \mathfrak{M}^a\dot{u}_a\,\dot{u}^b)\mathfrak{f}_b\, d\sigma dt \quad (a,b = 1,2,3). \tag{6.15}$$

The remaining part is

$$\begin{aligned} &(-\mathfrak{E}^b + \mathfrak{E}\dot{u}^b + \mathfrak{T}^{ab}\dot{u}_a + \mathfrak{M}^a\dot{u}_a\,\dot{u}^b)\mathfrak{f}_b\, d\sigma dt\\ &\qquad = (-c\mathfrak{P}^{4b} + \mathfrak{P}^{44}\dot{u}^b - \mathfrak{P}^{ab}\dot{u}_a + \frac{1}{c}\mathfrak{P}^{a4}\dot{u}_a\,\dot{u}^b)\mathfrak{f}_b\, d\sigma dt\\ &\qquad\qquad (a,b = 1,2,3), \end{aligned} \tag{6.16}$$

and this is the flow of heat in the time dt by conduction through $\mathfrak{f}_b\, d\sigma$

in a sense opposite to its orientation. Hence the current density vector of heat is

$$\mathfrak{Q}^a = c\mathfrak{P}^{4a} - \mathfrak{P}^{44}\dot{u}^a + \mathfrak{P}^{ba}\dot{u}_b - \frac{1}{c}\mathfrak{P}^{b4}\dot{u}_b\,\dot{u}^a \quad (a,b = 1,2,3). \qquad (6.17)$$

In order to draw further conclusions from (6.17) we go back to the equations (6.5) that can now be written in the form

$$p^a\,d\sigma = -\left(\mathfrak{P}^{ab} - \frac{1}{c}\mathfrak{P}^{a4}w^b\right)\mathfrak{f}_b\,d\sigma,$$

$$\frac{1}{c}E\,d\sigma = -\left(\mathfrak{P}^{4b} - \frac{1}{c}\mathfrak{P}^{44}w^b\right)\mathfrak{f}_b\,d\sigma \quad (a,b = 1,2,3). \qquad (6.18)$$

According to the theorem of Stokes we have

$$\int_\sigma p^a\,d\sigma = -\int_\tau \left(\partial_b\,\mathfrak{P}^{ab} - \frac{1}{c}\partial_b\,\mathfrak{P}^{a4}w^b\right)d\tau$$

$$\frac{1}{c}\int_\sigma E\,d\sigma = -\int_\tau \left(\partial_b\,\mathfrak{P}^{4b} - \frac{1}{c}\partial_b\,\mathfrak{P}^{44}w^b\right)d\tau \quad (a,b = 1,2,3). \qquad (6.19)$$

Now the integrals on the left-hand side represent the increase in τ of momentum and energy divided by c per unit time

$$\int_\sigma p^a\,d\sigma = \frac{d}{dt}\int_\tau \mathfrak{M}^a\,d\tau = \int_\tau \left(\frac{d}{dt}\mathfrak{M}^a\right)d\tau + \int_\tau \mathfrak{M}^a\frac{d}{dt}d\tau,$$

$$\frac{1}{c}\int E\,d\sigma = \frac{1}{c}\frac{d}{dt}\int_\tau \mathfrak{E}\,d\tau = \frac{1}{c}\int_\tau \left(\frac{d}{dt}\mathfrak{E}\right)d\tau + \frac{1}{c}\int_\tau \mathfrak{E}\frac{d}{dt}d\tau$$

$$(a = 1,2,3), \qquad (6.20)$$

and

$$\frac{d}{dt}d\tau = \partial_b\,w^b\,d\tau. \qquad (6.21)$$

Hence

$$\int_\sigma p^a\,d\sigma = \int_\tau \left(\frac{\partial\mathfrak{M}^a}{\partial t} + (\partial_b\,\mathfrak{M}^a)w^b + \mathfrak{M}^a\partial_b\,w^b\right)d\tau$$

$$= \int_\tau \left(\partial_4\,\mathfrak{P}^{a4} + \frac{1}{c}\partial_b\,\mathfrak{P}^{a4}w^b\right)d\tau$$

$$\frac{1}{c}\int_\sigma E\,d\sigma = \frac{1}{c}\int_\tau \left(\frac{\partial\mathfrak{E}}{\partial t} + (\partial_b\,\mathfrak{E})w^b + \mathfrak{E}\partial_b\,w^b\right)d\tau$$

$$= \int_\tau (\partial_4\,\mathfrak{P}^{44} + \partial_b\,\mathfrak{P}^{44}w^b)\,d\tau$$

$$(a,b = 1,2,3). \qquad (6.22)$$

Comparing (6.19) and (6.22) we see that

$$\partial_b \mathfrak{P}^{ab} + \partial_4 \mathfrak{P}^{a4} = 0$$

$$\partial_b \mathfrak{P}^{4b} + \partial_4 \mathfrak{P}^{44} = 0 \quad (a, b = 1, 2, 3), \tag{6.23}$$

or

$$\partial_i \mathfrak{P}^{hi} = 0 \quad (h, i = 1, 2, 3, 4). \tag{6.24}$$

If there are external influences (for instance from an electromagnetic field), instead of (6.20) we get

$$\int_\sigma p^a \, d\sigma + \int_\tau \mathfrak{k}^a \, d\tau = \frac{d}{dt} \int_\tau \mathfrak{M}^a \, d\tau$$

$$\frac{1}{c} \int E \, d\sigma + \int_\tau \mathfrak{k}^4 \, d\tau = \frac{1}{c} \frac{d}{dt} \int_\tau \mathfrak{E} \, d\tau \quad (a = 1, 2, 3), \tag{6.25}$$

where $\mathfrak{k}^a$ represents the increase of momentum and $c\mathfrak{k}^4$ the increase of energy per unit volume and per unit time due to external causes. Then instead of (6.24) we have

$$\mathfrak{k}^h = \partial_i \mathfrak{P}^{hi} \quad (h, i = 1, 2, 3, 4), \tag{6.26}$$

and this is the equation (4.30) which we had already found for the special case of continuous distribution of matter.

In (6.22) a differentiation appears which occurs very often in investigations of this kind. It is convenient to introduce a special symbol for it. If f is a function of x, y, z, t we write

$$\frac{\bar{d}f}{dt} \overset{\text{def}}{=} \frac{\partial f}{\partial t} + w^b \partial_b f + f \partial_b w^b = \frac{df}{dt} + f \partial_b w^b = \frac{\partial f}{\partial t} + \partial_b f w^b$$

$$(b = 1, 2, 3). \tag{6.27}$$

Then we always have

$$\frac{d}{dt}(f \, d\tau) = \frac{\bar{d}f}{dt} d\tau. \tag{6.28}$$

Taking w^a once more equal to $\dot{u}^a$, that is, taking the boundary σ in such a way that the total amount of matter passing through each of its elements is zero, the only energy passing through the boundary is heat and the total amount of heat passing through σ from the outside is (cf. (6.23))

$$-\int_\sigma \mathfrak{Q}^b \mathfrak{f}_b \, d\sigma = -\int_\tau \partial_b \mathfrak{Q}^b \, d\tau$$

$$= -\int (c\partial_b \mathfrak{P}^{4b} - \partial_b \mathfrak{P}^{44} \dot{u}^b + \partial_b \mathfrak{P}^{ab} \dot{u}_a - \frac{1}{c} \partial_b \mathfrak{P}^{a4} \dot{u}_a \dot{u}^b) \, d\tau$$

$$= \int\limits_\tau \Big\{\partial_t \mathfrak{P}^{44} + \partial_b \mathfrak{P}^{44}\dot{u}^b + \frac{1}{c}(\partial_t \mathfrak{P}^{a4})\dot{u}_a - \mathfrak{P}^{ab}\partial_b \dot{u}_a + \frac{1}{c}(\partial_b \mathfrak{P}^{a4})\dot{u}_a \dot{u}^b +$$
$$+ \frac{1}{c}\mathfrak{P}^{a4}(\partial_b \dot{u}_a)\dot{u}^b + \frac{1}{c}\mathfrak{P}^{a4}\dot{u}_a \partial_b \dot{u}^b\Big\}\, d\tau$$
$$= \int\limits_\tau \left(\frac{\bar{d}\mathfrak{P}^{44}}{dt} + \frac{1}{c}\frac{\bar{d}\mathfrak{P}^{a4}}{dt}\dot{u}_a - \mathfrak{P}^{ab}\partial_b \dot{u}_a + \frac{1}{c}\mathfrak{P}^{a4}\dot{u}^b \partial_b \dot{u}_a\right) d\tau$$
$$= \int\limits_\tau \frac{d}{dt}(\mathfrak{P}^{44}\, d\tau) + \frac{1}{c}\frac{d}{dt}(\mathfrak{P}^{a4}\, d\tau)\dot{u}_a + \left(-\mathfrak{P}^{ab} + \frac{1}{c}\mathfrak{P}^{a4}\dot{u}^b\right)(\partial_b \dot{u}_a)\, d\tau$$
$$= \int\limits_\tau \frac{d}{dt}(\mathfrak{E}\, d\tau) + \dot{u}_a \frac{d}{dt}(\mathfrak{M}^a\, d\tau) + (\mathfrak{T}^{ab} + \mathfrak{M}^a \dot{u}^b)(\partial_b \dot{u}_a)\, d\tau$$
$$(a, b = 1, 2, 3). \quad (6.29)$$

Hence, if $\mathfrak{Q}$ is the heat per unit volume, this integral has to be equal to

$$\frac{d}{dt}\int\limits_\tau \mathfrak{Q}\, d\tau = \int\limits_\tau \frac{d}{dt}(\mathfrak{Q}\, d\tau). \quad (6.30)$$

It follows from (6.29) and (6.30) that

$$d(\mathfrak{Q}\, d\tau) = d(\mathfrak{E}\, d\tau) + \dot{u}_a d(\mathfrak{M}^a\, d\tau) + (\mathfrak{T}^{ab} + \mathfrak{M}^a \dot{u}^b)(\partial_b du_a)\, d\tau$$
$$(a, b = 1, 2, 3) \quad (6.31)$$

or
$$d(\mathfrak{E}\, d\tau) = -\dot{u}_a d(\mathfrak{M}^a\, d\tau) + (\mathfrak{T}^{ab} + \mathfrak{M}^a \dot{u}^b)(-\partial_b du_a)\, d\tau + d(\mathfrak{Q}\, d\tau)$$
$$(a, b = 1, 2, 3) \quad (6.32)$$

for a small displacement u^a, and this equation expresses the fact that the increase of total energy in $d\tau$ is the sum of the increase of the kinetic energy of the matter in $d\tau$ moving as a whole with the momentum $\mathfrak{M}^a\, d\tau$, the increase of the potential elastic energy, and the increase of heat due to the heat flowing in from outside. If $\dot{u}^b = 0$, we get the expression

$$-\mathfrak{T}^{ab}(\partial_b du_a)\, d\tau = \mathfrak{T}^{ab}\, dS_{ab}\, d\tau \quad (a, b = 1, 2, 3) \quad (6.33)$$

for the increase of the elastic energy in accordance with (VII.1.6), upper sign, and (VII.3.2).

$\mathfrak{Q}^b \mathfrak{f}_b\, d\sigma dt$ is the heat flowing through $\mathfrak{f}_b\, d\sigma$ in the time dt. In order to obtain the transformation formula for an amount of heat we start from (6.17) and use the equations (6.6, 8):

$$d\mathfrak{Q} = \mathfrak{Q}^b \mathfrak{f}_b\, d\sigma dt = c\mathfrak{P}^{4b}\mathfrak{B}_b\, d\omega + c\mathfrak{P}^{44}\mathfrak{B}_4\, d\omega + \mathfrak{P}^{ab}\dot{u}_a \mathfrak{B}_b\, d\omega + \mathfrak{P}^{a4}\dot{u}_a \mathfrak{B}_4\, d\omega$$
$$= c\mathfrak{P}^{4i}\mathfrak{B}_i\, d\omega + \mathfrak{P}^{ai}\dot{u}_a \mathfrak{B}_i\, d\omega$$
$$= c\sqrt{(1-\beta^2)}(\mathfrak{P}^{4i}\underset{I}{u_4} + \mathfrak{P}^{ai}\underset{I}{u_a})\mathfrak{B}_i\, d\omega$$
$$(a, b = 1, 2, 3), \quad (6.34)$$

where $\underset{I}{u}^h$ is the four-dimensional velocity belonging to $\dot{u}^a$ and

$$\beta^2 c^2 = -\dot{u}^a \dot{u}_a;$$

hence

$$\frac{dQ}{\sqrt{(1-\beta^2)}} = c\mathfrak{P}^{hi}\underset{I}{u}_h \mathfrak{B}_i\,d\omega \quad (h,i = 1,2,3,4). \tag{6.35}$$

The right-hand side of this equation is a scalar. Hence

$$dQ = \underset{0}{dQ}\sqrt{(1-\beta^2)}, \tag{6.36}$$

where $\underset{0}{dQ}$ is the amount of heat from the point of view of an observer moving with the velocity $\dot{u}^a$. Because $d\tau = \underset{0}{d\tau}\sqrt{(1-\beta^2)}$ it follows that an amount of heat per unit volume is an invariant.

The temperature is the heat energy per particle multiplied by a scalar factor depending on the choice of the temperature scale. Now all observers agree about the average amount of particles in τ, and this proves that the transformation formulae of dQ and T are the same, i.e.

$$T = \underset{0}{T}\sqrt{(1-\beta^2)}, \tag{6.37}$$

It follows immediately from this that the entropy S defined by

$$\frac{dQ}{T} = \frac{\underset{0}{dQ}}{\underset{0}{T}} = dS \tag{6.38}$$

is an invariant and that the entropy per unit volume $\mathfrak{S}$ transforms in the following way

$$\underset{0}{\mathfrak{S}} = \mathfrak{S}\sqrt{(1-\beta^2)}, \tag{6.39}$$

Since $ds = dt\sqrt{(1-\beta^2)}$ the product $\underset{0}{T}\,ds = T\,dt$ is an invariant and the expression

$$T^h = \frac{1}{k\underset{0}{T}}\frac{dx^h}{ds} = \frac{1}{kT}\frac{dx^h}{dt}; \quad k = \text{Boltzmann constant}$$

$$(h = 1,2,3,4) \tag{6.40}$$

is a four-dimensional vector, the *temperature vector*.

A fluid is called *perfect* if $p^a\,d\sigma$ is always perpendicular to $\mathfrak{f}_b\,d\sigma$, and this is the case if and only if (cf. (6.5, 18))

$$-\mathfrak{P}^{ab} + \frac{1}{c}\mathfrak{P}^{a4}\dot{u}^b = \mathfrak{p}g^{ab}, \quad (a,b = 1,2,3) \tag{6.41}$$

where $\mathfrak{p}$ is a suitable scalar density. Then

$$p^a\,d\sigma = \mathfrak{p}\mathfrak{f}^a\,d\sigma \quad (a = 1,2,3). \tag{6.42}$$

Now if $\mathfrak{f}_b\,d\sigma$ is perpendicular to the x-axis and has an orientation in its $+$-direction, $\mathfrak{f}^1$ is negative and the force is directed towards the inside. Hence $\mathfrak{p}$ is the pressure.

If the definition of a perfect fluid is to have a meaning which is independent of the coordinate system, (6.41) and the equation for the flow of heat that follows from (6.17) and (6.41)

$$\mathfrak{Q}^a = c\mathfrak{P}^{4a} - \mathfrak{P}^{44}\dot{u}^a - \mathfrak{p}\dot{u}^a \quad (a = 1,2,3) \tag{6.43}$$

must be invariant. The equations of the principal directions of $\mathfrak{P}^{hi}$ are

$$\mathfrak{P}^{hi}r_i - \lambda g^{hi}r_i = 0 \quad (h,i = 1,2,3,4) \tag{6.44}$$

and the eigenvalues are the roots of the equation

$$\begin{vmatrix} \mathfrak{p}+\frac{1}{c}\mathfrak{P}^{14}\dot{u}^1+\lambda & \frac{1}{c}\mathfrak{P}^{14}\dot{u}^2 & \frac{1}{c}\mathfrak{P}^{14}\dot{u}^3 & \mathfrak{P}^{14} \\ \frac{1}{c}\mathfrak{P}^{24}\dot{u}^1 & \mathfrak{p}+\frac{1}{c}\mathfrak{P}^{24}\dot{u}^2+\lambda & \frac{1}{c}\mathfrak{P}^{24}\dot{u}^3 & \mathfrak{P}^{24} \\ \frac{1}{c}\mathfrak{P}^{34}\dot{u}^1 & \frac{1}{c}\mathfrak{P}^{34}\dot{u}^2 & \mathfrak{p}+\frac{1}{c}\mathfrak{P}^{34}\dot{u}^3+\lambda & \mathfrak{P}^{34} \\ -\frac{1}{c}\mathfrak{Q}^1+\frac{1}{c}\mathfrak{P}^{44}\dot{u}^1-\frac{1}{c}\mathfrak{p}\dot{u}^1 & -\frac{1}{c}\mathfrak{Q}^2+\frac{1}{c}\mathfrak{P}^{44}\dot{u}^2-\frac{1}{c}\mathfrak{p}\dot{u}^2 & -\frac{1}{c}\mathfrak{Q}^3+\frac{1}{c}\mathfrak{P}^{44}\dot{u}^3-\frac{1}{c}\mathfrak{p}\dot{u}^3 & \mathfrak{P}^{44}-\lambda \end{vmatrix} = 0 \tag{6.45}$$

or, by elementary transformations,

$$\begin{vmatrix} \mathfrak{p}+\lambda & 0 & 0 & \mathfrak{P}^{14} \\ 0 & \mathfrak{p}+\lambda & 0 & \mathfrak{P}^{24} \\ 0 & 0 & \mathfrak{p}+\lambda & \mathfrak{P}^{34} \\ -\frac{1}{c}\mathfrak{Q}^1 & -\frac{1}{c}\mathfrak{Q}^2 & -\frac{1}{c}\mathfrak{Q}^3 & \mathfrak{P}^{44}+\frac{1}{c}\mathfrak{P}^{a4}\dot{u}_a-\lambda \end{vmatrix} = 0 \tag{6.46}$$

or

$$(\mathfrak{p}+\lambda)^2\left\{(\mathfrak{p}+\lambda)\left(\mathfrak{P}^{44}+\frac{1}{c}\mathfrak{P}^{a4}\dot{u}_a-\lambda\right)\right\} - (\mathfrak{p}+\lambda)^2\mathfrak{P}^{b4}\mathfrak{Q}_b = 0$$
$$(a,b = 1,2,3). \tag{6.47}$$

Since this equation is invariant, a direction in space-time belongs to each root. These directions must have a physical meaning. Now the only invariant direction in space-time in this case is the direction of $\underset{I}{u}^h$. This implies firstly that the first three roots have to be equal and the corresponding R_3 must be perpendicular to $\underset{I}{u}^h$ and secondly that the direction corresponding to the fourth root must be the direction of $\underset{I}{u}^h$. Hence $\mathfrak{Q}^a = 0$ and $\mathfrak{P}^{hi}$ has the form

$$\mathfrak{P}^{hi} = \pi\underset{I}{u}^h\underset{I}{u}^i - \mathfrak{p}g^{hi}. \tag{6.48}$$

It follows from (6.48), (6.41), and (6.43) that

$$c\mathfrak{P}^{4a} - \mathfrak{P}^{44}\dot{u}^a - \mathfrak{p}\dot{u}^a = 0 \tag{6.49}$$

and

$$\pi = (1-\beta^2)(\mathfrak{E}+\mathfrak{p}). \tag{6.50}$$

Consequently, if $\underset{0}{\mathfrak{E}}$ is the mean energy per measuring parallelepiped for an observer moving with the same velocity we have

$$\pi = (\underset{0}{\mathfrak{E}} + \mathfrak{p}) \qquad (6.51)$$

and

$$\mathfrak{P}^{hi} = (\underset{0}{\mathfrak{E}} + \mathfrak{p})\underset{I}{u^h}\underset{I}{u^i} - \mathfrak{p}g^{hi}. \qquad (6.52)$$

If $\mathfrak{Q}^a = 0$ v. Dantzig calls the fluid *perfectly perfect.* Hence a relativistically perfect fluid is always perfectly perfect as was pointed out by v. Dantzig.†

See Note J, p. 273.

EXERCISES

IX.1. If an Einstein-V_n, $n > 3$, can be transformed into an R_n by a conformal transformation (cf. V, Ex. 7) it is a space of constant curvature, i.e. its curvature affinor has the form

$$K_{\nu\mu\lambda\kappa} = -\frac{2}{n(n-1)} K g_{[\nu[\lambda} g_{\mu]\kappa]}.‡ \qquad (1\,\alpha)$$

It follows from (1 α) that K is constant.

IX.2. A V_4 which cannot be transformed into an R_4 by a conformal transformation (cf. V, Ex. 7) can be mapped conformally on at most one Einstein space, and the mapping can be accomplished (if at all) in one way only, provided we neglect a change of scale.§ This means physically that the linear element of the empty part of space-time is known if the world-lines of the light rays are known.||

IX.3. In a homogeneous medium moving with a constant velocity v^a with respect to any rectilinear coordinate system the following equations hold:

$$(\mathfrak{F}_{\lambda\kappa} - \epsilon c F_{\lambda\kappa})\underset{I}{v^\kappa} = 0, \qquad (3\,\alpha)$$

$$\left(\mathfrak{F}^{[\kappa\lambda} - \frac{1}{\mu c} F^{[\kappa\lambda}\right)\underset{I}{v^{\mu]}} = 0.†† \qquad (3\,\beta)$$

IX.4. If $\mathfrak{L} \overset{\text{def}}{=} \frac{1}{2}\epsilon c^2 \mathfrak{g}^{\frac{1}{2}} F_{\kappa\lambda} F_{\mu\nu} g^{\kappa\mu} g^{\lambda\nu} \quad (\kappa, \lambda, \mu, \nu = 1, 2, 3, 4)$

is considered as a function of the $g_{\lambda\kappa}$, prove that

$$[\mathfrak{L}]^{\kappa\lambda} = \underset{e}{\mathfrak{P}^{\kappa\lambda}}. \qquad (4\,\alpha)$$

† 1939. 3, p. 688. ‡ Schouten and Struik 1921. 3. § Brinkmann 1923. 2.
|| Cf. Kasner 1921. 4; 1922. 2. †† Pauli 1921. 1, p. 658.

X

DIRAC'S MATRIX CALCULUS†

1. Introduction

THE great difficulty in matrix calculus when applied to quantum mechanics is to find an efficient system of labelling the quantities because every label must contain a lot of information. Our chief aim in this chapter is to show clearly the great merits of Dirac's methods. Therefore we develop in § 2 and § 3 the tensor calculus in U_n and in §§ 4 and 5 a matrix calculus in U_n based on the classical methods described in II, § 13. After considering the algebra of general vector-like sets it is then possible to show that Dirac's bra–ket symbolism is just what we need for labelling purposes. But in order to find the most satisfactory way of labelling, it is necessary to consider the physical interpretation and to find out what kinds of quantities really occur. It appears that we have to deal with states and with observables and that in order to be able to represent each quantity by a set of numbers, it is necessary to start from a complete set of commuting observables and to construct the set of simultaneous eigenstates. These eigenstates can all be labelled by the eigenvalues to which they belong, and to every state or observable there now belongs a 'representation', i.e. a set of numbers analogous to the orthogonal components in U_n. A symbolism arises that is closely analogous to matrix calculus in U_n. Moreover, the orthogonal components are intimately connected with the probabilities of finding certain results if an observable is measured in a certain state.

When a complete set of commuting observables has been given it is possible to define functions of these observables that are observables themselves. Using this new conception it is possible to fix every state uniquely by such a function, and this has a consequence that this function can be used for labelling states. After the introduction of some abbreviations this second method of labelling leads immediately to the equations known as wave equations.

2. Quantities of the second kind and hybrid quantities‡

If the group G_a of an E_n (cf. I, § 1) is supposed to contain not only real but also complex transformations, besides the x^κ we have also the complex conjugates

$$x^{\bar{\kappa}} \stackrel{\text{def}}{=} \overline{x^\kappa} \qquad (2.1)$$

† General references: Jordan 1936. 4; Kramers 1937. 4; Dirac 1947. 5.

‡ Cf. E. I, 1935. 1, p. 8 ff.

subject to the transformations

$$\bar{x}^{\kappa'} = \bar{A}^{\bar{\kappa}'}_{\bar{\kappa}} \bar{x}^{\bar{\kappa}}, \qquad \bar{A}^{\bar{\kappa}'}_{\bar{\kappa}} \overset{\text{def}}{=} \overline{A^{\kappa'}_{\kappa}}. \tag{2.2}$$

Using these transformations we may define vectors, affinors, etc., *of the second kind*, for instance,

$$\begin{aligned} \bar{v}^{\bar{\kappa}'} &= \bar{A}^{\bar{\kappa}'}_{\bar{\kappa}} \bar{v}^{\bar{\kappa}}, \\ \bar{w}_{\bar{\lambda}'} &= \bar{A}^{\bar{\lambda}}_{\bar{\lambda}'} \bar{w}_{\bar{\lambda}}. \end{aligned} \tag{2.3}$$

Since the complex conjugates of the components of any ordinary quantity (e.g. $P^{\kappa}_{\cdot\lambda}$) are the components of a quantity of the second kind,† we have not introduced something really new but only emphasized the fact that these conjugates transform in a different way. But something new does arise if we consider quantities with indices of both kinds, e.g. $Q^{\bar{\kappa}}_{\cdot\lambda}$ with the transformation

$$Q^{\bar{\kappa}'}_{\cdot\lambda'} = \bar{A}^{\bar{\kappa}'}_{\bar{\kappa}} A^{\lambda}_{\lambda'} Q^{\bar{\kappa}}_{\cdot\lambda}. \tag{2.4}$$

Such quantities we call *hybrid quantities*. The complex conjugates of the components of a hybrid quantity $Q^{\bar{\kappa}}_{\cdot\lambda}$ are components of another hybrid quantity $\bar{Q}^{\kappa}_{\cdot\bar{\lambda}}$ called the *complex conjugate* of the first one.

A co- or contravariant hybrid quantity with a valence > 2, e.g. $P^{\bar{\kappa}\lambda\bar{\mu}}$, cannot be subjected to the operation of mixing or alternation because $P^{\bar{\kappa}\bar{\lambda}\mu}$ has no meaning. But if the valence is 2 the analogues of the tensor and the bivector exist. We call a quantity $P^{\bar{\kappa}\lambda}$ which satisfies the condition

$$P^{\bar{\kappa}\lambda} = \bar{P}^{\lambda\bar{\kappa}} \tag{2.5}$$

hermitian (*symmetric*) or a *hermitian tensor*. It would be possible to define a *hermitian bivector* $Q^{\bar{\kappa}\lambda}$ in the same way by means of the equation

$$Q^{\bar{\kappa}\lambda} = -\bar{Q}^{\lambda\bar{\kappa}}. \tag{2.6}$$

But, as $iQ^{\bar{\kappa}\lambda}$ is hermitian symmetric, we need only to consider the symmetric case.

For co- or contravariant hermitian tensors the following theorem holds. It is analogous to the theorem for ordinary tensors of valence 2 (cf. II, § 10).

If r is the rank of $P_{\bar{\kappa}\lambda} = \bar{P}_{\lambda\bar{\kappa}}$, the coordinate system (κ) can always be chosen in such a way that

$$P_{\bar{\kappa}\lambda} = \begin{cases} -1 \text{ for } \kappa = \lambda \leqslant s \\ +1 \text{ for } s < \kappa = \lambda \leqslant r \\ \;\;\,0 \text{ for } r < \kappa = \lambda \text{ and for } \kappa \neq \lambda, \end{cases} \tag{2.7}$$

or

$$P_{\bar{\kappa}\lambda} = -\overset{\bar{1}}{e}_{\bar{\kappa}} \overset{1}{e}_{\lambda} - \ldots - \overset{\bar{s}}{e}_{\bar{\kappa}} \overset{s}{e}_{\lambda} + \overset{\overline{s+1}}{e}_{\bar{\kappa}} \overset{s+1}{e}_{\lambda} + \ldots + \overset{\bar{r}}{e}_{\bar{\kappa}} \overset{r}{e}_{\lambda}. \tag{2.8}$$

† Written $\bar{P}^{\bar{\kappa}}_{\cdot\bar{\lambda}}$.

Here is another form of the same theorem:

If r is the rank of $P_{\bar{\kappa}\lambda} = \bar{P}_{\lambda\bar{\kappa}}$, there is always a transformation $T^{\kappa}{}_{.\lambda}$ such that the components with respect to (κ) of $\bar{T}^{\bar{\rho}}{}_{.\bar{\kappa}} T^{\sigma}{}_{.\lambda} P_{\bar{\rho}\sigma}$ satisfy conditions of the form (2.7).

s is an invariant of $P_{\bar{\kappa}\lambda}$, called the *index*. $P_{\bar{\kappa}\lambda}$ is said to be *positive definite* if $s = 0$, *negative definite* if $s = n$, and *indefinite* if $0 < s < n$. The sequence $---...+++...$ in (2.8) is called the *signature* of $P_{\bar{\kappa}\lambda}$. The signature is said to be *even* if s is even and *odd* if s is odd.

If $r = n$ an inverse $\overset{-1}{P}{}^{\kappa\bar{\lambda}} = \overset{-1}{\bar{P}}{}^{\bar{\lambda}\kappa}$ exists. It has the same index and the same signature as $P_{\bar{\kappa}\lambda}$ as can be seen from the formula

$$\overset{-1}{P}{}^{\kappa\bar{\lambda}} = -\underset{1}{e}{}^{\kappa}\underset{1}{\bar{e}}{}^{\bar{\lambda}} - \dots - \underset{s}{e}{}^{\kappa}\underset{s}{\bar{e}}{}^{\bar{\lambda}} + \underset{s+1}{e}{}^{\kappa}\underset{s+1}{\bar{e}}{}^{\bar{\lambda}} + \dots + \underset{r}{e}{}^{\kappa}\underset{r}{\bar{e}}{}^{\bar{\lambda}}.$$

In a centred E_n a hermitian tensor is represented by one of the figures

$$P_{\bar{\kappa}\lambda}\bar{x}^{\bar{\kappa}}x^{\lambda} = \pm 1$$

that can be considered as a hypersurface in the E_{2n} in which the x^{κ} and $\bar{x}^{\bar{\kappa}}$ and also, for instance, the real and imaginary parts of the x^{κ} are $2n$ *independent* coordinates.†

3. The fundamental tensor; the U_n

If a hermitian tensor $a_{\bar{\lambda}\kappa} = \bar{a}_{\kappa\bar{\lambda}}$ of rank n is introduced in an E_n the E_n is called a U_n and $a_{\bar{\lambda}\kappa}$ its *fundamental tensor*. According to § 2 there is at least one coordinate system (h) $(h = 1,\dots, n)$ with measuring vectors $\underset{i}{i}{}^{\kappa}$, $\overset{h}{i}_{\lambda}$ such that

$$a_{\bar{\lambda}\kappa} = -\overset{\bar{1}}{i}_{\bar{\lambda}}\overset{1}{i}_{\kappa} - \dots - \overset{\bar{s}}{i}_{\bar{\lambda}}\overset{s}{i}_{\kappa} + \overset{\overline{s+1}}{i}_{\bar{\lambda}}\overset{s+1}{i}_{\kappa} + \dots + \overset{\bar{n}}{i}_{\bar{\lambda}}\overset{n}{i}_{\kappa}. \tag{3.1}$$

Let the inverse of $a_{\bar{\lambda}\kappa}$ be written $a^{\kappa\bar{\lambda}}$ (cf. II, § 12).

The fundamental tensor establishes a one-to-one correspondence between co-(contra-)variant vectors of the first kind and contra-(co-)variant vectors of the second kind

$$v_{\bar{\lambda}} \overset{\text{def}}{=} a_{\bar{\lambda}\kappa}v^{\kappa}, \qquad \bar{w}_{\lambda} \overset{\text{def}}{=} \bar{w}^{\bar{\kappa}}a_{\bar{\kappa}\lambda}. \tag{3.2}$$

In the same way as in an R_n, corresponding quantities are given the same kernel. Then v^{κ} and $v_{\bar{\lambda}}$, and in the same way $\bar{w}_{\lambda}$ and $\bar{w}^{\bar{\kappa}}$, are two different sets of components for the same quantity. But, unlike in the R_n, in a U_n two different kinds of vectors remain. They cannot be distinguished either by the names first and second kind or by the

† Cf. E I, p. 60.

adjectives co- and contravariant. Therefore, anticipating § 7 we introduce here the names proposed by Dirac† for a much more general case:

ket = contravariant vector of the first kind = covariant vector of the second kind: v^{κ}, $v_{\bar{\lambda}}$,

bra = covariant vector of the first kind = contravariant vector of the second kind: $\bar{w}_{\lambda}$, $\bar{w}^{\bar{\kappa}}$,

and remark that the conjugate complex of a ket is a bra and vice versa. It is convenient to write the kernel of a ket without a bar and that of a bra with a bar. Then the *kernel* and the *upper* index both have a bar or both have no bar. Applying the raising and lowering of indices to $A^{\kappa}_{\bar{\lambda}}$, ambiguity can only be avoided by writing $A^{\kappa}{}_{.\lambda}$ (which is inconvenient) or by making the rule that the upper index is first and the lower second. Then

$$A_{\bar{\kappa}\lambda} = a_{\bar{\kappa}\lambda}, \qquad A^{\kappa\bar{\lambda}} = a^{\kappa\bar{\lambda}}. \tag{3.3}$$

After the introduction of $a_{\bar{\lambda}\kappa}$ to every set of two kets u^{κ}, v^{κ} or two bras $\bar{u}_{\lambda}$, $\bar{v}_{\lambda}$ belong a number and its complex conjugate and we have

$$\begin{aligned} \bar{u}_{\lambda} v^{\lambda} &= \bar{u}^{\bar{\kappa}} a_{\bar{\kappa}\lambda} v^{\lambda} = v^{\lambda} \bar{a}_{\lambda\bar{\kappa}} \bar{u}^{\bar{\kappa}} = v_{\bar{\kappa}} \bar{u}^{\bar{\kappa}}, \\ u_{\bar{\lambda}} \bar{v}^{\bar{\lambda}} &= u^{\kappa} \bar{a}_{\kappa\bar{\lambda}} \bar{v}^{\bar{\lambda}} = \bar{v}^{\bar{\lambda}} a_{\bar{\lambda}\kappa} u^{\kappa} = \bar{v}_{\kappa} u^{\kappa}. \end{aligned} \tag{3.4}$$

$\bar{u}_{\lambda} v^{\lambda}$ is called the *scalar product of* u^{κ} *and* v^{κ} *in this order. If the order is reversed the product is replaced by its complex conjugate.* This product depends linearly on the second factor and antilinearly (i.e. with complex conjugate coefficients) on the first one. The scalar product of a vector with itself is called its *norm*. The norm is always *real*.

Two vectors are said to be *unitary orthogonal* or *orthogonal* (if no ambiguity can arise) if their scalar product vanishes. A vector with norm zero is orthogonal to itself and is called a (unitary) *null vector*. All null vectors fill the (unitary) *null cone* with the equation

$$\bar{x}^{\bar{\lambda}} a_{\bar{\lambda}\kappa} x^{\kappa} = 0 \tag{3.5}$$

If $0 < s < n$, vectors exist which have positive or negative norms, just as in an R_n. But if the index is 0 or n no null cone exists. As in an R_n, an E_m in U_n is a U_m if and only if the E_m is not in a singular position with respect to the null cone.

A vector is called a *unit vector* (*unit ket, unit bra*) if its norm is $+1$ or -1. It remains a unit vector if it is multiplied by a scalar of the form $e^{i\phi}$ (*phase factor*). Obviously the measuring vectors $\underset{i}{i}^{\kappa}$ in (3.1) are unit kets and the $\overset{h}{i}_{\lambda}$ are unit bras. They form a (*unitary*) *cartesian system*.

† 1947. 5, p. 18 ff.

Cartesian systems are transformed into each other by all *unitary* transformations, i.e. the transformations which leave $a_{\bar{\lambda}\kappa}$ invariant. $U^{\kappa}_{.\lambda}$ is a unitary transformation if and only if

$$U^{\kappa}_{.\rho}\,\overline{U}^{\bar{\lambda}}_{.\bar{\sigma}}\,a^{\rho\bar{\sigma}} = a^{\kappa\bar{\lambda}}, \tag{3.6}$$

or in another form

$$\overline{U}^{\bar{\kappa}\lambda} = \overset{-1}{U}{}^{\lambda\bar{\kappa}}. \tag{3.7}$$

The geometry in U_n is based on the *unitary group* G_{un}, consisting of all unitary transformations.

If the fundamental tensor $a_{\bar{\lambda}\kappa}$ is *definite* ($s = 0$ or $s = n$) the following theorem holds for hermitian tensors in U_n (cf. II, § 14):

THEOREM OF PRINCIPAL AXES OF A HERMITIAN TENSOR.

If the hermitian fundamental tensor is definite and if $T_{\bar{\mu}\lambda}$ is a hermitian tensor, it is always possible to find a unitary cartesian system (h) such that

$$T_{\bar{j}i} = 0 \quad (i,j = 1,\ldots,n;\quad j \neq i). \tag{3.8}$$

If the fundamental tensor is not definite the theorem of principal axes holds only for those tensors that are not in a singular position with respect to the null cone (3.5).

4. Matrix calculus in E_n and U_n

An abbreviated calculus could be constructed in E_n (cf. II, § 13) so that it contained not only quantities of the first but also of the second kind and also hybrid quantities. Now since there are four kinds of vectors and sixteen kinds of quantities of valence 2, such an abbreviation cannot be considered seriously. But in a U_n we have only two kinds of vectors, kets and bras, and four different kinds of quantities of valence 2 that can be distinguished by their characters: ket-ket ($f^{\kappa\lambda}$), ket-bra ($P^{\kappa}_{.\lambda}$), bra-ket ($Q_{\lambda}^{.\kappa}$), and bra-bra ($h_{\lambda\kappa}$). Moreover, the complex conjugate of a ket is a bra and this reduces to three the number of quantities required for purposes of symbolization. Every vector has two kinds of components and every quantity of valence 2 has four, e.g. $f^{\kappa\lambda}$, $f^{\kappa}_{.\bar{\lambda}}$, $f_{\bar{\kappa}}^{.\lambda}$, $f_{\bar{\kappa}\bar{\lambda}}$. But if the fundamental tensor is *positive definite* and if only unitary cartesian systems of coordinates are used all those different components are numerically equal.

As in II, § 13, we use the sign | to denote the isomer. It will appear convenient to use it also for scalars and vectors although the isomer of a scalar or vector is of course the quantity itself. By — we indicate the operation of forming the complex conjugate, i.e. the quantity with the complex conjugate components. The new sign + may symbolize the combination of these two (always commuting) operations.

The scalar product of a bra $\bar{u}$ and a ket v is written $\bar{u}v$ or $v\bar{u}$. Hence

$$\bar{u}v \text{ and } v\bar{u} \quad \text{stand for} \quad \bar{u}_\lambda v^\lambda = \bar{u}^{\bar{\kappa}} v_{\bar{\kappa}}, \tag{4.1}$$

but if we wish to consider this product as the scalar product of the kets u, v or of the bras $\bar{u}$, $\bar{v}$ we use a dot as the multiplication sign:

$$\begin{aligned} u.v &= \bar{u}v = v\bar{u} = \overline{\bar{v}u} = \overline{v.u}, \\ \bar{u}.\bar{v} &= u\bar{v} = \bar{v}u = \overline{v\bar{u}} = \overline{\bar{v}.\bar{u}}. \end{aligned} \tag{4.2}$$

In writing out products without multiplication signs we have to take care that every factor with ket-(bra-)character to the right is followed by a factor with bra-(ket-)character to the left. Hence the characters of all quantities concerned must be memorized. To help their memory beginners may at first take different sets of kernels for quantities of different kinds, e.g. p, q, r, s, t for scalars, a, b, c, u, v, w for vectors, capitals for quantities with bra-ket or ket-bra character, and f, g, h, k, l for those with ket-ket or bra-bra character. But when some skill has been obtained† such devices should be dropped because in physical applications there is not much freedom left for the choice of letters. Besides, as we will see in Dirac's calculus, there are still other ways to aid the memory.

We have the following rules for the application of $|$, $-$, and $+$ on products:

$|$ is effected by applying $|$ to every factor and reversing the order of the factors, for instance

$$\begin{aligned} (Pv)' &= \overset{|}{v}\overset{|}{P} = v\overset{|}{P} = Pv, \\ (PQ)' &= \overset{|}{Q}\overset{|}{P}, \\ (\bar{u}Pv)' &= \overset{|}{v}\overset{|}{P}\overset{+}{u} = v\overset{|}{P}\bar{u} = \bar{u}Pv. \end{aligned} \tag{4.3}$$

$-$ is effected by applying $-$ to every factor and leaving the order of the factors unchanged.

$+$ is effected by applying $+$ to every factor and reversing the order of the factors, e.g.

$$\begin{aligned} (\bar{u}PQv)^+ &= \overset{+}{v}\overset{+}{Q}\overset{+}{P}\overset{|}{u} = \bar{v}\overset{+}{Q}\overset{+}{P}u, \\ (hu)^+ &= \overset{+}{u}\overset{+}{h} = \bar{u}\overset{+}{h} = \bar{h}\bar{u}, \\ (h\bar{f}\bar{u})^+ &= \overset{|}{u}\overset{|}{f}\overset{+}{h} = u\overset{|}{f}\overset{+}{h} = \bar{h}fu, \\ (P\bar{h}fu)^+ &= \overset{+}{u}\overset{+}{f}\overset{|}{h}\overset{+}{P} = \bar{u}\overset{+}{f}\overset{|}{h}\overset{+}{P} = \bar{P}h\bar{f}\bar{u}. \end{aligned} \tag{4.4}$$

† In general physicists get this skill much sooner than pure mathematicians, perhaps because they are eager to get physical results and are not at all interested in the formal side of the calculus.

The simplicity of the rules for | and + is of course due to our agreement about the application of | to scalars and vectors.

5. Linear operators

There are two kinds of linear operators, the ket-bra operators $T^{\kappa}{}_{.\lambda}$ and their complex conjugate bra-ket operators $\bar{T}^{\bar{\kappa}}{}_{.\bar{\lambda}}$. Eigenvalues and eigenvectors (kets and bras) are defined as in II, § 14. But here we are especially interested in the case when there is a positive definite hermitian fundamental tensor and when the operator is *hermitian*, i.e. when $T_{\bar{\mu}\lambda}$ is a hermitian tensor (cf. (2.5))

$$T = \overset{+}{T}. \tag{5.1}$$

In this case the eigenvalues are the roots of the equation

$$\mathrm{Det}(T_{\bar{\mu}\lambda} - \sigma a_{\bar{\mu}\lambda}) = 0. \tag{5.2}$$

If $\underset{1}{\sigma}$ is an eigenvalue there exists a ket v^{κ} such that

$$Tv = \underset{1}{\sigma} v. \tag{5.3}$$

Applying the operator + we get

$$\overset{+}{v}\overset{+}{T} = \bar{v}T = \underset{1}{\bar{\sigma}}\bar{v}, \tag{5.4}$$

hence

$$\bar{v}Tv = \underset{1}{\sigma}\bar{v}v = \underset{1}{\bar{\sigma}}\bar{v}v, \tag{5.5}$$

and this is only possible if $\underset{1}{\sigma} = \underset{1}{\bar{\sigma}}$. Hence the eigenvalues of a hermitian operator are all *real*.

If $\underset{1}{\sigma}$ and $\underset{2}{\sigma}$ are two *different* eigenvalues and $\underset{1}{v}$ and $\underset{2}{v}$ are eigenkets belonging to them, we have

$$\underset{1}{\bar{v}}T\underset{2}{v} = \underset{2}{\sigma}\underset{1}{\bar{v}}\underset{2}{v} = \underset{1}{\sigma}\underset{1}{\bar{v}}\underset{2}{v}, \tag{5.6}$$

and this is only possible if $\underset{1}{v}$ and $\underset{2}{v}$ are *unitary orthogonal*.

According to the theorem of principal axes there exists a unitary cartesian coordinate system (h) such that

$$T_{\bar{\lambda}\kappa} \overset{*}{=} T_{\bar{i}h}\overset{\bar{i}}{i}_{\bar{\lambda}}\overset{h}{i}_{\kappa} = T_{\bar{1}1}\overset{\bar{1}}{i}_{\bar{\lambda}}\overset{1}{i}_{\kappa} + \dots + T_{\bar{n}n}\overset{\bar{n}}{i}_{\bar{\lambda}}\overset{n}{i}_{\kappa}, \tag{5.7}$$

from which we see that the components $T_{\bar{1}1}, \dots, T_{\bar{n}n}$ are all eigenvalues. They need not be different and some of them may be zero. If $P_{\bar{1}1}$ belongs to a set of z equal eigenvalues, for instance,

$$P_{\bar{1}1} = \dots = P_{\bar{z}z}, \tag{5.8}$$

the vectors $\underset{1}{i}^{\kappa}, \dots, \underset{z}{i}^{\kappa}$ span an E_z in which each vector is an eigenvector belonging to the same eigenvalue $P_{\bar{1}1}$. Since $\underset{1}{i^{\kappa}},\ldots,\underset{z}{i^{\kappa}}$ form a unitary orthogonal set this E_z is a U_z. In this way a set of flat subspaces $U_{z_1}, U_{z_2},\ldots$ $(z_1+z_2+\ldots = n)$ can be found, orthogonal to each other and each belonging to one definite eigenvalue. Just z linearly independent eigenkets (and bras) belong to an eigenvalue if its corresponding subspace is z-dimensional.

Let P and Q be two hermitian operators. We wish to know whether the product PQ is hermitian as well. From

$$PQ = (PQ)^{+} = \overset{+}{Q}\overset{+}{P} = QP \tag{5.9}$$

we see that the necessary and sufficient condition is that P and Q *commute.*

The following theorem holds for commuting operators:

A set of hermitian operators P, Q,... commutes, if and only if they have at least one set of n linearly independent eigenkets (bras) in common.

PROOF.†

We give the proof here for two commuting operators. The generalization is obvious. Let P and Q be commuting and let v be an eigenket of Q belonging to the eigenvalue $\underset{1}{\mu}$. Now v can be expressed linearly in eigenkets of P. Let this expression be

$$v = \underset{1}{\alpha}\underset{1}{v}+\ldots+\underset{z}{\alpha}\underset{z}{v}, \tag{5.10}$$

$\underset{1}{v},\ldots,\underset{z}{v}$ being eigenkets of P belonging to the eigenvalues $\underset{1}{\lambda},\ldots,\underset{z}{\lambda}$ of P. Without loss of generality we may now assume that *all these eigenvalues are different.* Then we have

$$(Q-\underset{1}{\mu})(\underset{1}{\alpha}\underset{1}{v}+\ldots+\underset{z}{\alpha}\underset{z}{v}) = 0 \tag{5.11}$$

and, for every choice of $\mathfrak{a}$ $(\mathfrak{a} = 1,\ldots,z)$,

$$P(Q-\underset{1}{\mu})\underset{\mathfrak{a}}{\alpha}\underset{\mathfrak{a}}{v} = (Q-\underset{1}{\mu})P\underset{\mathfrak{a}}{\alpha}\underset{\mathfrak{a}}{v} = \underset{\mathfrak{a}}{\lambda}(Q-\underset{1}{\mu})\underset{\mathfrak{a}}{\alpha}\underset{\mathfrak{a}}{v}. \tag{5.12}$$

Hence the z kets $(Q-\underset{1}{\mu})\underset{\mathfrak{a}}{\alpha}\underset{\mathfrak{a}}{v}$ are all eigenkets of P. But then (5.11) expresses the fact that a sum of z eigenkets of P, belonging to z different eigenvalues is zero. This is only possible if every one of these eigenkets is zero, hence

$$Q\underset{\mathfrak{a}}{v} = \underset{1}{\mu}\underset{\mathfrak{a}}{v} \quad (\mathfrak{a} = 1,\ldots,z), \tag{5.13}$$

and this means that every $\underset{\mathfrak{a}}{v}$ $(\mathfrak{a} = 1,\ldots,z)$ is a simultaneous eigenket of P and Q. We have now proved that every eigenket of Q can be expressed linearly in simultaneous eigenkets of P and Q. Accordingly,

† Cf. Dirac 1947. 5, p. 49.

since there are n linearly independent eigenkets of Q, there must be n linearly independent simultaneous eigenkets of P and Q. It follows immediately from (5.7) that P and Q commute if they have n linearly independent simultaneous eigenkets.

A set of n linearly independent simultaneous eigenkets (bras) of the commuting hermitian operators P, Q,... is called a *complete set* belonging to these operators. Every ket (bra) of a complete set belongs to one definite combination of eigenvalues of P, Q,.... Conversely to every combination of eigenvalues of P, Q,... there belongs *at least* one ket (bra) of the complete set. As we will see later, this is important for the labelling of the kets (bras) of a complete set.

Any operator P is transformed into TPT^{-1} by any other (not necessarily hermitian) operator T. If P is hermitian, we may wish to know whether the transform is hermitian too for every choice of P. From

$$TPT^{-1} = (TPT^{-1})^{+} = \overset{+}{T}{}^{-1}\overset{+}{P}\overset{+}{T} = \overset{+}{T}{}^{-1}P\overset{+}{T} \tag{5.14}$$

it follows that

$$\overset{+}{T}TP = P\overset{+}{T}T. \tag{5.15}$$

Hence $\overset{+}{T}T$ commutes with every hermitian operator and accordingly with every operator. But this is only possible if $\overset{+}{T}T$ is equal to the identical operator to within a scalar factor. This factor must be real because

$$\overset{+}{T}T = (\overset{+}{T}T)^{+} \tag{5.16}$$

Hence the necessary and sufficient condition is that T is a *unitary* transformation ($T = \overset{+}{T}{}^{-1}$, cf. (3.7)) to within a scalar factor.

6. The algebra of vector-like sets

Consider a set of elements A, B,... subjected to the three operations of the addition and subtraction of elements and the multiplication of an element with a complex number. Suppose that the operations satisfy the following conditions:

$$\begin{aligned} A+B &= B+A \\ (A+B)+C &= A+(B+C) \\ B = C-A \quad &\text{if} \quad A+B = C \\ (\alpha+\beta)A &= \alpha A+\beta A \\ \alpha(A+B) &= \alpha A+\alpha B \\ \alpha(\beta A) &= (\alpha\beta)A \\ 1.A &= A \end{aligned} \tag{6.1}$$

$(\alpha, \beta$ — complex numbers).

Such a set may be called *vector-like* because the set of all contravariant vectors in E_n satisfies the conditions. But all affinors in E_n with the same valences and the same places for the indices also form a vector-like set. Now in all these special sets all elements can be expressed linearly in a finite number out of them. We do not, however, make this assumption for the more general sets considered here.

Let $a, b, \ldots$ be another vector-like set. Then we may consider a linear correspondence between the two sets such that to every element A there corresponds just one element a

$$a = F(A), \tag{6.2}$$

and that for every choice of the elements A and B

$$F(A+B) = F(A)+F(B), \qquad F(\alpha A) = \alpha F(A) \tag{6.3}$$

holds. Such a correspondence may be called a *linear transformation of the type* $A \to a$. Let $G(A)$ be another linear transformation of the type $A \to a$. Then we may define another transformation $\alpha F+\beta G$, of the same type, by the equation

$$(\alpha F+\beta G)(A) \overset{\text{def}}{=} \alpha F(A)+\beta G(A), \tag{6.4}$$

and from this it follows that all these linear transformations form a new vector-like set. Hence

If two vector-like sets A and a are given, all linear transformations of the type $A \to a$ form another vector-like set and this also holds for all linear transformations of the type $a \to A$.

As an example we take all tensors $h^{\kappa\lambda}$ in E_n for the set A and all vectors w_λ for the set a. Then the transformations of the type $A \to a$ correspond to the affinors $P_{\kappa\lambda\mu}$ which are symmetric in $\kappa\lambda$ and the transformations of the type $a \to A$ to the affinors $Q^{\kappa\lambda\mu}$ which are symmetric in $\kappa\lambda$.

Obviously the complex numbers form a vector-like set by themselves. If for any correspondence

$$a = {}'F(A) \tag{6.5}$$

and if

$${}'F(A+B) = {}'F(A)+{}'F(B), \qquad {}'F(\alpha A) = \bar{\alpha}\{{}'F(A)\} \tag{6.6}$$

holds instead of (6.3), this correspondence is called an *antilinear transformation of the type* $A \to a$. If $\alpha\,{}'F+\beta\,{}'G$ is defined by

$$(\alpha\,{}'F+\beta\,{}'G)(A) \overset{\text{def}}{=} \bar{\alpha}\{{}'F(A)\}+\bar{\beta}\{{}'F(B)\} = {}'F(\alpha A)+{}'G(\beta B), \tag{6.7}$$

the antilinear transformations form another vector-like set.

In the special case of complex numbers an antilinear transformation is given *a priori*, viz. $\alpha \to \bar{\alpha}$. Now if $a, b,\ldots$ is any vector-like set and $\alpha, \beta,\ldots$ the set of complex numbers, let us consider simultaneously the linear and the antilinear transformations of the type $\alpha \to a$. Every one of these transformations can be determined by giving only the transform of 1. Let T_1, T_2 and $'T_1$, $'T_2$ be the transformations for which

$$\begin{aligned} T_1\colon\ & 1 \to a_1, \quad \alpha \to \alpha a_1; \qquad & 'T_1\colon\ & 1 \to a_1, \quad \alpha \to \bar{\alpha} a_1, \\ T_2\colon\ & 1 \to a_2, \quad \alpha \to \alpha a_2; \qquad & 'T_2\colon\ & 1 \to a_2, \quad \alpha \to \bar{\alpha} a_2 \end{aligned} \tag{6.8}$$

then we have

$$\begin{aligned} \lambda_1 T_1 + \lambda_2 T_2\colon\ & 1 \to \lambda_1 a_1 + \lambda_2 a_2; \quad \alpha \to \lambda_1 \alpha a_1 + \lambda_2 \alpha a_2 = \alpha(\lambda_1 a_1 + \lambda_2 a_2), \\ \bar{\lambda}_1{}'T_1 + \bar{\lambda}_2{}'T_2\colon\ & 1 \to \lambda_1 a_1 + \lambda_2 a_2; \quad \alpha \to \lambda_1 \bar{\alpha} a_1 + \lambda_2 \bar{\alpha} a_2 = \bar{\alpha}(\lambda_1 a_1 + \lambda_2 a_2) \end{aligned} \tag{6.9}$$

for the transformations $\lambda_1 T_1 + \lambda_2 T_2$ and $\bar{\lambda}_1{}'T_1 + \bar{\lambda}_2{}'T_2$, where λ_1 and λ_2 are complex numbers. Hence there exists a linear one-to-one correspondence between the set $a, b,\ldots$ and the set $T_1, T_2,\ldots$. In fact T_1 and T_2 could be identified with a_1 and a_2, etc. But between the set $a, b,\ldots$ and the set $'T_1$, $'T_2$ there exists an antilinear one-to-one correspondence. We write $\bar{a}$ for the transformation $\alpha \to \bar{\alpha} a$ and call $\bar{a}, \bar{b},\ldots$ the *conjugates* of $a, b,\ldots$ and vice versa. The term complex conjugate would be deceptive because a and $\bar{a}$ cannot be added, and accordingly a cannot be split up into a real and an imaginary part.† It is easily verified that $\bar{\lambda}\bar{a} + \bar{\mu}\bar{b}$ and $\lambda a + \mu b$ are conjugate to each other.

In this way, by starting from one given vector-like set, a whole system of vector-like sets can be constructed. If the first set is the set of all contravariant vectors in E_n we find not only all covariant, contravariant, and mixed affinors of the first kind but also those of the second kind and all hybrid quantities. The general process is quite analogous to this but much less special because here we do not assume anywhere that in any set all elements can be expressed linearly in a finite number of them.

7. Dirac's kets and bras‡

Following Dirac we start from one given vector-like set and call its elements *kets*. These kets are not yet analogous to the kets of § 3 in U_n but only to contra-(co-)variant vectors in E_n. Vectors in E_n can be denoted by their components, for instance v^κ. But this notation is intimately connected with the possibility of expressing a vector linearly

† For the same reason Dirac uses the expression 'conjugate imaginary' instead of 'conjugate complex' in a similar case (1947. 5, p. 21).

‡ Cf. Dirac 1947. 5, p. 18 ff.

in terms of a finite number of measuring vectors. Since this is no longer possible we have to look for another method of labelling kets. This labelling is no simple matter because, as we shall see later, the label must contain a lot of information. So, labelling by one kernel letter as in the matrix calculus of § 4 would not be satisfactory. Hence the only way is to give a ket a symbol in the form of a box into which all information can be put. This is just what Dirac does. For a ket he writes a box of the form $|\,\rangle$ with a label within which consists of as many letters and ciphers as are necessary to express all necessary information. According to § 6 a new vector-like set can be deduced from kets and complex numbers. This set is called *bra* and denoted by a box of the form $\langle\,|$ with a label within. They do not correspond yet to the bras of § 3 in U_n but only to co-(contra-)variant vectors in E_n. According to § 6, to every bra $\langle A|$ and ket $|B\rangle$ there belongs one and only one complex number, which will be called the *scalar product* of $\langle A|$ and $|B\rangle$ and be denoted by the *bra-c-ket symbol* $\langle A|B\rangle$. This unveils the mystery of the names bra and ket and shows in the most pleasant way that there is still some humour in mathematics!

In many cases some restrictions are necessary. If we take, for instance, the functions of a variable x in a certain interval as kets the bras are so-called *interval functions*, that is a certain kind of additive set functions.† They are best symbolized by F^X where X denotes an interval. The meaning of such a function is that to every interval X there belongs a number F^X such that $F^{X+Y} = F^X+F^Y$ always if X and Y have no point in common and if $X+Y$ is the interval consisting of all points that are either in X or in Y. The product of a bra F^X and a ket $f(x)$ is then the integral

$$\int F^{dx} f(x) \tag{7.1}$$

taken over some definitely chosen interval. Hence the restriction has to be made that only such functions and interval functions may occur for which these integrals exist.

Every integrable function $f(x)$ may be looked upon as an interval function because to every interval where $f(x)$ is integrable it makes the integral $\int f(x)\,dx$ over the interval correspond. But not every interval function corresponds to some function of x. As an example we mention the interval function $\delta(x-a)$ defined by

$$\int \delta(x-a)\,dx = \begin{cases} 1 \text{ if the interval contains } x = a, \\ 0 \quad ,, \quad ,, \quad \text{ does not contain } x = a. \end{cases} \tag{7.2}$$

† Cf. v. Dantzig 1936. 3.

This function could be approximated by some function taking very high positive values in the neighbourhood of $x = a$ and values very near to zero at all other points. But no function of x exists that could replace $\delta(x-a)$ exactly. The interval function $\delta(x-a)$ satisfies a number of elementary equations, for instance

$$\begin{aligned}
&(a) & \delta(-x) &= \delta(x), \\
&(b) & x\,\delta(x) &= 0, \\
&(c) & \delta(ax) &= a^{-1}\delta(x) \quad (a > 0), \qquad (7.3)\\
&(d) & \int \delta(a-x)\,dx\,\delta(x-b) &= \delta(a-b), \\
&(e) & f(x)\,\delta(x-a) &= f(a)\,\delta(x-a), \\
&(f) & \int f(x)\,\delta(x-a) &= f(a),
\end{aligned}$$

that can be proved easily.† Instead of $\delta(x-a)$, v. Dantzig‡ introduces a point-interval function defined by the equation

$$E_x^X = \begin{cases} 1 \text{ if } x \text{ is in the interval } X, \\ 0 \text{ if } x \text{ is not in } X, \end{cases} \qquad (7.4)$$

and gets the equations

$$\begin{aligned}
&(a) & \int E_x^{dy} f_y &= f_x, \qquad f_x \stackrel{\text{def}}{=} f(x), \\
&(b) & \int F^{dy} E_x^X &= F^X
\end{aligned} \qquad (7.5)$$

in strict analogy to the symbolism of transvection in tensor calculus. (7.5 a) replaces the fundamental relation (7.3 f).

In order to make the kets and bras defined here correspond to the kets and bras of § 3 we could now introduce as in § 6 the vector-like sets corresponding to the vectors of the second kind $\bar{v}^{\bar{\kappa}}$, $\bar{w}_{\bar{\lambda}}$, after that the set corresponding to the hybrid quantities $P_{\bar{\kappa}\lambda}$, and finally one of the elements of this last set could be proclaimed a fundamental tensor. But if we only have in mind the construction of an abbreviated calculus for the group G_{un} with a positive definite fundamental tensor it can be done in a shorter way by introducing immediately one definite anti-linear one-to-one correspondence between kets and bras. Dirac uses this opportunity. He denotes the conjugate bra of the ket $|A\rangle$ by $\langle A|$, hence if $|A\rangle$ corresponds to v^κ, $\langle A|$ corresponds to $\bar{v}_\lambda = \bar{v}^{\bar{\kappa}} a_{\bar{\kappa}\lambda}$. The changing of $|\ \rangle$ into $\langle\ |$ making sufficiently clear what is meant, it is

† Cf. Dirac 1947. 5, p. 60.

‡ 1936. 3. This point of view is more general as his point functions are functions defined on a general separable topological space. Then the F^X are absolutely additive set functions which determine a real or complex number with respect to each Borel's subset.

not necessary to write $\langle \bar{A}|$.† The properties of the fundamental tensor $a_{\bar{\lambda}\kappa}$ that it is hermitian and positive definite are expressed here in the axioms:

$$\begin{aligned} &(a) \qquad \langle A|B\rangle = \overline{\langle B|A\rangle} \quad (\bar{u}_\lambda v^\lambda = \overline{\bar{v}_\lambda u^\lambda}), \\ &(b) \qquad \langle A|A\rangle = 0 \text{ only if } |A\rangle = 0 \quad (\bar{v}_\lambda v^\lambda = 0). \end{aligned} \tag{7.6}$$

After these steps the kets and bras correspond to the kets and bras of § 3, but they are much more general. So we see that, following the methods of § 6, we first get four kinds of quantities that have to be identified two by two after the introduction of the fundamental tensor. Using Dirac's short cut we immediately get only two kinds of quantities whilst the fundamental tensor does not yet appear explicitly.

The *length* of a ket (bra) can now be defined as the square root of $\langle A|A\rangle$ and *orthogonality* of kets (bras) by the equation

$$\langle A|B\rangle = 0. \tag{7.7}$$

The method of § 6 applied to the linear transformations ket → ket or bra → bra leads to two different kinds of linear operators, the ket-bra and the bra-ket type. Now, returning to the notation of § 4, a product of the form $\bar{u}Tv$ can be written in the forms

$$\bar{u}Tv = \overset{|}{v}\overset{|+}{T}u = v\overset{|}{T}\bar{u}, \tag{7.8}$$

and from this we see that such an expression can also be written with the bra-ket operator T instead of the ket-bra operator T by merely changing the order of the factors. This fact can be used to simplify the notation again by only using ket-bra operators and denoting them by some letter without any brackets. The following is a list of some products and some of their corresponding forms in kernel-index notation

$$\begin{aligned} \alpha|A\rangle&: \quad \alpha^{\kappa}_{\cdot\lambda}A^\lambda, \quad \alpha^{\kappa\bar{\lambda}}A_{\bar{\lambda}}, \quad \alpha_{\bar{\kappa}}^{\cdot\bar{\lambda}}A_{\bar{\lambda}}; \\ \langle B|\alpha&: \quad B_\kappa \alpha^{\kappa}_{\cdot\lambda}, \quad B^{\bar{\kappa}}\alpha_{\bar{\kappa}\lambda}, \quad B^{\bar{\kappa}}\alpha_{\bar{\kappa}}^{\cdot\bar{\lambda}}; \\ \langle B|\alpha|A\rangle&: \quad B_\kappa \alpha^{\kappa}_{\cdot\lambda}A^\lambda = B^{\bar{\kappa}}\alpha_{\bar{\kappa}\lambda}A^\lambda = B_\kappa\alpha^{\kappa\bar{\lambda}}A_{\bar{\lambda}} = B^{\bar{\kappa}}\alpha_{\bar{\kappa}}^{\cdot\bar{\lambda}}A_{\bar{\lambda}}; \\ \langle B|\alpha\beta|A\rangle&: \quad B_\kappa \alpha^{\kappa}_{\cdot\rho}\beta^{\rho}_{\cdot\lambda}A^\lambda. \end{aligned} \tag{7.9}$$

If the ket-bra operator α has been given we may require the operator β such that

$$\overline{\alpha|A\rangle} = \langle A|\beta \tag{7.10}$$

corresponding to

$$\bar{\alpha}^{\bar{\kappa}}_{\cdot\bar{\lambda}}\bar{A}^{\bar{\lambda}} = \bar{A}_\rho \beta^{\rho}_{\cdot\lambda}a^{\lambda\bar{\kappa}} = \bar{A}^{\bar{\lambda}}\beta_{\bar{\lambda}}^{\cdot\bar{\kappa}}. \tag{7.11}$$

Hence β corresponds to the $\overset{+}{\alpha}$ of § 4. But in Dirac's calculus, $\overset{|}{\alpha}$, that is

† But we have an objection here that will be dealt with in § 12.

$\alpha_{\dot{\lambda}}^{\kappa}$, since it is a bra-ket operator, never occurs. Hence it is practical to denote by $\bar{\alpha}$ what was written $\overset{+}{\alpha}$ in § 4 and to call $\bar{\alpha}$ the *conjugate*† of α. From the rule for the application of $+$ to products we get the rule:

The conjugate of a product is the product of the conjugate of the factors in the reversed order.

The identical operator, corresponding to $A_{\bar{\lambda}}^{\kappa}$ and also to $a_{\bar{\lambda}\kappa}$ and $a^{\kappa\bar{\lambda}}$ in § 3 is now merely the number *1*. Here the fundamental tensor appears at last!

Vector-like sets corresponding to co- or contravariant affinors of any given valence could be derived by the methods of § 6. Now every quantity of this kind is a sum of general products of co- or contravariant vectors. Hence expressions

$$|A_1\rangle\,|A_2\rangle ... |A_u\rangle \tag{7.12}$$

and sums of these expressions may be considered as symbols for elements of a vector-like set corresponding to (ket-)affinors of valence u. This also holds for bras. If no ambiguity can arise the abbreviation $|A_1 ... A_u\rangle$ may be used. This development of the calculus is of great importance for dealing with complicated dynamical systems such as systems of particles with Bose statistics (corresponding to tensors) or Fermi statistics (corresponding to multivectors).

Eigenvalues of an operator are defined as in § 5. If we have

$$\alpha|A\rangle = \alpha'|A\rangle \tag{7.13}$$

for some operator α, where α' is a scalar, then α' is called a *right-hand eigenvalue* of α and $|A\rangle$ an *eigenket* belonging to this eigenvalue. *Left-hand eigenvalues* and *eigenbras* are defined in the same way. If α is hermitian, $\alpha = \bar{\alpha}$, (7.13) can be written

$$\langle A|\alpha = \bar{\alpha}'\langle A|. \tag{7.14}$$

But it follows from (7.13) and (7.14) that

$$\langle A|\alpha|A\rangle = \alpha' = \bar{\alpha}', \tag{7.15}$$

and this proves that every eigenvalue is *real* and at the same time right-hand and left-hand, and that the conjugate of every eigenket (bra) is an eigenbra (ket) belonging to the same eigenvalue.

If α' and α'' are two *different* eigenvalues of α, and $|A\rangle$ and $|B\rangle$ are two eigenkets belonging to them, we have

$$\langle A|\alpha|B\rangle = \alpha'\langle A|B\rangle = \alpha''\langle A|B\rangle, \tag{7.16}$$

† Dirac uses the expressions 'adjoint' and 'conjugate complex'. We prefer to use the term 'conjugate complex' for numbers only and the term 'conjugate' for all other objects.

and consequently $$\langle A|B\rangle = 0. \tag{7.17}$$

Hence *eigenkets (bras) belonging to different eigenvalues of a hermitian operator are mutually orthogonal.*

If $|A\rangle$ is an eigenket, every product of $|A\rangle$ with a scalar is an eigenket as well, belonging to the same eigenvalue. This can be used to *normalize* the eigenket (bra) by choosing the scalar factor in such a way that the norm takes any suitable value r. We call this '*normalizing to r*'.† r can take any real value, for instance, 1 or ∞ (of some kind). The scalar factor necessary for the normalization of a ket (bra) is only fixed to within a scalar factor of the form $e^{i\phi}$ (*phase factor*). But we assume that *the process of normalization includes the choice of the phase factor.* So a *normalized* ket (bra) is *absolutely fixed.*

8. Physical interpretation‡

In order to understand the necessity of the further developments we need to know something about the physical interpretation of kets, bras, and linear operators. According to quantum mechanics a dynamical system can be in different *states*, and the result of the measurement of some dynamical variable (viz. a component of some physical quantity) depends on the state the system is in. But, contrary to classical mechanics, if the system is in an arbitrary state the result of the measurement cannot be predicted. If a dynamical variable can be measured at all, it is only possible to predict the set of possible values and for each of them the probability that it may be the result of the measurement. After the measurement it looks just as if the system had 'jumped' into a state of special kind with respect to *this* variable, because in this new state the result of a second measurement of the *same* variable will always be the same as that of the first measurement. For example, in classical mechanics, the place of a point on the x-axis is given by one dynamical variable x and there are as many states as there are possible values of x. If the point is in the state $x = 3$ every measurement in this state gives with certainty the result $x = 3$. But in quantum mechanics other states exist (and we know how to produce them) where there is only a certain probability for x, to be found by measurement for instance between 3 and $3\frac{1}{2}$.

Now in matrix calculus every linear operator is connected with a series of numerical values, its eigenvalues. Dynamical variables are always

† We use the term 'normalize' in a more general sense than Dirac. His normalizing is always normalizing to 1.

‡ Cf. Dirac 1947. 5, p. 45 ff.

real, and we know that hermitian operators have real eigenvalues only. Hence it seems opportune to connect every dynamical variable with a hermitian operator (also called *real* operator) and to assume that the result of a measurement (if possible at all) can only be one of the eigenvalues. This leads to the assumption that every state corresponds to a ket (and its conjugate bra), determined to within a scalar factor. As we know that every eigenvalue corresponds to an eigenket or a set of eigenkets, we may expect that the *eigenstates*, i.e. the states corresponding to these eigenkets, are the states in which the result of the measurement is, with certainty, this eigenvalue. As to any other state, it will be necessary to assume that its corresponding ket can be expressed linearly in terms of those eigenkets whose eigenvalues can be possible results of a measurement in this state. But since this holds for every other state, this means that the operator *must have so many eigenkets that every ket can be expressed linearly in them.* A dynamical variable corresponding to such a hermitian operator is called an *observable.* Hence every *measurable* dynamical variable is an *observable.* The states of an observable and also the corresponding kets (bras) are said to form a *complete set.*

As we have seen in § 5, a hermitian operator in U_n has always just n mutually orthogonal eigenkets. But in the general case the number of eigenvalues may be enumerable or non-enumerable infinite and these values may be *discrete,* i.e. distributed discretely, or *continuous,* i.e. consisting of all numbers in one or more intervals. So we have the following possibilities:

only discrete: . . ..

only continuous: ——— —— —————

discrete and continuous: . .. ——— . . —— . . . . .

If ξ is a hermitian operator we denote the discrete eigenvalues by ξ^r ($r = 1, 2,\ldots$). If we happen to use for some purpose from the continuous eigenvalues an enumerable set only, we denote the eigenvalues of this set by ξ^s ($s = 1, 2,\ldots$). Finally ξ', ξ'',..., each denote a real variable that can take all the values of the range of continuous eigenvalues.† The eigenkets corresponding to them are denoted by $|\xi^r\rangle$, $|\xi^s d\rangle$, $|\xi' c\rangle$, $|\xi'' c\rangle$,.... This is an example of Dirac's putting information into the boxes.‡

Now suppose ξ to be an observable. Then every ket $|P\rangle$ can be expressed as a sum of discrete eigenkets, together with an enumerable set

† We differ slightly here from Dirac, so as to clarify things for the beginner.

‡ c and d are not factors but stand for 'continuous' and 'discrete'.

of eigenkets from the range of continuous eigenkets and an integral of continuous eigenkets. Hence there is an equation of the form

$$|P\rangle = \sum_r c^r|\xi^r\rangle + \sum_s c^s|\xi^s d\rangle + \int c(\xi')|\xi' c\rangle\, d\xi', \tag{8.1}$$

where the integral has to be taken over the range, where c^r, c^s, and $c(\xi')$ are finite, $|\xi^r\rangle$ and $|\xi^s d\rangle$ are normalized to *1* and where $|\xi' c\rangle$ has to be normalized in some suitable way.

If two kets $|P\rangle$ and $|Q\rangle$ can both be written as an integral over the range of the same continuous eigenvalues ξ', i.e.

$$|P\rangle = \int \underset{P}{c}(\xi')|\xi' c\rangle\, d\xi', \qquad |Q\rangle = \int \underset{Q}{c}(\xi')|\xi' c\rangle\, d\xi', \tag{8.2}$$

we have

$$\langle P|Q\rangle = \iint \underset{P}{c}(\xi'')\underset{Q}{c}(\xi')\langle \xi'' c|\xi' c\rangle\, d\xi' d\xi''. \tag{8.3}$$

Now in the single integral

$$\int \underset{P}{c}(\xi'')\underset{Q}{c}(\xi')\langle \xi'' c|\xi' c\rangle\, d\xi'' \tag{8.4}$$

$|\xi'' c\rangle$ is orthogonal to $|\xi' c\rangle$ over the whole range except at the point $\xi'' = \xi'$ and $\underset{P}{c}$ and $\underset{Q}{c}$ are finite. But this means that the integral would be zero if $\langle \xi' c|\xi' c\rangle$ were finite. Now $\langle P|Q\rangle$ cannot be zero always, and accordingly $|\xi' c\rangle$ cannot be normalized to any finite number r. The integral (8.4) can be written

$$\underset{P}{c}(\xi')\underset{Q}{c}(\xi') \int \langle \xi'' c|\xi' c\rangle\, d\xi'', \tag{8.5}$$

and this integral can only be finite if $\langle \xi'' c|\xi' c\rangle$ is equal to $\delta(\xi''-\xi')$ to within a finite scalar factor. Hence the finite scalars $\underset{P}{c}(\xi')$ and $\underset{Q}{c}(\xi')$ can always be transformed in such a way that

$$\langle \xi'' c|\xi' c\rangle = \delta(\xi''-\xi'). \tag{8.6}$$

This we call *normalizing to* δ. It is an infinite normalization ($r = \infty$) but a very special one.

It can be proved that the expansion (8.1) of $|P\rangle$ is unique *provided that no two terms of the sums belong to the same eigenvalue.*† This will always be the case if to every eigenvalue of ξ there belongs only one eigenstate (i.e. ∞^1 eigenkets differing only by a scalar factor).

9. Functions of observables‡

By a function $f(\xi)$ of an observable, we understand an observable such that the measurement of $f(\xi)$ in any state gives for certain the result $f(\xi')$ if and only if the measurement of ξ in that same state gives

† Cf. Dirac 1947. 5, p. 40.

‡ Ibid. pp. 41 ff.

for certain the result ξ', where ξ' now denotes any eigenvalue. This leads to the mathematical definition

$$f(\xi)|\xi'\rangle = f(\xi')|\xi'\rangle \tag{9.1}$$

which is valid for every eigenvalue ξ' under the condition that the function $f(x)$ of a real variable x is defined in a domain containing all eigenvalues of ξ and that for each of these eigenvalues $f(x)$ is single-valued. From this we see that the values of $f(x)$ for values of x which are not eigenvalues of ξ do not affect the function $f(\xi)$. It can be proved that a sum and a product of functions of ξ and a function of a function of ξ are all functions of ξ.

The physical interpretation of a result of a measurement must now be completed by an expression for the probability of such a result. *We assume that for any state $|X\rangle$, the real number $\langle X|\xi|X\rangle$ represents the average value of the result of measuring ξ for this state, provided that $|X\rangle$ is normalized to 1*: $\langle X|X\rangle = 1$. This can be said in another way. Let δ_{xa} be the function of x which takes the value 1 for $x = a$ and the value 0 for $x \neq a$. Then we can form the function $\delta_{\xi a}$ of the observable ξ and the scalar variable a. For all values of x which are not eigenvalues of ξ the values of δ_{xa} do not affect the function $\delta_{\xi a}$. Hence, if a is not an eigenvalue of ξ we have $\delta_{\xi a} = 0$, but if a is an eigenvalue of ξ the result of the measurement of $\delta_{\xi a}$ is zero for all states that are not eigenstates belonging to the eigenvalue a and is 1 for an eigenstate belonging to a. Now, according to our assumption, for every value of a,

$$P_a \overset{\text{def}}{=} \langle X|\delta_{\xi a}|X\rangle \tag{9.2}$$

is the average value of the result of the measurement of $\delta_{\xi a}$ in the state $|X\rangle$. $P_a = 0$ if a is not an eigenvalue of ξ, because $\delta_{\xi a}$ vanishes.

If a is an eigenvalue of ξ, which does not lie in a range of eigenvalues, every time that the measurement of ξ gives a (not a), the measurement of $\delta_{\xi a}$ gives one (zero). If now a great number N of measurements of ξ and $\delta_{\xi a}$ is made, and if the measurement of ξ gives the value a N' times and another eigenvalue $N-N'$ times, the measurement of $\delta_{\xi a}$ gives one N' times and zero $N-N'$ times. Hence the average value $P_a = \lim_{N\to\infty} N'/N$ is the probability that ξ has the value a in the state X.

If a happens to lie in a range of eigenvalues, the probability of finding a exactly is of course zero. But here we are interested in the probability $P(a)\,da$ that ξ has a value between a and $a+da$. $P(a)\,da$ is really an interval function. Now let E_x^{da} be the point-interval function of v. Dantzig which takes the value 1 if x lies in the interval between a and $a+da$ and zero otherwise (cf. § 7). Then E_ξ^{da} is the corresponding observable-interval function.† If the interval da does not contain an eigenvalue of ξ we have $E_{\xi}^{da} = 0$. But if the interval contains one or more eigenvalues, the result of the measurement of E_{ξ}^{da} is zero for all states that are not eigenstates belonging to an eigenvalue of ξ lying in the interval and is *1* for an eigenstate whose eigenvalue lies in the interval. With the same reasoning as above we see that now

$$P(a)\,da = \langle X|E_{\xi}^{da}|X\rangle \tag{9.3}$$

provided that $|X\rangle$ is normalized to *1*.

If $|X\rangle$ is not normalized to one, P_a and $P(a)\,da$ will still be *proportional* to the probability that ξ has the value a and lies in the interval respectively.

Obviously we have

$$\delta_{\xi a}|\xi'\rangle = \delta_{\xi' a}|\xi'\rangle = \begin{cases} |\xi'\rangle \text{ for } \xi' = a \\ 0 \quad ,, \quad \xi' \neq a, \end{cases} \tag{9.4}$$

because $|\xi'\rangle$ is an eigenstate of $\delta_{\xi a}$ belonging to the eigenvalue zero and in the same way

$$E_{\xi}^{da}|\xi'\rangle = E_{\xi'}^{da}|\xi'\rangle = \begin{cases} |\xi'\rangle \text{ for } \xi' \text{ lying between } a \text{ and } a+da, \\ 0 \text{ for } \xi' \text{ not lying between } a \text{ and } a+da. \end{cases} \tag{9.5}$$

10. Representations and matrices

As we have seen in § 5, every set of hermitian operators in E_n has at least one complete set of simultaneous eigenvalues if and only if it is commuting. It can be proved that in the general case this is also true for observables:

A set of observables has at least one complete set of simultaneous eigenstates if and only if it is commuting.‡

But in order to get representations we need something more. The expansion (8.1) is unique if and only if no two terms belong to the same eigenvalue, and this can be made certain only by the assumption that the complete set of eigenstates of the observable is such that to every eigenvalue there belongs one and only one eigenstate. If an observable or a set of commuting observables satisfies this condition it is said to form a *complete commuting set.* In general an observable or a commuting set of observables does not form a complete set but there is a very remarkable theorem:

Every set of commuting observables can be made into a complete commuting set by adding certain observables to it.§

† We differ slightly here from Dirac (1947. 5, p. 48) because he does not mention interval functions explicitly.

‡ Cf. for the proof Dirac 1947. 5, p. 49 f.

§ Ibid. pp. 53 ff.

When a complete commuting set is given, we call the eigenkets (bras) belonging to the simultaneous eigenvalues, normalized in some suitable way, a system of (always orthogonal) *basic kets* (bras). A convenient way of labelling them is by means of the eigenvalues belonging to them. Let $\xi_1, \ldots, \xi_u$ be the complete commuting set of observables and let $\xi'_1, \ldots, \xi'_u$ be the eigenvalues (discrete or continuous) belonging to some eigenket. Then this ket (normalized in some suitable way) will be denoted by $|\xi'_1 \ldots \xi'_u\rangle$ and the conjugate bra by $\langle \xi'_1 \ldots \xi'_u|$. The basic kets are a kind of measuring kets like the orthogonal measuring vectors $\underset{i}{i}^{\kappa}$ of tensor calculus, and the systems of numbers $\xi'_1 \ldots \xi'_u$ are marks to distinguish them from other kets and from each other, just as the kernel $\underset{i}{i}$ serves to distinguish the measuring vectors $\underset{i}{i}^{\kappa}$ from other vectors and from each other. But it would not be satisfactory always to mark our more general kets by a kernel containing one index because we often want a notation showing clearly the eigenvalues to which the basic ket belongs. Here is the reason why in this special case, where only observables are concerned and basic kets are always derived from complete commuting sets, and where there are moreover enumerable or non-enumerable infinite basic kets, it is so efficient to introduce a notation with boxes $\langle\ |, |\ \rangle$ in which all this valuable information can be stored.

Having thus introduced a set of orthogonal basic kets and their conjugate bras, an *orthogonal representation* can be set up, i.e. every ket, bra, or linear operator can be *represented by a set of real or complex* numbers. For a ket $|P\rangle$ or a bra $\langle Q|$ we get the representations

$$\langle \xi'_1 \ldots \xi'_u | P\rangle, \qquad \langle Q|\xi'_1 \ldots \xi'_u\rangle. \tag{10.1}$$

and for an operator α the representation

$$\langle \xi'_1 \ldots \xi'_u|\alpha|\xi''_1 \ldots \xi''_u\rangle. \tag{10.2}$$

The numbers of a representation correspond to the orthogonal components in U_n and there is no reason why we should not call them *orthogonal components with respect to the given set of orthogonal basic kets.*

Conversely we wish now to work out $|P\rangle$, $\langle Q|$, and α if their orthogonal components are given. Let us suppose first that there is only one observable ξ forming a complete set. Then the expression (8.1)

$$|P\rangle = \sum_r c^r|\xi^r\rangle + \sum_s c^s|\xi^s d\rangle + \int c(\xi')|\xi' c\rangle\, d\xi' \tag{10.3}$$

is unique if $|\xi^r\rangle$, $|\xi^s d\rangle$, and $|\xi' c\rangle$ are normalized. The representation of P consists of the orthogonal components $\langle \xi^r|P\rangle$ and $\langle \xi' c|P\rangle$. Let $|\xi^r\rangle$ and $|\xi^s d\rangle$ be normalized to *1* and $|\xi' c\rangle$ to $\rho^{-1}\delta$ where ρ is a function of

ξ^r, ξ', called the *weight function* of the representation. Now the second term on the right-hand side of (10.3) can be written

$$\sum_s c^s|\xi^s d\rangle = \sum_s \int c^s \alpha^s \delta(\xi^s - \xi')|\xi' c\rangle\, d\xi' \tag{10.4}$$

where the α^s are scalar factors such that

$$\alpha^s|\xi^s c\rangle = |\xi^s \alpha\rangle, \tag{10.5}$$

where $|\xi^s d\rangle$ is normalized to *1* and $|\xi^s d\rangle$ is normalized to $\rho^{-1}\delta$. Hence $|P\rangle$ can be written in the form

$$|P\rangle = \sum_r c^r|\xi^r\rangle + \int c(\xi')|\xi'\rangle\, d\xi', \tag{10.6}$$

where $|\xi'\rangle$ is now written instead of $|\xi' c\rangle$ and $c(\xi')$ is no longer a finite ordinary function but a sum of a finite ordinary function and an interval function connected with a δ-function. Multiplying (10.6) by $\langle\xi^r|$ and by $\rho''\langle\xi''|$ we get

$$\begin{aligned} &(a) \qquad \langle\xi^r|P\rangle = c^r, \\ &(b) \qquad \rho''\langle\xi''|P\rangle = \int c(\xi')\rho''\langle\xi''|\xi'\rangle\, d\xi' = c(\xi''), \end{aligned} \tag{10.7}$$

and if these values are substituted into (10.6) we get $|P\rangle$ expressed in terms of its orthogonal components

$$|P\rangle = \sum_r |\xi^r\rangle\langle\xi^r|P\rangle + \int |\xi'\rangle\rho'\, d\xi'\langle\xi'|P\rangle. \tag{10.8}$$

This can be expressed also in the formula

$$\sum_r |\xi^r\rangle\langle\xi^r| + \int |\xi'\rangle\rho'\, d\xi'\langle\xi'| = 1. \tag{10.9}$$

If there is a complete set of commuting observables $\xi_1,\ldots, \xi_u$ we will deal here with the case where $\xi_1,\ldots, \xi_v$ have discrete eigenvalues only and $\xi_{v+1},\ldots, \xi_u$ continuous ones only. Then the general simultaneous eigenket $|\xi'_1 \ldots \xi'_u\rangle$ can be normalized by the equation

$$\langle\xi'_1 \ldots \xi'_u|\xi''_1 \ldots \xi''_u\rangle = \rho'^{-1}\delta_{\xi'_1\xi''_1}\ldots\delta_{\xi'_v\xi''_v}\delta(\xi'_{v+1} - \xi''_{v+1})\ldots\delta(\xi'_u - \xi''_u), \tag{10.10}$$

and instead of (10.8) and (10.9) we get

$$|P\rangle = \sum_{\xi'_1\ldots\xi'_v} \int |\xi'_1\ldots\xi'_u\rangle\rho'\, d\xi'_{v+1}\ldots d\xi'_u\langle\xi'_1\ldots\xi'_u|P\rangle \tag{10.11}$$

and

$$\sum_{\xi'_1\ldots\xi'_v} \int |\xi'_1\ldots\xi'_u\rangle\rho'\, d\xi'_{v+1}\ldots d\xi'_u\langle\xi'_1\ldots\xi'_u| = 1. \tag{10.12}$$

In this equation we see that by using ρ' we introduce some kind of measuring volume in the space which has coordinates $\xi'_{v+1},\ldots, \xi'_u$.

The general case when all observables have both discrete and continuous eigenvalues can be treated in the same way.

If there is only one observable and a finite number of eigenstates the orthogonal components (10.2) of an operator form an ordinary matrix. First we consider now the case of a complete set of commuting observables $\xi_1, \ldots, \xi_u$ and an *enumerable* infinite number of simultaneous eigenstates, hence only discrete eigenvalues. Then the orthogonal components (10.2) may again be looked upon as elements of a matrix although it is an infinite one. Numbering the eigenvalues $\xi^1, \xi^2, \ldots$, every element of the matrix is of the form $\langle \xi^r | \alpha | \xi^s \rangle$; $r, s = 1, 2, \ldots$. We see immediately that $\xi_1, \ldots, \xi_u$ and all observables that are functions of them are represented by *diagonal matrices*, i.e. matrices with non-zero elements only in the main diagonal. The identical operator is represented by the *unit matrix*, i.e. the diagonal matrix with elements $+1$ only.

If α is hermitian, $\alpha = \bar{\alpha}$, it follows that

$$\overline{\langle \xi_1' \ldots \xi_u' | \alpha | \xi_1'' \ldots \xi_u'' \rangle} = \langle \xi_1'' \ldots \xi_u'' | \alpha | \xi_1' \ldots \xi_u' \rangle. \tag{10.13}$$

Hence the matrix of a hermitian operator is a hermitian matrix.

From (10.10) it follows for normalization to 1 that

$$\langle \xi_1' \ldots \xi_u' | \alpha\beta | \xi_1'' \ldots \xi_u'' \rangle = \sum \langle \xi_1' \ldots \xi_u' | \alpha | \xi_1''' \ldots \xi_u''' \rangle \langle \xi_1''' \ldots \xi_u''' | \beta | \xi_1'' \ldots \xi_u'' \rangle, \tag{10.14}$$

and this means that the matrix of a product can be derived from the matrices of the factors by the ordinary rules of matrix multiplication.

We see in the same way from

$$\langle \xi_1' \ldots \xi_u' | \alpha | P \rangle = \sum \langle \xi_1' \ldots \xi_u' | \alpha | \xi_1'' \ldots \xi_u'' \rangle \langle \xi_1'' \ldots \xi_u'' | P \rangle \tag{10.15}$$

that, if $|P\rangle$ is considered as a matrix with an enumerable infinite number of rows and one column, the ordinary rules of matrix multiplication hold.

Now we consider the more general case when $\xi_1, \ldots, \xi_v$ have discrete eigenvalues only and $\xi_{v+1}, \ldots, \xi_u$ continuous ones only. Again we have a set of orthogonal components of α but this set is non-enumerable infinite. Nevertheless the whole set of numbers is considered as a matrix representing α. Instead of (10.14) we get according to (10.12) for $\rho' = 1$

$$\langle \xi_1' \ldots \xi_u' | \alpha\beta | \xi_1'' \ldots \xi_u'' \rangle = \sum_{\xi_1' \ldots \xi_v'} \int \langle \xi_1' \ldots \xi_u' | \alpha | \xi_1''' \ldots \xi_u''' \rangle \, d\xi_{v+1}''' \ldots d\xi_u''' \langle \xi_1''' \ldots \xi_u''' | \beta | \xi_1'' \ldots \xi_u'' \rangle \tag{10.16}$$

as a rule for the multiplication of these generalized matrices. Instead of (10.15) we get

$$\langle \xi_1' \ldots \xi_u' | \alpha | P \rangle = \sum_{\xi_1' \ldots \xi_v'} \int \langle \xi_1' \ldots \xi_u' | \alpha | \xi_1'' \ldots \xi_u'' \rangle \, d\xi_{v+1}'' \ldots d\xi_u'' \langle \xi_1'' \ldots \xi_u'' | P \rangle, \tag{10.17}$$

showing that P may be considered as a generalized matrix with a non-enumerable infinite number of rows and one column.

For $\alpha = \xi_1$ and $\rho' = 1$ we get

$$\langle \xi'_1 \ldots \xi'_u | \xi_1 | \xi''_1 \ldots \xi''_u \rangle = \xi'_1 \delta_{\xi'_1 \xi''_1} \ldots \delta_{\xi'_v \xi''_v} \delta(\xi'_{v+1} - \xi''_{v+1}) \ldots \delta(\xi'_u - \xi''_u), \tag{10.18}$$

and this means that non-zero elements only are in the main diagonal and that these elements have a very particular form, containing a product of $u-v$ δ-functions.

Now let α be an operator commuting with ξ_1. Then we have according to (10.16) and (10.18)

$$\int \langle \xi'_1 \ldots \xi'_u | \alpha | \xi'''_1 \ldots \xi'''_u \rangle \, d\xi'''_{v+1} \ldots d\xi'''_u \, \xi'''_1 \delta_{\xi'''_1 \xi''_1} \ldots \delta_{\xi'''_v \xi''_v} \delta(\xi'''_{v+1} - \xi''_{v+1}) \ldots \delta(\xi'''_u - \xi''_u)$$
$$= \int \xi'_1 \delta_{\xi'_1 \xi'''_1} \ldots \delta_{\xi'_v \xi'''_v} \delta(\xi'_{v+1} - \xi'''_{v+1}) \ldots \delta(\xi'_u - \xi'''_u) \, d\xi'''_{v+1} \ldots d\xi'''_u \langle \xi'''_1 \ldots \xi'''_u | \alpha | \xi''_1 \ldots \xi''_u \rangle \tag{10.19}$$

or, making use of the properties of the δ-functions,

$$(\xi'_1 - \xi''_1) \langle \xi'_1 \ldots \xi'_u | \alpha | \xi''_1 \ldots \xi''_u \rangle = 0. \tag{10.20}$$

ξ_1 has only discrete eigenvalues. Hence it follows from (10.20) that $\langle \xi'_1 \ldots \xi'_u | \alpha | \xi''_1 \ldots \xi''_u \rangle$ must have $\delta_{\xi'_1 \xi''_1}$ as a factor. But if we take ξ_{u+1} instead of ξ_1, the same reasoning leads to the conclusion that $\langle \xi'_1 \ldots \xi'_u | \alpha | \xi''_1 \ldots \xi''_u \rangle$ contains $\delta(\xi'_{u+1} - \xi''_{u+1})$ as a factor. Hence, if α commutes with all observables $\xi_1, \ldots, \xi_u$ all elements outside the main diagonal vanish and the elements in the main diagonal have the same particular form as the matrices of $\xi_1, \ldots, \xi_u$. All matrices with this particular form of elements commute. Now commuting is a very important property of matrices and therefore, for matrices with a non-enumerable number of rows and columns, we use the term *diagonal matrix* only if the elements have this particular form. Non-diagonal matrices with a non-enumerable number of rows and columns can have the property that all elements outside the main diagonal vanish. They occur frequently in quantum mechanics but they do not commute with the diagonal matrices.

11. Probabilities and orthogonal components†

If, from the complete set of commuting observables $\xi_1, \ldots, \xi_u$, the first v have only discrete eigenvalues and the other $u-v$ only continuous ones, we wish to know the probability that the result of a measurement of all these observables in a state $|S\rangle$ is $\xi'_1, \ldots, \xi'_v$ for the first v observables and lies in the intervals ξ'_{v+1} to $\xi'_{v+1} + d\xi'_{v+1}, \ldots, \xi'_u$ to $\xi'_u + d\xi'_u$ for the other $u-v$ observables. According to our assumptions in § 9 this probability is

$$P_{\xi'_1 \ldots \xi'_u} d\xi'_{v+1} \ldots d\xi'_u = \langle S | \delta_{\xi_1 \xi'_1} \ldots \delta_{\xi_v \xi'_v} E^{d\xi'_{v+1}}_{\xi_{v+1}} \ldots E^{d\xi'_u}_{\xi_u} | S \rangle \tag{11.1}$$

if $|S\rangle$ is normalized to 1.

† Cf. Dirac 1947. 5, pp. 45 ff., 72 ff.

Taking $u = 2$, $v = 1$ for convenience and making use of (10.12) for $\rho' = 1$ we get according to (9.4, 5)

$$P_{\xi_1'\xi_2'}d\xi_2' = \sum_{\xi_1''}\int \langle S|\delta_{\xi_1\xi_1'}E_{\xi_2}^{d\xi_2}|\xi_1''\xi_2''\rangle d\xi_2''\langle\xi_1''\xi_2''|S\rangle$$

$$= \langle S|\xi_1'\xi_2'\rangle\langle\xi_1'\xi_2'|S\rangle d\xi_2' = |\langle\xi_1'\xi_2'|S\rangle|^2 d\xi_2'. \quad (11.2)$$

In the same way we get in the more general case (11.1)

$$P_{\xi_1'\ldots\xi_u'}d\xi_{v+1}'\ldots d\xi_u' = |\langle\xi_1'\ldots\xi_u'|S\rangle|^2 d\xi_{v+1}'\ldots d\xi_u', \quad (11.3)$$

and this means that the probability distribution for the measurement of the complete set of observables $\xi_1,\ldots, \xi_u$ is given by the squares of the absolute values of the orthogonal components of the state concerned with respect to the complete set of simultaneous eigenstates.

The orthogonal components $\langle\xi_1'\ldots\xi_u'|S\rangle$ need not be real but they may contain a factor of the form $e^{i\phi}$. They are called the *probability amplitudes* and the factor is called the *phase factor*. The phase factor does not affect the probability.

If $|S\rangle$ is not normalized to 1 but to δ, the $|\langle\xi_1'\ldots\xi_u'|S\rangle|^2 d\xi_{v+1}'\ldots d\xi_u'$ are not probabilities but they are *proportional to probabilities*. In this case the orthogonal components $\langle\xi_1'\ldots\xi_u'|S\rangle$ are called the *relative probability amplitudes*.

We are often interested in two different representations of the same dynamical system. Let $\eta_1,\ldots, \eta_w$ be another complete set of commuting observables (in general not commuting with the ξ) and let $\eta_1,\ldots, \eta_x$ have only discrete eigenvalues and $\eta_{x+1},\ldots, \eta_w$ only continuous ones. Then, from (10.12) and the corresponding formula in η, we get the equations

$$\langle\eta_1'\ldots\eta_w'|P\rangle = \sum_{\xi_1'\ldots\xi_v'}\int \langle\eta_1'\ldots\eta_w'|\xi_1'\ldots\xi_u'\rangle d\xi_{v+1}'\ldots d\xi_u'\langle\xi_1'\ldots\xi_u'|P\rangle, \quad (11.4)$$

$$\langle\xi_1'\ldots\xi_u'|P\rangle = \sum_{\eta_1'\ldots\eta_x'}\int \langle\xi_1'\ldots\xi_u'|\eta_1'\ldots\eta_w'\rangle d\eta_{x+1}'\ldots d\eta_w'\langle\eta_1'\ldots\eta_w'|P\rangle, \quad (11.5)$$

which express the orthogonal components of P with respect to the set of simultaneous eigenstates of $\xi_1,\ldots, \xi_u$ and those with respect to the set of simultaneous eigenstates of $\eta_1,\ldots, \eta_w$ in terms of each other. We have seen in § 5 that transformations which transform a complete orthogonal set of unit kets into a system of the same kind are unitary transformations. Now here we have complete sets consisting of an infinite number of kets and kets normalized to δ instead of unit kets. Nevertheless, (11.4) and (11.5) are unitary transformations, as is to be seen from (cf. (3.7) and (7.6a))

$$\overline{\langle\eta_1'\ldots\eta_w'|\xi_1'\ldots\xi_u'\rangle} = \langle\xi_1'\ldots\xi_u'|\eta_1'\ldots\eta_w'\rangle. \quad (11.6)$$

The coefficients $\langle \eta'_1 \dots \eta'_w | \xi'_1 \dots \xi'_u \rangle$ may be looked upon as a kind of probabilities. If the ξ's and η's all have discrete eigenvalues only, all eigenkets are normalized to *1*. Then $\langle \eta'_1 \dots \eta'_w | \xi'_1 \dots \xi'_u \rangle$ is one of the orthogonal components of the ket $|\xi'_1 \dots \xi'_u\rangle$ with respect to the complete set of simultaneous eigenkets of $\eta_1, \dots, \eta_w$ and thus its absolute value represents the probability that the measurement of $\eta_1, \dots, \eta_w$ leads to $\eta'_1, \dots, \eta'_w$ if this measurement is done in the state $|\xi'_1 \dots \xi'_u\rangle$, that is, in the state in which the measurement of $\xi_1, \dots, \xi_u$ leads for certain to the result $\xi'_1, \dots, \xi'_u$. From this point of view, (11.6) can be interpreted as a theorem of reciprocity.

12. The labelling by means of functions†

In § 9 we have seen that a function of the complete set of commuting observables $\xi_1, \dots, \xi_u$ is a correspondence between the simultaneous eigenstates and the complex numbers such that to every one of these eigenstates there belongs one and only one number. Hence the representatives $\langle \xi'_1 \dots \xi'_u | P \rangle$ of any ket form a function of $\xi_1, \dots, \xi_u$ according to this definition. Let $\psi(\xi_1, \dots, \xi_u)$, or $\psi(\xi)$ for short, be this function

$$\langle \xi'_1 \dots \xi'_u | P \rangle = \psi(\xi'_1, \dots, \xi'_u) = \psi(\xi'). \tag{12.1}$$

Then we may use $\psi(\xi)$ to label the ket $|P\rangle$:

$$|P\rangle = |\psi(\xi)\rangle. \tag{12.2}$$

In order to see what is really going on from the point of view of tensor calculus, let us suppose that the number of simultaneous eigenvalues is n and that $\underset{1}{i^{\kappa}}, \dots, \underset{n}{i^{\kappa}}$ are the orthogonal simultaneous eigenkets normalized to *1*. Let $|P\rangle$ be the ket v^{κ}, $v_{\bar{\lambda}}$. Then the representatives are the orthogonal components $v^1, \dots, v^n$, which are numerically equal to $v_{\bar{1}}, \dots, v_{\bar{n}}$. But the function $\psi(\xi)$ is now the operator

$$v^1 \overset{1}{i^{\kappa}} \overset{1}{i_{\lambda}} + \dots + v^n \overset{n}{i^{\kappa}} \overset{n}{i_{\lambda}}. \tag{12.3}$$

This is not a concomitant of the vector v^{κ}. But it is a concomitant of v^{κ} and the set of operators corresponding to $\xi_1, \dots, \xi_n$, because this set is complete and determines uniquely the eigenstates corresponding to $\underset{1}{i^{\kappa}}, \dots, \underset{n}{i^{\kappa}}$.

Using this new method of labelling, every ket is labelled with a function of $\xi_1, \dots, \xi_u$. Now consider the ket $|P\rangle = |\psi(\xi)\rangle$ and another function $f(\xi)$. Then we wish to determine the label of the product $f(\xi)|P\rangle$ (note that $f(\xi)$ is an observable, not a scalar!). If $\xi'_1 \dots \xi'_u$, or shortly ξ', is a simultaneous eigenvalue we have

$$\langle \xi' | f(\xi) = \langle \xi' | f(\xi') \tag{12.4}$$

† Cf. Dirac 1947. 5, pp. 79 ff.

and $$\langle \xi' | f(\xi) | P \rangle = f(\xi') \langle \xi' | P \rangle = f(\xi') \psi(\xi'), \tag{12.5}$$

hence $$f(\xi) | \psi(\xi) \rangle = | f(\xi) \psi(\xi) \rangle. \tag{12.6}$$

But this equation shows that with this new labelling the vertical bar is superfluous and that we may write

$$| P \rangle = \psi(\xi) \rangle. \tag{12.7}$$

When we have done this we may look upon $\psi(\xi)\rangle$ as the product of the observable $\psi(\xi)$ and the ket $1\rangle$ or shortly $\rangle$. $\rangle$ is called the *standard ket.* Obviously it always belongs to a definitely chosen complete set of commuting observables, and its representative numbers or orthogonal components are all equal to $+1$. In tensor calculus it corresponds to the ket with n orthogonal components all equal to 1:

$$\underset{1}{i^\kappa} + \ldots + \underset{n}{i^\kappa}. \tag{12.8}$$

Transvection of (12.8) with (12.3) gives the vector v^κ.

The same can be done with bras. If

$$\langle Q | \xi'_1 \ldots \xi'_u \rangle = \phi(\xi'_1, \ldots, \xi'_u) = \phi(\xi'), \tag{12.9}$$

the bra $\langle Q|$ can be written $\langle \phi(\xi)|$ or $\langle \phi(\xi)$ and can be looked upon as the product of the observable $\phi(\xi)$ and the *basic bra* $\langle$, corresponding to the bra $\overset{1}{i_\lambda} + \ldots + \overset{n}{i_\lambda}$ in tensor calculus. Obviously $\langle P|$ must be written $\langle \bar{\psi}(\xi)$.†

Now kets and bras are both labelled by functions of $\xi_1, \ldots, \xi_u$ and the only difference exists in the signs $\rangle$ and $\langle$, denoting the multiplication of a ket (bra) by any operator on the left (right). We may even drop these signs and leave it to memory to remember whether a function $f(\xi)$ stands for a ket or for a bra or for an operator. This is, in fact, what many authors have done from the very beginning of quantum mechanics. They work with some hermitian operator, for instance H, which is a function of a set of commuting observables and a *wave function,* for instance ψ, of the same observables. This wave function can be multiplied by H from the left: $H\psi$, standing for $H|\psi\rangle$. Then the conjugate $\bar{\psi}$ can only be multiplied from the right: $\bar{\psi}H$, standing for $\langle \bar{\psi}|H$. The product $\bar{\psi}\psi$ stands for $\langle \bar{\psi}(\xi) | \psi(\xi) \rangle$.

Using these abbreviations Dirac returns to a well-known form of wave calculus. But the abbreviations are now not introduced *ad hoc* but form a part of a theoretically well-established calculus that can be

† According to the abbreviation introduced in § 7 Dirac writes $\langle P|$ for the conjugate of $|P\rangle$, but with this new labelling we have to write $\langle \bar{\psi}|$ for the conjugate of $|\psi\rangle$. Here is an inconsequence in the notation that could only be avoided by always writing $\langle \bar{P}|$ for the conjugate of $|P\rangle$ independent of the kind of labelling used.

used in all cases, even in those cases where too much abbreviation would lead to ambiguity. With this last finishing touch Dirac's calculus is a highly adapted instrument and a very good illustration of the thesis that mathematical methods which are really good for practical use cannot be found by theoretical considerations only but by collaboration between theoretical and practical investigators.

EXERCISES

X.1. If P and Q are hermitian, prove that $PQ+QP$ and $iPQ-iQP$ are both hermitian.

X.2. If U is a unitary operator, prove that

$$(Uu).(Uv) = u.v. \tag{2α}$$

X.3. Let ξ be a linear operator, m a positive integer, and $\xi^m\,|P\rangle = 0$. Prove that $\xi|P\rangle = 0$.†

X.4. Prove that
$$\frac{d}{dx}\log x = \frac{1}{x} - i\pi\,\delta(x)$$
holds for all real values of x.‡

X.5. A linear operator that commutes with an observable ξ also commutes with any function of ξ.§

X.6. A linear operator that commutes with each of a complete set of commuting observables is a function of these observables.‖

† Cf. Dirac 1947. 5, p. 29. ‡ Ibid. p. 61.
§ Ibid. p. 77. ‖ Ibid. p. 78.

BIBLIOGRAPHY

1872. 1. F. KLEIN, *Vergleichende Betrachtungen über neuere geometrische Forschungen.* Erlangen. Reprinted *Math. Ann.* **43** (1893), 63–100.

1877. 1. L. B. CHRISTOFFEL, 'Ueber die Fortpflanzung von Stössen durch elastische feste Körper.' *Ann. di Mat.* II **8**, 193–243.

1880. 1. A. VOSS, 'Zur Theorie der Transformation quadratischer Differentialausdrücke und der Krümmung höherer Mannigfaltigkeiten.' *Math. Ann.* **16**, 129–78.

1894. 1. L. KRONECKER, *Vorlesungen über die Theorie der einfachen und der vielfachen Integrale.* Leipzig, Teubner.

1898. 1. W. VOIGT, *Die fundamentalen physikalischen Eigenschaften der Kristalle.* Leipzig, Teubner.

1900. 1. E. v. WEBER, *Vorlesungen über das Pfaffsche Problem.* Leipzig, Teubner.

1900. 2. H. BURKHARDT and W. F. MEYER, 'Potentialtheorie.' *Enc. d. m. Wiss.* ii A. 7b. Leipzig, Teubner.

1901. 1. G. RICCI and T. LEVI CIVITA, 'Méthodes de calcul différentiel absolu et leurs applications.' *Math. Ann.* **54**, 125–201; errata, p. 608; reprinted 1923 Paris, Blanchard.

1904. 1. G. HAMEL, 'Die Lagrange–Eulerschen Gleichungen der Mechanik', *Z. Math. Phys.* **50**, 1.

1904. 2. A. E. H. LOVE, 'Wave Motions with Discontinuities at Wave Fronts.' *Proc. Lond. Math. Soc.* ii. **1**, 37–62.

1907. 1. O. TEDONE, 'Allgemeine Theoreme der mathematischen Elastizitätslehre.' *Enc. d. m. Wiss.* iv. 24. Leipzig, Teubner.

1908. 1. R. WEITZENBÖCK, *Komplex-Symbolik.* Leipzig, 1908.

1913. 1. E. B. WILSON, *Vector Analysis.* New Haven, Yale University Press.

1916. 1. A. EINSTEIN, *Die Grundlagen der allgemeinen Relativitätstheorie.* Leipzig, Barth.

1917. 1. E. T. WHITTAKER, *Analytical Dynamics.* Cambridge University Press.

1917. 2. T. LEVI CIVITA, 'Nozione di parallelismo in una varietà qualunque.' *Rend. Circ. Mat. Pal.* **42**, 173–205.

1918. 1. J. A. SCHOUTEN, 'Die direkte Analysis zur neueren Relativitätstheorie.' *Verh. Kon. Akad. v. Wet.* **12**, No. 6, 95 pp.

1920. 1. M. v. LAUE, *Die Relativitätstheorie*, 4th edition. Braunschweig, Vieweg.

1921. 1. W. PAULI, 'Relativitätstheorie'. *Enc. d. m. Wiss.* v. 12. Leipzig–Berlin, Teubner.

1921. 2. H. WEYL, *Raum, Zeit, Materie*, 4th edition. Berlin, Springer.

1921. 3. J. A. SCHOUTEN and D. J. STRUIK, 'On some Properties of general Manifolds relating to Einstein's theory of Gravitation.' *Am. J. of Math.* **43**, 213–16.

1921. 4. E. KASNER, 'Geometrical Theorems on Einstein's Cosmological Equations.' Ibid. p. 217–21.

1921. 5. J. A. SCHOUTEN, 'Über die konforme Abbildung n-dimensionaler Mannigfaltigkeiten mit quadratischer Massbestimmung, auf eine Mannigfaltigkeit mit euklidischer Massbestimmung.' *Math. Zeitschr.* **11**, 58–88.

1922. 1. O. VEBLEN, 'Normal Coordinates for the Geometry of Paths.' *Proc. Nat. Acad.* **8**, 192–7.

1922. 2. E. KASNER, 'The Solar Gravitational Field completely determined by its Light Rays.' *Math. Ann.* **85**, 227–36.

1923. 1. A. S. EDDINGTON, *The Mathematical Theory of Relativity.* Cambridge University Press.

1923. 2. H. W. BRINKMANN, 'On Riemann Spaces conformal to Einstein Spaces.' *Proc. Nat. Acad.* **9**, 172–4.

1923. 3. R. WEITZENBÖCK, *Invariantentheorie.* Groningen, Noordhoff.

1924. 1. J. A. SCHOUTEN, *Der Ricci-Kalkül.* Berlin, Springer. *Here referred to as* R.K. Second unrevised edition N.Y. Edward Bros. Third revised edition in preparation.

1925. 1. L. T. OKAYA, 'L'extrémal dans un champ gravifique à pseudo-orthogonalité.' *Proc. Physicomath. Soc. Tokyo*, **3**, 51–8.

1926. 1. L. P. EISENHART, *Riemannian Geometry.* Princeton University Press, Oxford University Press. Second edition 1949.

1926. 2. J. L. SYNGE, 'On the Geometry of Dynamics.' *Trans. Roy. Soc. London* A **226**, 31–106.

1926. 3. G. VRANCEANU, 'Sopra le equazioni del moto di un sistema anolonome.' *Rend. Acc. Linc.* (6) **4**, 508–11.

1927. 1. G. D. BIRKHOFF, 'Dynamical Systems.' *Amer. Math. Soc. Coll. Publ.* New York.

1927. 2. T. LEVI CIVITA, *The Absolute Differential Calculus.* London, Blackie & Son.

1927. 3. L. P. EISENHART, 'Non-Riemannian Geometry.' *Amer. Math. Soc. Coll. Publ.* New York.

1927. 4. J. A. SCHOUTEN, 'Über n-fache Orthogonalsysteme in V_n.' *Math. Zeitschr.* **26**, 706–30.

1928. 1. Z. HOŘÁK, 'Sur les systèmes non holonomes.' *Bull. Int. Ac. Sc. de Bohème*, **29**, 1–18.

1929. 1. A. D. FOKKER, *Relativiteitstheorie.* Groningen, Noordhoff.

1931. 1. T. Y. THOMAS, *The Elementary Theory of Tensors.* New York–London. McGraw-Hill.

1931. 2. W. SLEBODZINSKI, 'Sur les équations de Hamilton.' *Bull. Acad. Roy. Belg.* (5) **17**, 864–70.

1932. 1. O. VEBLEN and J. H. C. WHITEHEAD, *The Foundations of Differential Geometry.* Cambridge Tracts No. 29, 96 pp.

1932. 2. A. WUNDHEILER, 'Rheonome Geometrie.' *Absolute Mechanik. Prac. Matematyczno-Fizycznych Warszawa*, **40**, 97–142.

1933. 1. J. A. SCHOUTEN and E. R. v. KAMPEN, 'Beiträge zur Theorie der Deformation.' *Prac. Matematyczno-Fizycznych Warszawa*, **41**, 1–19.

1933. 2. I. KOGA, 'Vibrations of Piezo-electric Oscillating Crystals.' *London, Edinburgh, and Dublin Phil. Mag. and J. of Sc.* (7) **12**, 275–83.

1933. 3. E. JAHNKE and F. EMDE, *Funktionentafeln.* Leipzig–Berlin, Teubner. New York, Dover Publications.

1933. 4. P. A. M. DIRAC, 'Homogeneous Variables in Classical Dynamics.' *Proc. Cambr. Phil. Soc.* **29**, 389–400.

1934. 1. G. PRANGE, 'Die allgemeinen Integrationsmethoden der analytischen Mechanik.' *Enc. d. m. Wiss.* iv. 12, 13. Leipzig, Teubner.

1934. 2. F. R. LACK, G. W. WILLARD, and J. E. FAIR, 'Some Improvements in Quartz Crystal Circuit Elements.' *Bell System Techn. Journ.* **13**, 453–63.

1934. 3, 4, 5, 6. D. v. DANTZIG, 'Electromagnetism, independent of metrical geometry.' *Proc. Kon. Ned. Akad. v. Wet.* **37**, 521–5, 526–31, 644–52, 825–36.

1934. 7. J. L. SYNGE, 'The Energy Tensor of a Continuous Medium.' *Trans. Roy. Soc. of Canada*, **3**, 28, 127–71.

1935. 1. J. A. SCHOUTEN and D. J. STRUIK, *Einführung in die neueren Methoden der Differentialgeometrie*, I. Groningen–Batavia, Noordhoff. *Here referred to as* E I.

1936. 1. B. v. DIJL, 'The Applications of Ricci Calculus to the Solutions of Vibrational Equations of Piezo-electric Quartz.' *Physica*, **3**, 317–26.

1936. 2. J. L. SYNGE, *Tensorial Methods in Dynamics*. Toronto University Press.

1936. 3. D. v. DANTZIG, 'Ricci Calculus and Functional Analysis.' *Proc. Kon. Ned. Akad. v. Wet.* **39**, 785–94.

1936. 4. P. JORDAN, *Anschauliche Quantentheorie*. Berlin, Springer.

1937. 1. J. W. GIVENS, 'Tensor Coordinates of Linear Spaces.' *Ann. of Math.* **38**, 355–85.

1937. 2. J. v. WEYSSENHOFF, 'Duale Grössen, Grossrotation, Grossdivergenz und die Stokes-Gaussischen Sätze in allgemeinen Räumen.' *Ann. Soc. Pol. de Math.* **16**, 127–44.

1937. 3. J. L. SYNGE, 'Relativistic Hydrodynamics.' *Proc. Lond. Math. Soc.* (2) **43**, 376–416.

1937. 4. H. A. KRAMERS, *Die Grundlagen der Quantentheorie*. Leipzig. Akad. Verl. Ges.

1938. 1. L. BRILLOUIN, *Les Tenseurs en Mécanique et en Élasticité*. Paris, Masson.

1938. 2. J. A. SCHOUTEN, 'Über die geometrische Deutung von gewöhnlichen p-Vektoren und W-p-Vektoren und den korrespondierenden Dichten.' *Proc. Kon. Ned. Akad. v. Wet.* **41**, 709–16.

1938. 3. J. A. SCHOUTEN and D. J. STRUIK, *Einführung in die neueren Methoden der Differentialgeometrie* II. Groningen–Batavia, Noordhoff. *Here referred to as* E II.

1939. 1. A. FOKKER, 'De beweging van golfgroepen volgens de kanonische vergelijkingen van Hamilton. *Hand. 27ste Ned. Nat. Gen. Congr.* 1–4.

1939. 2. A. FOKKER, 'Hamilton's Canonical equations for the Motion of Wave Groups.' *Physica*, **6**, 785–90.

1939. 3. D. v. DANTZIG, 'On the Phenomenological Thermodynamics of Moving Matter.' *Physica*, **6**, 673–704.

1939. 4. D. v. DANTZIG, 'On Relativistic Thermodynamics.' *Proc. Kon. Ned. Akad. v. Wet.* **42**, 601–7.

1939. 5. D. v. DANTZIG, 'Stress Tensor and Particle Density in special Relativity Theory.' *Nature*, **143**, 855.

1939. 6. D. v. DANTZIG, 'On Relativistic Gas Theory.' *Proc. Kon. Ned. Akad. v. Wet.* **42**, 608–25.

1940. 1. J. A. SCHOUTEN and D. v. DANTZIG, 'On Ordinary Quantities and W-quantities.' *Comp. Math.* **7**, 447–73.

1940. 2. D. v. DANTZIG, 'On the Thermo-hydrodynamics of Perfectly Perfect Fluids.' *Proc. Kon. Ned. Akad. v. Wet.* **43**, 387–402, 609–18.

1944. 1. A. E. H. LOVE, *A Treatise on the Mathematical Theory of Elasticity*. New York, Dover Publications.

1945. 1. D. E. LITTLEWOOD, 'Invariant Theory under Orthogonal Groups.' *Proc. Lond. Math. Soc.* (2) **50**, 349–79.

1946. 1. H. Dorgelo and J. A. Schouten, 'On Units and Dimensions.' *Proc. Kon. Ned. Akad. v. Wet.* **48**, 124–31, 282–91, 393–403.

1946. 2. W. G. Cady, *Piezo-electricity.* Intern. Ser. in Pure and Applied Physics. New York–London, McGraw-Hill.

1947. 1. A. Lichnerowicz, *Algèbre et Analyse Linéaire.* Paris, Masson.

1947. 2. L. Brandt, *Vector and Tensor Analysis.* New York, Wiley & Sons; London, Chapman & Hall.

1947. 3. A. D. Michal, *Matrix and Tensor Calculus.* Galcit Aeronautic Series. New York, John Wiley & Sons; London, Chapman & Hall.

1947. 4. W. P. Mason, 'First and Second Order Equations for Piezo-electric Crystals expressed in Tensor Form.' *Bell System Techn. Journal,* **26,** 80–138.

1947. 5. P. A. M. Dirac, *The Principles of Quantum Mechanics.* Oxford, Clarendon Press.

1947. 6. A. Duschek, 'Matrizen, Vektoren und Tensoren.' *Ing. Arch.* **1,** 371–82.

1948. 1. E. J. Post, 'Reciprocal Properties of Elastic Waves in Anisotropic Media.' *Proc. Kon. Ned. Akad. v. Wet.* **51,** 65–72.

1949. 1. J. A. Schouten and W. v. d. Kulk, *Pfaff's Problem and its Generalizations.* Oxford, Clarendon Press. *Here referred to as* P.P.

1951. 1. J. A. Schouten, *Regular Systems of Equations and Supernumerary Coordinates.* Scriptum 6 of the Mathematical Centre, Amsterdam.

1951. 2. F. G. Fumi, 'Third order elastic coefficients of crystals.' *Phys. Rev.* **83,** 1274–5.

1952. 1. F. G. Fumi, 'Physical properties of crystals: the direct inspection method; the direct inspection method in systems with a principal axis of symmetry.' *Acta Cryst.* **5,** 44–48, 691–4.

1952. 2. F. G. Fumi, 'Third order elastic coefficients in trigonal and hexagonal crystals.' *Phys. Rev.* **86,** 561–2.

1952. 3. F. G. Fumi, 'Matter tensors in symmetrical systems.' *Nuovo Cimento,* **9,** 739–56.

1952. 4. H. Rund, 'Die Hamiltonsche Funktion bei allgemeinen dynamischen Systemen.' *Arch. d. Math.* **3,** 207–15.

1953. 1. R. Fieschi and F. G. Fumi, 'High order matter tensors in symmetrical systems.' *Nuovo Cimento,* **10,** 865–82.

1954. 1. J. A. Schouten, *Ricci Calculus.* London, Lange, Maxwell, and Springer. Third revised edition of R.K. 1924. 1. *Here referred to as* R.C.

ADDITIONAL NOTES

NOTE A, p. 15.

The use of $\stackrel{*}{=}$ is not always obligatory. Of course it must be used in an equation like (2.16). But if in an investigation a special coordinate system is used with respect to which certain equations get a simple form, it may be so obvious that this simple form only occurs with respect to this special coordinate system, that no doubt can arise and then the sign $\stackrel{*}{=}$ can be omitted.

NOTE B, p. 21.

The rank can, in the same way, be defined for every expression with indices that is not an affinor. But then the rank defined in this way has a sense only with respect to the coordinate system that has been used in the definition.

NOTE C, p. 45.

For $n = 3$ we show here how the figure of $\mathfrak{v}^\kappa$ arises from the figure of v^κ by means of a volume, and similar correspondences in three other cases. In all these cases the $\pm$ sign depends on the sign of $\mathfrak{q}$.

NOTE D, p. 84.

Many authors use the term parallel instead of pseudo-parallel.

NOTE E, p. 93.

This method is called the *transposition* of a problem. Cf. R.C. III § 3 also for problems that need other methods.

NOTE F, p. 100.

If $K_{\mu\lambda} = cg_{\mu\lambda}$ it can be proved that c is a constant. The space is then called an *Einstein space* and for $c = 0$ a *special Einstein space.*

NOTE G, p. 161.

The same method can be used for affinors and W-affinors of all valences. Fumi did this (1951. 2, 1952. 1, 2, 3) for the valences 3, 4, and 6. Cf. Fieschi and Fumi 1953. 1.

NOTE H, p. 167.

It is interesting to compare these tables and those for the piezo-magnetic quantity on page 168 with the tables of Fumi (1952. 3, pp. 12, 13) for general quantities of valence 3 and for the trigonal and hexagonal symmetry.

NOTE, I, p. 201.

Rund 1952. 4 has given another method for the homogenization of every problem that can be put in the form of a variational equation. He uses a so-called Finsler geometry in which the fundamental tensor not only depends on the ξ^κ but also on a direction at the point considered, and he gets quite another Hamiltonian function that is uniquely determined. There is some kind of duality between this new function and $\mathfrak{L}$. In this remarkable theory that seems important especially for more general problems, momenta of another kind arise and the author claims and promises to prove that with these new momenta many simplifications can be obtained.

NOTE, J, p. 239.

§ 7. *The field equations derived from a variational equation*

In V § 5 we have proved that $-\mathfrak{G}^{\kappa\lambda}$ is the Lagrange derivative of the scalar density of weight $+1$

$$\underset{g}{\mathfrak{L}} \overset{\text{def}}{=} \frac{c^2}{\kappa}\mathfrak{g}^{\frac{1}{2}}K; \tag{7.1}$$

we call $\underset{g}{\mathfrak{L}}$ the *world-function of gravitation* in space-time. The momentum-energy tensor density $\underset{e}{\mathfrak{P}}^{\kappa\lambda}$ defined in (4.36) is the Lagrange derivative of the scalar density of weight $+1$

$$\underset{e}{\mathfrak{L}} \overset{\text{def}}{=} \tfrac{1}{2}\epsilon c^2\mathfrak{g}^{\frac{1}{2}}F_{\kappa\lambda}F_{\mu\nu}g^{\kappa\mu}g^{\lambda\nu}, \tag{7.2}$$

called the *electromagnetic world-function*.† Let further $\underset{m}{\mathfrak{L}}$ be a scalar density of weight $+1$ such that (cf. 4.31)‡

$$\overset{v}{d}\int \underset{m}{\mathfrak{L}}\, d\xi^1...d\xi^4 = \int \underset{m}{\mathfrak{P}}^{\kappa\lambda}\, \overset{v}{d}g_{\lambda\kappa}\, d\xi^1...d\xi^4. \tag{7.3}$$

We call $\underset{m}{\mathfrak{L}}$ the material world function. The equations (5.10), written in full,

$$\underset{m}{\mathfrak{P}}^{\kappa\lambda}+\underset{e}{\mathfrak{P}}^{\kappa\lambda}-\frac{c^2}{\kappa}\mathfrak{G}^{\kappa\lambda} = 0, \tag{7.4}$$

can now be derived from the variational equation $\overset{v}{d}\int \mathfrak{L}\, d\xi^1...d\xi^4 = 0$ with the total world function

$$\mathfrak{L} = \underset{m}{\mathfrak{L}}+\underset{e}{\mathfrak{L}}+\underset{g}{\mathfrak{L}}. \tag{7.5}$$

This world function consists now of three parts that are not connected in any way. It would certainly be more satisfactory if there were only one world function and if material and electromagnetic phenomena could be considered as geometric properties of time-space just as gravitational phenomena. The material part would certainly require quantum-mechanical considerations, but without these it could be tried to establish a field theory unifying gravitational and electromagnetic equations. Of course, the new geometry necessary for such a unified field theory could only be a generalization of the Riemannian geometry in V_4. Weyl tried a geometry with $\nabla_\mu g_{\lambda\kappa} = Q_\mu g_{\lambda\kappa}$ instead of $\nabla_\mu g_{\lambda\kappa} = 0$ and connected Q_μ with the electromagnetic potential vector. After him many attempts were made in several directions. An interesting unified field theory with a non-symmetric linear connexion and a non-symmetric fundamental tensor was first given by Einstein and later developed by Einstein, Strauss, Schrödinger and other authors.§

† Exercise IX. 4.

‡ For the construction of this integral in a very simple case see for instance Weyl 1921. 2, p. 209. Of course, in the general case the function $\underset{m}{\mathfrak{L}}$ cannot be dealt with without quantum mechanics.

§ For a detailed discussion and also for literature see R.C. 1954, 1, I § 11 and V. Hlavaty, *Geometry of Einstein's Unified Field Theory.* (Noordhoff, Groningen, 1958).

INDEX

A CATALOG OF SELECTED

DOVER BOOKS

IN SCIENCE AND MATHEMATICS

THE ELECTROMAGNETIC FIELD, Albert Shadowitz. Comprehensive undergraduate text covers basics of electric and magnetic fields, builds up to electromagnetic theory. Also related topics, including relativity. Over 900 problems. 768pp. 5⅝ × 8¼. 65660-8 Pa. $17.95

FOURIER SERIES, Georgi P. Tolstov. Translated by Richard A. Silverman. A valuable addition to the literature on the subject, moving clearly from subject to subject and theorem to theorem. 107 problems, answers. 336pp. 5⅜ × 8½. 63317-9 Pa. $7.95

THEORY OF ELECTROMAGNETIC WAVE PROPAGATION, Charles Herach Papas. Graduate-level study discusses the Maxwell field equations, radiation from wire antennas, the Doppler effect and more. xiii + 244pp. 5⅜ × 8½. 65678-0 Pa. $6.95

DISTRIBUTION THEORY AND TRANSFORM ANALYSIS: An Introduction to Generalized Functions, with Applications, A.H. Zemanian. Provides basics of distribution theory, describes generalized Fourier and Laplace transformations. Numerous problems. 384pp. 5⅜ × 8½. 65479-6 Pa. $9.95

THE PHYSICS OF WAVES, William C. Elmore and Mark A. Heald. Unique overview of classical wave theory. Acoustics, optics, electromagnetic radiation, more. Ideal as classroom text or for self-study. Problems. 477pp. 5⅜ × 8½. 64926-1 Pa. $11.95

CALCULUS OF VARIATIONS WITH APPLICATIONS, George M. Ewing. Applications-oriented introduction to variational theory develops insight and promotes understanding of specialized books, research papers. Suitable for advanced undergraduate/graduate students as primary, supplementary text. 352pp. 5⅜ × 8½. 64856-7 Pa. $8.95

A TREATISE ON ELECTRICITY AND MAGNETISM, James Clerk Maxwell. Important foundation work of modern physics. Brings to final form Maxwell's theory of electromagnetism and rigorously derives his general equations of field theory. 1,084pp. 5⅜ × 8½. 60636-8, 60637-6 Pa., Two-vol. set $19.90

AN INTRODUCTION TO THE CALCULUS OF VARIATIONS, Charles Fox. Graduate-level text covers variations of an integral, isoperimetrical problems, least action, special relativity, approximations, more. References. 279pp. 5⅜ × 8½. 65499-0 Pa. $7.95

HYDRODYNAMIC AND HYDROMAGNETIC STABILITY, S. Chandrasekhar. Lucid examination of the Rayleigh-Benard problem; clear coverage of the theory of instabilities causing convection. 704pp. 5⅝ × 8¼. 64071-X Pa. $14.95

CALCULUS OF VARIATIONS, Robert Weinstock. Basic introduction covering isoperimetric problems, theory of elasticity, quantum mechanics, electrostatics, etc. Exercises throughout. 326pp. 5⅜ × 8½. 63069-2 Pa. $7.95

DYNAMICS OF FLUIDS IN POROUS MEDIA, Jacob Bear. For advanced students of ground water hydrology, soil mechanics and physics, drainage and irrigation engineering and more. 335 illustrations. Exercises, with answers. 784pp. 6⅛ × 9¼. 65675-6 Pa. $19.95

ROTARY-WING AERODYNAMICS, W.Z. Stepniewski. Clear, concise text covers aerodynamic phenomena of the rotor and offers guidelines for helicopter performance evaluation. Originally prepared for NASA. 537 figures. 640pp. 6⅛ × 9¼.
64647-5 Pa. $14.95

DIFFERENTIAL GEOMETRY, Heinrich W. Guggenheimer. Local differential geometry as an application of advanced calculus and linear algebra. Curvature, transformation groups, surfaces, more. Exercises. 62 figures. 378pp. 5⅜ × 8½.
63433-7 Pa. $7.95

INTRODUCTION TO SPACE DYNAMICS, William Tyrrell Thomson. Comprehensive, classic introduction to space-flight engineering for advanced undergraduate and graduate students. Includes vector algebra, kinematics, transformation of coordinates. Bibliography. Index. 352pp. 5⅜ × 8½. 65113-4 Pa. $8.95

A SURVEY OF MINIMAL SURFACES, Robert Osserman. Up-to-date, in-depth discussion of the field for advanced students. Corrected and enlarged edition covers new developments. Includes numerous problems. 192pp. 5⅜ × 8½.
64998-9 Pa. $8.95

ANALYTICAL MECHANICS OF GEARS, Earle Buckingham. Indispensable reference for modern gear manufacture covers conjugate gear-tooth action, gear-tooth profiles of various gears, many other topics. 263 figures. 102 tables. 546pp. 5⅜ × 8½. 65712-4 Pa. $11.95

SET THEORY AND LOGIC, Robert R. Stoll. Lucid introduction to unified theory of mathematical concepts. Set theory and logic seen as tools for conceptual understanding of real number system. 496pp. 5⅜ × 8¼. 63829-4 Pa. $10.95

A HISTORY OF MECHANICS, René Dugas. Monumental study of mechanical principles from antiquity to quantum mechanics. Contributions of ancient Greeks, Galileo, Leonardo, Kepler, Lagrange, many others. 671pp. 5⅜ × 8½.
65632-2 Pa. $14.95

FAMOUS PROBLEMS OF GEOMETRY AND HOW TO SOLVE THEM, Benjamin Bold. Squaring the circle, trisecting the angle, duplicating the cube: learn their history, why they are impossible to solve, then solve them yourself. 128pp. 5⅜ × 8½. 24297-8 Pa. $3.95

MECHANICAL VIBRATIONS, J.P. Den Hartog. Classic textbook offers lucid explanations and illustrative models, applying theories of vibrations to a variety of practical industrial engineering problems. Numerous figures. 233 problems, solutions. Appendix. Index. Preface. 436pp. 5⅜ × 8½. 64785-4 Pa. $9.95

CURVATURE AND HOMOLOGY, Samuel I. Goldberg. Thorough treatment of specialized branch of differential geometry. Covers Riemannian manifolds, topology of differentiable manifolds, compact Lie groups, other topics. Exercises. 315pp. 5⅜ × 8½. 64314-X Pa. $8.95

HISTORY OF STRENGTH OF MATERIALS, Stephen P. Timoshenko. Excellent historical survey of the strength of materials with many references to the theories of elasticity and structure. 245 figures. 452pp. 5⅜ × 8½. 61187-6 Pa. $10.95

CATALOG OF DOVER BOOKS

CHALLENGING MATHEMATICAL PROBLEMS WITH ELEMENTARY SOLUTIONS, A.M. Yaglom and I.M. Yaglom. Over 170 challenging problems on probability theory, combinatorial analysis, points and lines, topology, convex polygons, many other topics. Solutions. Total of 445pp. 5⅜ × 8½. Two-vol. set.
Vol. I 65536-9 Pa. $6.95
Vol. II 65537-7 Pa. $6.95

FIFTY CHALLENGING PROBLEMS IN PROBABILITY WITH SOLUTIONS, Frederick Mosteller. Remarkable puzzlers, graded in difficulty, illustrate elementary and advanced aspects of probability. Detailed solutions. 88pp. 5⅜ × 8½.
65355-2 Pa. $3.95

EXPERIMENTS IN TOPOLOGY, Stephen Barr. Classic, lively explanation of one of the byways of mathematics. Klein bottles, Moebius strips, projective planes, map coloring, problem of the Koenigsberg bridges, much more, described with clarity and wit. 43 figures. 210pp. 5⅜ × 8½. 25933-1 Pa. $5.95

RELATIVITY IN ILLUSTRATIONS, Jacob T. Schwartz. Clear nontechnical treatment makes relativity more accessible than ever before. Over 60 drawings illustrate concepts more clearly than text alone. Only high school geometry needed. Bibliography. 128pp. 6⅛ × 9¼. 25965-X Pa. $5.95

AN INTRODUCTION TO ORDINARY DIFFERENTIAL EQUATIONS, Earl A. Coddington. A thorough and systematic first course in elementary differential equations for undergraduates in mathematics and science, with many exercises and problems (with answers). Index. 304pp. 5⅜ × 8½. 65942-9 Pa. $7.95

FOURIER SERIES AND ORTHOGONAL FUNCTIONS, Harry F. Davis. An incisive text combining theory and practical example to introduce Fourier series, orthogonal functions and applications of the Fourier method to boundary-value problems. 570 exercises. Answers and notes. 416pp. 5⅜ × 8½. 65973-9 Pa. $9.95

THE THEORY OF BRANCHING PROCESSES, Theodore E. Harris. First systematic, comprehensive treatment of branching (i.e. multiplicative) processes and their applications. Galton-Watson model, Markov branching processes, electron-photon cascade, many other topics. Rigorous proofs. Bibliography. 240pp. 5⅜ × 8½. 65952-6 Pa. $6.95

AN INTRODUCTION TO ALGEBRAIC STRUCTURES, Joseph Landin. Superb self-contained text covers "abstract algebra": sets and numbers, theory of groups, theory of rings, much more. Numerous well-chosen examples, exercises. 247pp. 5⅜ × 8½. 65940-2 Pa. $6.95